Advances in

ECOLOGICAL RESEARCH

VOLUME 18

Advances in

ECOLOGICAL RESEARCH

Edited by
M. BEGON
Department of Zoology, University of Liverpool, Liverpool, L69 3BX, UK

A. H. FITTER
Department of Biology, University of York, York, YO1 5DD, UK

E. D. FORD
Center for Quantitative Science, University of Washington, 3737 15th Avenue, Seattle, WA 98195, USA

A. MACFADYEN
23 Mountsandel Road, Coleraine, Northern Ireland

VOLUME 18

1988

[redacted]

ACADEMIC PRESS
Harcourt Brace Jovanovich, Publishers
London
San Diego New York Boston
Sydney Tokyo Toronto

ACADEMIC PRESS LIMITED
24/28 Oval Road
London NW1

United States Edition published by
ACADEMIC PRESS INC.
San Diego, CA 92101

British Library Cataloguing in Publication Data

Advances in ecological research.
Vol. 18
1. Ecology
I. Begon, Michael
574.5

ISBN 0-12-013918-9

Typeset by Latimer Trend & Company Ltd, Plymouth
Printed in Great Britain by the St Edmundsbury Press, Bury St Edmunds

Contributors to Volume 18

R. L. CHAZDON, *Department of Botany, University of California, Berkeley, California 94720, USA.*

E. A. LIVETT, *Department of Environmental Biology, University of Manchester, Manchester M13, 9PL, England.*

J. H. MYERS, *Department of Zoology, Oxford University, South Parks Road, Oxford OX1 3PS, England.* Present Address: *Department of Zoology and Plant Science, University of British Columbia, 6270 University Boulevard, Vancouver, British Columbia, Canada V6T 2A9.*

E. I. NEWMAN, *Department of Botany, University of Bristol, Woodland Road, Bristol BS8 1UG, England.*

I. C. PRENTICE, *Institute of Ecological Botany, Uppsala University, Box 559, S-751 22 Uppsala, Sweden.*

C. J. F. TER BRAAK, *Agricultural Mathematics Group, Box 100, 6700 AC Wageningen, The Netherlands.*

Preface

The five papers in this volume cover a very wide range of ecology, from environmental science through to population, mathematical and physiological ecology. Dr Robin Chazdon's paper on 'Sunflecks and their importance to forest understorey plants' covers a neglected but newly exciting field. Most of our understanding of photosynthetic ecology is based on measurements under continuous illumination, yet most leaves grow in shade and as Dr Chazdon shows, shade is a peculiarly difficult phenomenon to define. Sunflecks are of immense importance to shaded leaves and the ability to utilise sunflecks may determine the success of plants growing in shade. In a sense, sunflecks can be perceived as a resource for which plants must forage and possibly compete.

Competition below ground between plant roots has long been recognised as an important force structuring plant communities. The discovery that roots can be connected by mycorrhizal hyphae and that materials can pass from plant to plant through these hyphae threatens to upset seriously the perception of the below-ground relations of plants. Dr Edward Newman surveys the evidence currently available on this controversial topic and comes up with an important cautionary message, which emphasises the technical difficulties of work in this field.

Cycles of population density are, perhaps, the single most striking phenomenon that population ecologists seek to explain. In so doing, they address questions – of how a complex of interacting causes lead to a single effect – which are central to all studies of population ecology. Dr Judy Myers examines the nature of population cycles in forest lepidoptera and assesses the variety of possible causes that have been proposed for them. It is particularly appropriate that she should do so, since in Volume 8 of Advances in Ecological Research she was co-author with Dr Charles Krebs of one of the seminal papers on population cycles in small mammals. In the present case, she pays particular attention to the interactions between the insects and their pathogens.

The distribution of organisms in space and time is of fundamental ecological importance. Gradient analysis has developed as a major technique for exploring such patterns, and Drs Ter Braak and Prentice provide a unifying framework for the classification of the many methods of gradient analysis that have been used in ecology. They review both linear and non-linear methods and give a guide both to the choice of appropriate techniques and to the interpretation of the results of such analyses.

The decomposition of atmospheric pollutants is of enormous economic and environmental importance, and understanding the historical background is essential both to assessing the significance of current rates of deposition and predicting future changes. Dr Elizabeth Livett reviews the evidence that can be gained for heavy metal deposition from geochemical studies, broadly defined. Since many older studies have recently been re-assessed in the light of modern techniques this is a particularly timely study.

Contents

Contributors to Volume 18 v
Preface vii

Sunflecks and Their Importance to Forest Understorey Plants

ROBIN L. CHAZDON

I. Summary 2
II. Introduction: Sunflecks as a Resource 3
III. Measurement of Sunfleck Activity Beneath Forest Canopies 8
A. Area-survey Techniques 8
B. Instantaneous Sensor Measurements 9
C. Photographic Techniques 11
IV. Sunfleck Activity in Temperate and Tropical Forests 12
A. Defining Sunflecks 12
B. Temperate Deciduous Forests 13
C. Coniferous Forests 14
D. Tropical Evergreen Forests 15
E. Summary: Generalizations About Sunfleck Activity 19
V. Photosynthetic Responses to Sunflecks 20
A. Sunflecks and Carbon Gain in Understorey Habitats 20
B. Determinants of Sunfleck Utilization 25
C. Constraints on Sunfleck Utilization in Understorey Habitats . . 32
D. Sunfleck Regimes and Light Acclimation 37
E. Photosynthesis in Understorey Plants Revisited 41
VI. Seed Germination, Establishment and Growth in Relation to Sunfleck Activity 42
A. Seed Germination and Establishment in Understorey Habitats . . 43
B. Growth of Understorey Plants 44
VII. The Influence of Sunflecks on Reproductive Behavior and Distributions of Understorey Species 48
A. Light Availability, Size Variation and Reproductive Behavior . . 49
B. Vegetative and Sexual Reproductive Effort 50
C. Sunflecks, Canopy Gaps and Species Distributions 50
D. Vertical Distribution of Understorey Species 51
VIII. Conclusions 52
A. The Importance of Sunflecks: Scaling Up from Leaves to Whole Plants 52

B. Directions for Future Research 53
Acknowledgements 54
References 54

Geochemical Monitoring of Atmospheric Heavy Metal Pollution: Theory and Applications

ELIZABETH A. LIVETT

I. Summary 65
II. Introduction 67
III. Theoretical Considerations 69
A. The Accumulation of Heavy Metals 70
B. Chronology and Dating 89
C. Interpretation 99
IV. Practical Applications 106
A. Historical Perspectives 107
B. Present-day Sources of Atmospheric Heavy Metal Pollutants . . 130
C. Present World-wide Occurrence of Atmospheric Heavy Metal Pollutants 143
V. Conclusions 154
Acknowledgements 157
References 157
Appendix 174

Can a General Hypothesis Explain Population Cycles of Forest Lepidoptera?

JUDITH H. MYERS

I. Summary 179
II. Introduction 181
III. Evidence for Population Cycles in Forest Lepidoptera 182
IV. Characteristics of Cyclic Populations of Forest Lepidoptera 188
A. Characteristics of Cyclic Species 188
B. Patterns of Population Change 189
C. The Beginning of the Decline 193
D. Insect Fecundity and Population Fluctuations 195
E. Parasitoids and Population Fluctuations 196
F. Cyclic and Non-cyclic Populations 199
G. The Impact of Forest Defoliators on the Forests 201
V. Hypotheses to Explain Population Cycles 202
A. Variation in Insect Quality 202
B. The Climatic Release Hypothesis 206
C. Variation in Plant Quality 208
D. Disease Susceptibility 212
E. Mathematical Models 216
VI. Evaluation of Hypotheses 223
VII. Population Cycles of Other Organisms 226
VIII. Conclusions and Speculations 228

Acknowledgements 231
References 232

Mycorrhizal Links Between Plants: Their Functioning and Ecological Significance

E. I. NEWMAN

I. Introduction 243
II. Evidence that Links Between Plants Occur 245
III. Possible Functioning of Mycorrhizal Links 249
A. More Rapid or Greater Mycorrhizal Infection 249
B. Transport of Substances Between Plants 250
IV. Possible Roles of Mycorrhizal Links in Ecosystems 261
V. Conclusions 265
Acknowledgements 266
References 266

A Theory of Gradient Analysis

CAJO J. F. TER BRAAK and I. C. PRENTICE

I. Introduction 272
II. Linear Models 273
A. Regression 277
B. Calibration 278
C. Ordination 278
D. The Environmental Interpretation of Ordination Axes (Indirect Gradient Analysis) 280
E. Constrained Ordination (Multivariate Direct Gradient Analysis) . 281
III. Non-linear (Gaussian) Methods 282
A. Unimodal Response Models 282
B. Regression 285
C. Calibration 285
D. Ordination 286
E. Constrained Ordination 286
IV. Weighted Averaging Methods 287
A. Regression 287
B. Calibration 288
C. Ordination 290
D. Constrained Ordination 293
V. Ordination Diagrams and Their Interpretation 295
A. Principal Components: Biplots 295
B. Correspondence Analysis: Joint Plots 296
C. Redundancy Analysis 299
D. Canonical Correspondence Analysis 300
VI. Choosing the Method 302
A. Which Response Model? 302
B. Direct or Indirect? 304

C. Direct Gradient Analysis: Regression or Constrained Ordination?. 305
VII. Conclusions 306
Acknowledgements 308
References 309
Appendix 313

Subject Index 319

Sunflecks and Their Importance to Forest Understorey Plants

ROBIN L. CHAZDON

I. Summary 2
II. Introduction: Sunflecks as a Resource 3
III. Measurement of Sunfleck Activity Beneath Forest Canopies 8
A. Area-survey Techniques 8
B. Instaneous Sensor Measurements 9
C. Photographic Techniques 11
IV. Sunfleck Activity in Temperate and Tropical Forests 12
A. Defining Sunflecks 12
B. Temperate Deciduous Forests 13
C. Coniferous Forests 14
D. Tropical Evergreen Forests 15
E. Summary: Generalizations About Sunfleck Activity 19
V. Photosynthetic Responses to Sunflecks 20
A. Sunflecks and Carbon Gain to Understorey Habitats 20
B. Determinants of Sunfleck Utilization 25
C. Constraints on Sunfleck Utilization in Understorey Habitats . . 32
D. Sunfleck Regimes and Light Acclimation 37
E. Photosynthesis in Understorey Plants Revisited 41
VI Seed Germination, Establishment and Growth in Relation to Sunfleck Activity 42
A. Seed Germination and Establishment in Understorey Habitats . . 43
B. Growth of Understorey Plants 44
VII. The Influence of Sunflecks on Reproductive Behavior and Distributions of Understorey Species 48
A. Light Availibility, Size Variations and Reproductive Behavior . . 49
B. Vegative and Sexual Reproductive Effort 50
C. Sunflecks, Canopy Gaps and Species Distributions 50
D. Vertical Distribution of Understorey Species 51
VIII. Conclusions 52

ADVANCES IN ECOLOGICAL RESEARCH Vol. 18
ISBN 0-12-013918-9

A. The Importance of Sunflecks: Scaling Up From Leaves to Whole Plants 52
B. Directions for Future Research 53
Acknowledgements 54
References 54

I. SUMMARY

Sunfleck activity has profound effects on ecological processes ranging from photosynthesis to microsite distributions. Sunflecks occur when predominantly direct-beam radiation passes through openings in the forest canopy. In the forest understorey, sunflecks foster a high degree of spatial and temporal variation in light availability. Sunflecks may contribute more than 50% of the daily photon flux density in the understorey of temperate and tropical forests. Although understorey species are usually able to maintain positive carbon balance in the absence of sunflecks, photosynthesis during sunflecks may account for 30–60% of daily carbon gain.

Photosynthetic responses to sunflecks involve both short-term (dynamic) and longer-term (induction) responses. Following induction, leaves respond more rapidly to sunflecks. Carbon gain and photosynthetic efficiency during sunflecks depend strongly on sunfleck duration. When sunflecks are shorter than 40 s, measured carbon gain is often greater than predicted carbon gain based on steady-state photosynthetic responses. This enhancement of photosynthesis has been attributed to post-illumination CO_2 fixation, which contributes a large proportion of total carbon gain during brief sunflecks, but only a small proportion during longer sunflecks. Under natural conditions, photosynthetic utilization of sunflecks may be hindered by a variety of factors including loss of induction during low-light periods, restricted stomatal opening to conserve water loss, photoinhibition, wilting, and high leaf temperatures. There is no evidence that the photosynthetic characteristics responsible for efficient utilization of sunflecks impose any constraint on efficient utilization of low light. Some evidence does indicate, however, that photosynthetic adaptation to high light limits photosynthetic efficiency during sunflecks.

At light levels below 20% of full sun, light usually limits growth of understorey species. Accordingly, variation in sunfleck activity among understorey microsites has been correlated with differences in plant growth rates, size, sexual reproduction, and vegetative reproduction. The patchy distribution of some understorey species has, in some cases, been linked to microsite differences in light availability. Integrated organismal responses to changing light conditions make it exceedingly difficult to quantify the significance of sunflecks *per se* for growth, survivorship, and reproduction of forest understorey plants.

II. INTRODUCTION: SUNFLECKS AS A RESOURCE

> The interest of light as an ecological factor arises partly from the great variety of influences which the light climate exercises upon individual organisms, and also from the very complexity of the light climate itself, which has for long exercised a fascination over those who have been working in this field (Evans, 1966).

Plants that live beneath forest shade inherit the remnants of the sun's rays after they have filtered through the forest canopy. Those wavelengths of shortwave radiation that are most strongly absorbed by canopy foliage layers are also the most highly prized by leaves in the understorey, for these are the photosynthetically active wavelengths. In many forest types, less than 2% of the photosynthetically active radiation (PAR) incident above the canopy may actually reach the forest floor. How is it, then, that so many species of plants flourish in the dark recesses of forest undergrowth?

Although we are only beginning to understand in sufficient detail the light relations of forest understorey species, it is clear that their ecological and evolutionary successes are largely due to their abilities to capitalize on patterns of variation in light, the major limiting resource in most forest types. These patterns come in many different forms, with differing degrees of predictability and exploitability. Spatial patterns of light availability on the forest floor are fundamentally determined by the three-dimensional structure of the forest, but spatial patterns continually change, according to time of day, season, and latitude. Superimposed on the highly complex spatial pattern are temporal patterns of light fluctuation, which are affected by cloud cover, atmospheric conditions, and wind. If we examine further the extent of light variation among leaves within a single crown, the added influences of crown structure and leaf display must be considered. Ecological studies of understorey plants need to be concerned with patterns of light variation in natural habitats and reponses by leaves, individuals, and populations to these patterns.

Perhaps the most striking patterns of light variation within the understorey are those created by sunflecks, shafts of sunlight that penetrate small openings in the canopy (Fig. 1). For many years, plant ecologists and physiologists have recognized the importance of sunflecks to plants in the forest understorey (Lundegardh, 1922; Evans, 1939). Only recently, however, have researchers been able to measure light variation with sufficient resolution to evaluate critically its impact on physiological and ecological processes. In this review, I summarize, and attempt to synthesize, what is now

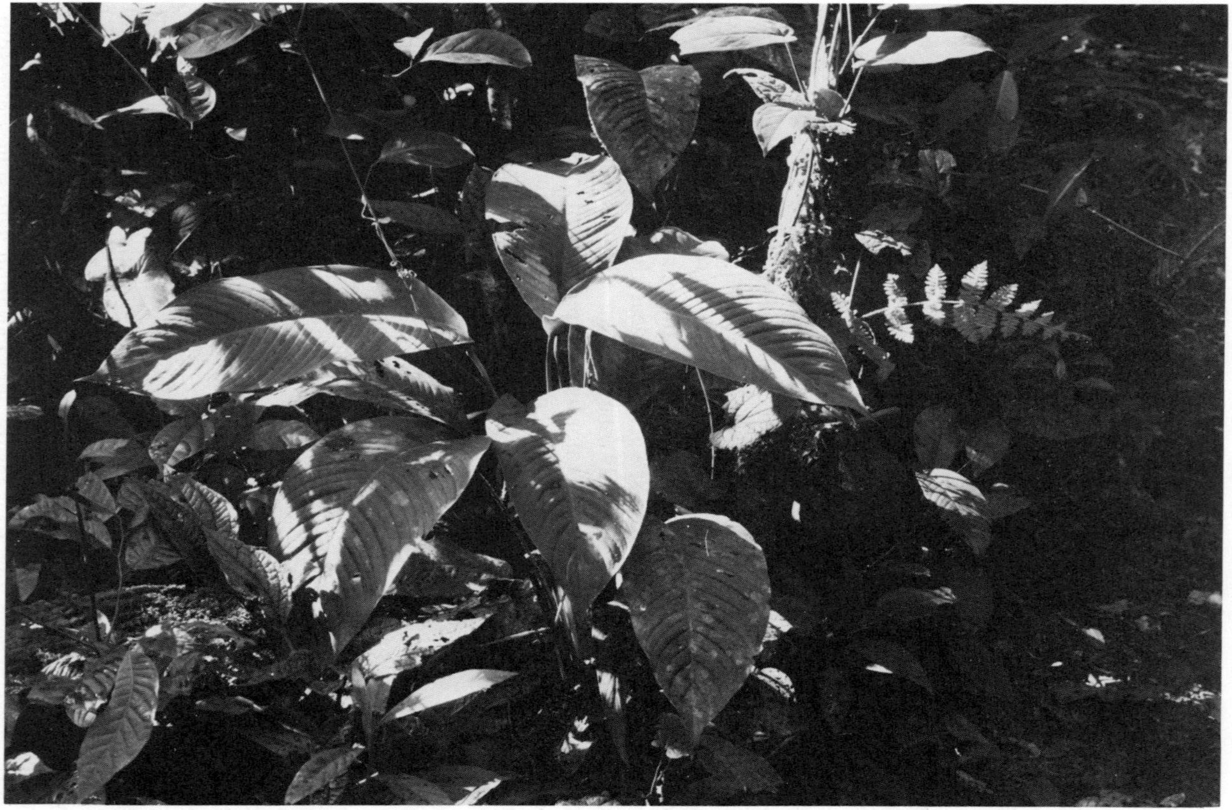

Fig. 1. Sunflecks on leaves of an understorey herb (Marantaceae) in a Costa Rican rainforest.

known about sunfleck activity and its importance to understorey plants in temperate and tropical forests.

The temporal and spatial scale of light variation are key factors in evaluating the range of physiological and ecological processes affected (Tables 1 and 2). Sunflecks lasting from a few seconds to several minutes may affect photosynthesis rates, stomatal responses, leaf temperature and morphogenesis, whereas variation in light availability on the scale of weeks to months may lead to differences in plant growth, morphology, survivorship, and reproduction (Table 1). Regardless of temporal scale, the spatial scale of light variation also has important consequences for the types of biological processes affected (Table 2). In this review, I examine the physiological and ecological consequences of sunfleck activity at different spatial and temporal scales. I consider two general classes of understorey plants: (1) species that complete their life-cycle in the forest understorey (shade tolerant herbs, shrubs, or small trees); and (2) seedlings and saplings of canopy tree species, which live in the understorey only during the early stages of their life-history. Mature canopy trees, non-forest species, and agricultural crops, have, in

Table 1

Physiological and ecological processes affected by light variation at different time-scales with spatial location maintained constant. Adapted from Chazdon (1987).

Time scale	Process affected
Seconds to minutes	Transient photosynthetic responses, stomatal responses, leaf temperature, seed germination, morphogenesis
Minutes to hours	Induction of photosynthetic apparatus, stomatal responses, chloroplast movements, leaf temperature, leaf movements, seed germination, morphogenesis
Hours to days	Changes in photosynthetic capacity, stomatal responses, leaf phenology, seed germination, morphogenesis, photoperiodic responses
Days to weeks	Changes in photosynthetic capacity and leaf biochemistry, leaf growth and morphology, plant growth, seedling establishment, survivorship, reproduction
Weeks to months	Photosynthetic acclimation, whole-plant growth, phenology, canopy structure, leaf morphology, biomass and nutrient allocation, seedling establishment, survivorship, reproduction
Months to years	Phenology, leaf turnover, whole-plant growth, plant architecture, survivorship, reproduction, nutrient cycling

Table 2

Physiological and ecological processes affected by light variation at different spatial scales with time maintained constant. Adapted from Chazdon (1987).

Spatial unit	Process affected
Cells within leaf	Light scattering and absorption, chloroplast movement, photosynthesis
Part of leaf	Photosynthesis, translocation, stomatal density, photomorphogenesis, herbivory, energy balance
Leaf	Photosynthetic capacity, energy balance, leaf movement, leaf morphology, leaf orientation, competition for light, photomorphogenesis, herbivory
Crown	Nutrient, water, and carbohydrate transport; shading among leaves, age structure of leaves, foliage distribution, branching pattern
Whole plant	Biomass allocation, establishment, growth, architecture, survival, reproduction, competition
Plant population	Ecotypic differentiation, age structure, population growth, recruitment
Communities	Succession, regeneration, vegetation structure, species diversity, nutrient cycling, hydrology

general, been excluded from this review, although light variation certainly influences these species as well.

Just as water and nutrients are environmental resources required in specific ways by different plants, so are sunflecks. If we consider sunflecks as a resource for understorey plants, we can then begin to assess the spatial and temporal distribution of the resource and physiological mechanisms of capture and utilization by different species. Different wavelengths within the radiation spectrum have unique realms of biological influence (Table 3). Total radiation affects leaf energy balance, whereas the ratio of red to far-red quanta controls photomorphogenetic processes mediated by phytochrome (Table 3). By studying sunflecks as a resource, their importance to specific physiological and ecological processes can be considered along with the effects of other critical plant resources. This is not to say that light is always perceived by understorey plants as either background (diffuse) light or as sunflecks. Furthermore, as discussed below, the utilization of sunflecks by leaves is not independent of leaf temperature, plant water status, nor plant nutrition. Rather, the value of regarding sunflecks as a resource lies in the usefulness of applying widely-accepted concepts of resource availability, utilization, and allocation, without implying any exclusive biological significance of sunflecks.

Table 3

Categories of radiation, their effects on physiological processes, instrumentation required for measurement, and units of measurement.

Category of radiation	Processes affected	Waveband	Instrument	Units
Total radiation	Energy balance	0·3–80 μm	Radiometer	Energy (J)
Net radiation	Energy balance	0·3–80 μm	Net radiometer	Radiant flux ($J\ s^{-1}$ or W)
Total short-wave radiation	Leaf energy balance, transpiration, leaf temperature	0·3–4 μm	Pyranometer	Radiant flux density or irradiance ($J\ m^{-2}\ s^{-1}$ or $W\ m^{-2}$)
Photosynthetically active radiation (PAR)	Photosynthesis	0·4–0·7 μm	Quantum sensor	Photon flux density (μmol photons $m^{-2}\ s^{-1}$)
			Spectral radiometer	Energy flux density or irradiance ($W\ m^{-2}$)
Red: far-red	Seed germination, photoperiodism, photomorphogenesis	Quanta 658–662 nm /quanta 728–732 nm	Spectral radiometer	Dimensionless
Blue light	Phototropism, stomatal movements, leaf movements	425–490 nm	Spectral radiometer	Quantum flux density (μmol quanta $m^{-2}\ s^{-1}$)

III. MEASUREMENT OF SUNFLECK ACTIVITY BENEATH FOREST CANOPIES

The solar radiation incident on the forest canopy is composed of two different forms: direct-beam radiation, and radiation diffused by the earth's atmosphere. These two forms penetrate the forest canopy in different ways (Anderson, 1964a; Reifsnyder *et al.*, 1971; Hutchison and Matt, 1977). When the sun is shining, predominantly direct-beam radiation passes through holes in the canopy, dappling the forest floor and leaves with sunflecks (Fig. 1). In contrast, the diffuse sky radiation incident on the forest canopy penetrates all canopy openings, although not always equally. As both forms of radiation pass through vegetation, spectral quality is altered by selective absorption, transmission, and reflection of wavelengths by foliage, branches and boles.

Sunflecks come in a wide variety of shapes, sizes, colors and durations; there is no such thing as a "typical" sunfleck, even within an intensely studied understorey microsite. The task of defining precisely what is and is not a sunfleck challenges even the most experienced ecologist. Contributing to the difficulty of describing sunfleck activity is the dual nature of sunflecks; they have both spatial and temporal dimensions. The spatial and temporal dimensions of sunflecks arriving at any particular location are related to the configuration of the forest canopy as seen from that location. Ideally, methods used to measure sunfleck activity should reflect the particular biological phenomenon being investigated. More often, however, definitions of the spatial and temporal dimensions of sunflecks have been limited by the techniques used to measure them. Three general methods have been used in the measurement of sunfleck activity in temperate and tropical forests: area-survey techniques, instantaneous sensor measurements, and analyses of hemispherical photographs.

A. Area-Survey Techniques

The first detailed studies of sunfleck activity in forest environments focused on describing spatial distributions based on continuous sampling of irradiance at points on a pre-determined grid within study plots. Evans (1956) described an area-survey technique for measuring the distribution of sunflecks in a Nigerian rainforest. This technique permitted the computation of the areas of sunflecks of different flux density, as measured by a galvanometer connected to a photoelectric cell. Evans then converted the area scale to an appropriate time scale to estimate the incidence of sunflecks of different flux density on an average day (Evans, 1956). The area-survey technique was subsequently used by Whitmore and Wong (1959) and Evans *et al.* (1960) in a Singapore rainforest understorey, and by Grubb and Whitmore (1967) in lowland and montane rainforests in Ecuador. As noted by Evans (1956), this technique is limited by the spectral response of the photocell and shading by the apparatus. Spatial distributions of sunflecks were also investigated by Miller and Norman (1971) and Norman *et al.* (1971), who proposed a technique for measuring sunfleck lengths using transect lines randomly drawn through a vegetation canopy. An area-survey technique was also used by Ustin *et al.* (1984) to measure sunfleck size and number in the understorey of a red fir forest.

B. Instantaneous Sensor Measurements

The use of photoelectric cells to measure instantaneous light conditions within forests has a long history (Atkins and Poole, 1926). It is perhaps within this realm of light measurement that the greatest progress has been made. Using portable data-loggers, it is now possible to record instantaneous light changes as frequently as every 0·01 s, obtain frequency distributions, averages, variances, and other statistical computations over any set of time intervals, and transfer these data directly to magnetic media, or to a microcomputer. A variety of light sensors have been used to measure the distribution of radiation within forests (Pearcy, 1988b; Table 3). These include photoelectric cells (Evans, 1939), net radiometers (Baldocchi *et al.*, 1984), pyranometers (Reifsnyder *et al.*, 1971), and PAR sensors (Biggs *et al.*, 1971; Gutschick *et al.*, 1985). Accordingly, radiation measurements may be made in units of energy (J), radiant flux ($J\ s^{-1}$ or W), flux density or irradiance ($J\ m^{-2}\ s^{-1}$ or $W\ m^{-2}$), or photon flux density (PFD; $\mu mol\ m^{-2}\ s^{-1}$; Table 3).

Several important considerations apply to the use of instantaneous sensor measurements for sunfleck descriptions. Each kind of sensor has a characteristic spectral and temporal response, which may or may not make it an appropriate choice for measuring sunfleck activity (Pearcy, 1988b). In studies of carbon gain during sunflecks, for example, sunfleck activity should be measured in units of photon flux density (Björkman and Ludlow, 1972), whereas studies of leaf temperature and energy balance require measurements of net radiation (Woodward, 1981; Table 3). Moreover, the sampling interval determines the level of sunfleck activity that can be measured. Sunflecks less than 10 s long are not accurately sampled using a sampling interval of 10 s (Chazdon and Fetcher, 1984a; Fig. 2). If significant sunfleck activity is overlooked because sampling intervals are long relative to sunfleck duration, substantial errors in calculating mean or integrated light levels can arise. These errors can be in the positive or negative direction. In the case of the 2-min sequence shown in Fig. 2, the average PFD computed from

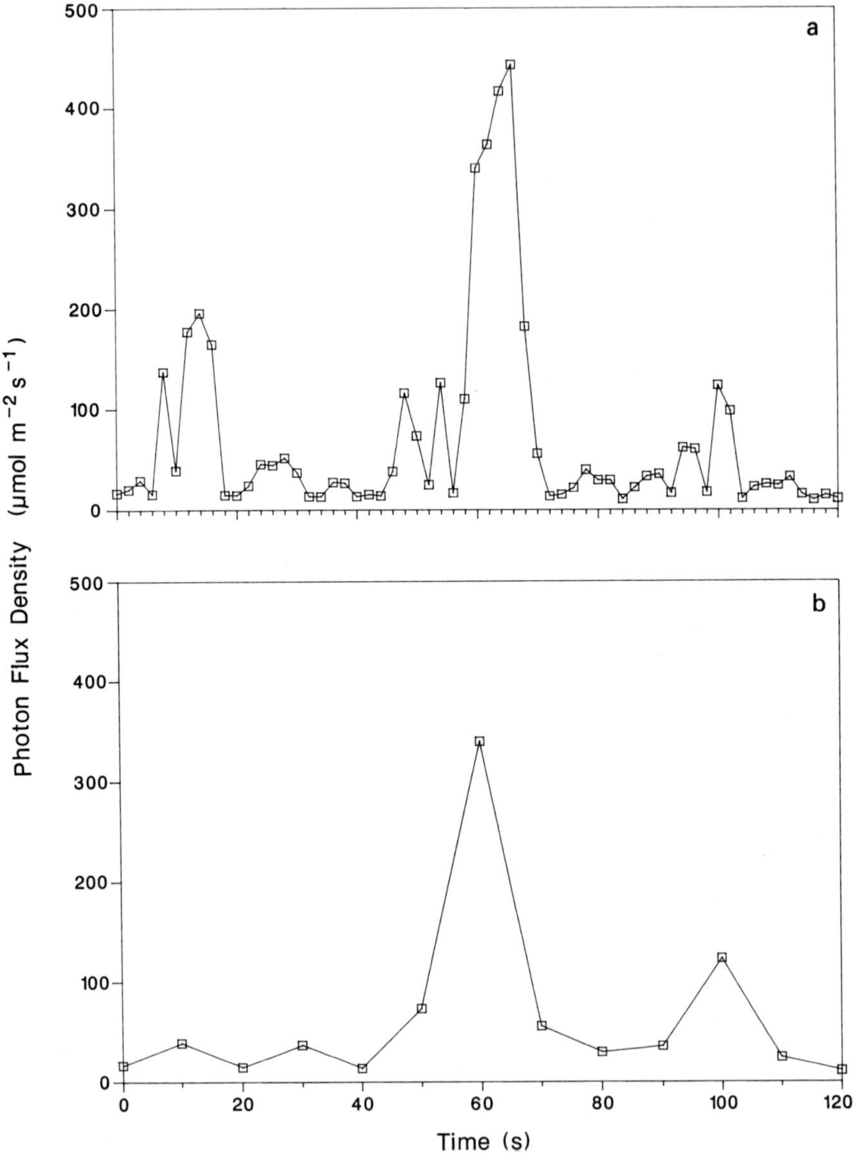

Fig. 2. Instantaneous measurements of photon flux density (PFD; μmol m^{-2} s^{-1}) during naturally-occurring sunflecks at (a) 2-s time intervals and (b) subsampled every 10 s. Two-min average PFD computed from 10-s readings was 10% lower than that computed from 2-s readings.

observations sampled every 10 s was 10% lower than the average computed from 2-s samples.

The exclusive use of instantaneous measurements to obtain integrated measurements over longer time periods leads to a substantial loss of information, such as the length and spacing of sunfleck intervals, and peak light intensities during sunflecks. Similarly, although light integrators (Woodward and Yaqub, 1979) provide a cumulative measure of PAR that is correlated with total sunfleck activity, they do not provide information on individual sunfleck periods. Furthermore, a single light sensor cannot provide information on the contribution of direct-beam radiation from sunflecks. To measure the relative contributions of direct and diffuse solar radiation during sunflecks, a pair of sensors is required, one with a shadow-band that obscures direct radiation (Horowitz, 1969).

C. Photographic Techniques

In 1924, Robin Hill invented and constructed a special camera to be used for observations of clouds. This "fish eye" camera was the first to be capable of recording an image of a complete hemisphere on a flat plate (Hill, 1924). The camera was borrowed by Evans and Coombe (1959), who found it most useful for photographing forest canopies. By orienting the photograph properly and then superimposing a transparency marked with solar tracks, they were able to see whether direct sunlight could reach the forest floor at any particular time. In 1964 hemispherical photographs were first used to estimate quantitatively the light conditions under forest canopies. Anderson (1964a) pioneered this technique, proposing a method for computing the percentage of diffuse light in the open received under the canopy (diffuse site factor) based on hemispherical photographs. She also estimated direct site factors using solar track diagrams to score, hour by hour, the percentage of direct light that could potentially reach the forest floor.

Hemispherical photographs have limited utility for predicting precise sunfleck activity because of variation in weather conditions, transmittance of light through foliage, and penumbral effects (Anderson, 1966; Norman *et al.*, 1971; Anderson and Miller, 1974; Salminen *et al.*, 1983; Chazdon and Field, 1987b). Nevertheless, as an indicator of potential sunfleck duration, they have proven very useful in a wide range of ecological studies of understorey plants (Pearcy, 1983; Ustin *et al.*, 1984; Orozco-Segovia, 1986: Walters and Field, 1987; Chazdon and Field, 1987a; Rich *et al.*, 1987). In contrast to sensor measurements, hemispherical photographs provide a means of assessing light conditions over a relatively long period of time (weeks to months) and for a large number of plants. They also offer the advantage of estimating diffuse and direct components of the light environment separately. Several computerized techniques are now being used to analyze hemispherical photographs (Jupp *et al.*, 1980; Chan *et al.*, 1986; Chazdon and Field, 1987b). Many types of "fish-eye" lenses are currently available (Evans *et al.*, 1975). Photographic analyses should account for lens distortion (if any), areal projection of a hemispherical image (Herbert, 1988), and leaf angles (Chazdon and Field, 1987a).

IV. SUNFLECK ACTIVITY IN TEMPERATE AND TROPICAL FORESTS

A. Defining Sunflecks

Sunflecks defy generalization. Nevertheless, in order to describe measurements of sunfleck activity, some general criteria must be established. Because sunflecks are caused by the penetration of predominantly direct-beam solar radiation through openings in the forest canopy, the spectral quality of sunflecks differs from that of diffuse shade light (Coombe, 1957; Federer and Tanner, 1966; Chazdon and Fetcher, 1984b; Lee, 1987). The degree to which sunflecks are composed of direct-beam radiation may vary greatly, however. The mean red:far-red ratio for sunflecks in a wheat canopy was only 15% lower than daylight values (Holmes and Smith, 1977a). In lowland rainforests of Panama and Costa Rica, the red:far-red ratio of sunflecks ranged from 0·37 to 1·30, compared to a mean of 1·22–1·33 in a clearing, and 0·35–0·40 in forest shade (Lee, 1987).

Variation in the relative proportions of diffuse and direct light in sunflecks can be partly explained by penumbral effects within forest canopies (Miller and Norman, 1971; Anderson and Miller, 1974). A penumbra is a partial shadow, an "edge effect" of a sunfleck. The probability of penumbras in a forest understorey depends primarily on the size of canopy openings and the canopy height (Norman *et al.*, 1971; Oker-Blom, 1984). The solar disk subtends an angle of $1/2°$ at the earth's surface, such that a canopy opening of at least this apparent size is required to transmit full-sun irradiance. In a tall forest with many small openings, sunfleck sizes will be small, and penumbras will be frequent (Anderson and Miller, 1974; Oker-Blom, 1984). In this case, irradiance during sunflecks will be considerably lower than that of direct beam solar radiation incident above the canopy, and the irradiance and spectral composition will more closely resemble those of shade light. When penumbras are combined with windy conditions, it is often difficult to differentiate sunflecks from fluctuations in background diffuse light. These patterns become even more complicated under partly cloudy conditions, because of cloud edge effects.

At the other extreme, canopy gaps or widely-spaced canopy trees create relatively large sunflecks, or sun patches, where full-sunlight irradiances are often observed (Young and Smith, 1979). Along the edges of gaps, and in small gaps formed by a branch fall, for example, the frequency of full-sun irradiance is low compared to conditions within larger gaps (Chazdon, 1986). In these intermediate habitats, sunflecks can often be easily distinguished from the slightly elevated levels of background diffuse light. In describing measurements of sunfleck activity, one must keep in mind that each investigator needs to adopt his or her own criteria for distinguishing sunflecks from shade light. For example, the PFD of diffuse light in a lodgepole pine forest in Wyoming (Young and Smith, 1979) is similar to the PFD during sunflecks in a Costa Rican rainforest (Chazdon and Fetcher, 1984a). If sunflecks were universally defined as periods when PFD exceeded 50 μmol m^{-2} s^{-1}, this definition would be useless for describing sunfleck activity in the Wyoming forest. Furthermore, investigators often have different concepts of what constitutes a forest understorey habitat relative to a gap. In the following sections, unless otherwise indicated, I have adopted the definitions used by each investigator. Although this approach limits direct comparisons of sunfleck activity among forest types, I believe it is the most reasonable way to describe the wide array of sunfleck measurements that have been described.

B. Temperate Deciduous Forests

The light regime in the understorey of temperate deciduous forests varies dramatically over the year, according to the timing of leaf production and leaf fall in canopy trees and changes in solar evaluation. During winter, when trees are leafless, branches and trunks alone can absorb 50–70% of the incoming solar radiation (Hutchison and Matt, 1977). In a New England hardwood forest, 30% of the daily photon flux incident on the canopy reached the forest floor in April, before the emergence of tree foliage (Curtis and Kincaid, 1984). Based on analyses of hemispherical photographs, Anderson (1964a) found that both diffuse and direct site factors (see p. 11) varied throughout the year in the understorey of a deciduous forest near Cambridge, UK. The diffuse site factor remained fairly constant at 30% from January to April, whereas the direct site factor increased from 3% in January to a maximum of 19% in April. Monthly irradiance in the understorey site reached maximum values in April for three consecutive years (Anderson, 1964b). Similarly, in a deciduous forest in Tennessee, maximum amounts of direct beam solar radiation reached the forest floor in early spring, accounting for over 90% of the total solar energy received in the understorey during this period (Hutchison and Matt, 1976, 1977).

In the summer, following tree leaf emergence, solar radiation in the understorey decreases to 1–5% of that available above the canopy, and remains low until autumn leaf fall (Hicks and Chabot, 1985). During this period, direct site factors range from 1–3%, and sunflecks become relatively infrequent (Anderson, 1964a,b). Nevertheless, penetrating direct beam solar radiation contributed over 50% of the total radiation budget during the summer months in a Tennessee deciduous forest understorey (Hutchison and Matt, 1976, 1977). Measurements of solar radiation penetration in a hardwood stand in Connecticut showed that only 21% of direct-beam radiation reached the forest floor during the summer (Reifsnyder *et al.*, 1971). Sunflecks, which were small and widely scattered, contributed to a high degree of spatial and temporal variation in solar radiation within the understorey; the coefficient of variation for individual 5-min observations was 225% (Reifsnyder *et al.*, 1971). In the understorey of a Michigan mixed hardwood forest during the summer, an estimated 45–55% of daily photon flux density (PFD) was contributed by sunflecks exceeding 100 $\mu mol\ m^{-2}\ s^{-1}$ (Weber *et al.*, 1985). Although PFD was below 50 $\mu mol\ m^{-2}\ s^{-1}$ more than 75% of the time, these readings contributed only 35–40% of daily PFD.

C. Coniferous forests

The light regime in the understorey of evergreen, coniferous forests also varies greatly during the year, but in this case seasonal variation is primarily due to changes in solar elevation rather than to changes in forest canopy cover. In general, a higher proportion of direct beam radiation reaches the understorey in coniferous forests compared to deciduous forests (Anderson, 1966; Reifsnyder *et al.*, 1971; Smith, 1985). Diffuse site factors in the understorey of a stand of *Pinus sylvestris* averaged 16·4%, whereas monthly direct site factors varied from 0 in January to a maximum of nearly 30% in June and July (Anderson, 1966). Direct site factors were higher in the summer, when there were more opportunities for direct sunlight to penetrate openings in the canopy. Sunfleck distributions did not vary in a consistent manner among sites during the year, however. Direct site factors within particular sites did not change in parallel throughout the seasons (Anderson, 1966).

Sunflecks accounted for over 50% of the total solar radiation beneath a *Pinus resinosa* stand in Connecticut during the summer (Reifsnyder *et al.*, 1971). In this forest, sunflecks were large and bright, and nearly 25% of direct-beam radiation incident above the canopy reached the forest floor. As in the hardwood forest, sunflecks contributed greatly to light variation within the understorey; the coefficient of variation for individual 5-min observations was 121% (Reifsnyder *et al.*, 1971).

Young and Smith (1979) made detailed observations on the frequency and duration of sunflecks received by two *Arnica* species in the understorey of two subalpine coniferous forests in Wyoming. In the lower-elevation lodgepole pine forest, 39% of the sunflecks received were less than 15 min long, and only 5% exceeded 60 min. The longest sunfleck (or sun patch) lasted 165 min. In contrast, sunflecks in the higher elevation spruce-fir forest were shorter, less frequent, and of lower flux density (Young and Smith, 1979). Both forests are characterized by the occurrence of large sunflecks, often lasting 20–30 min, that receive 40–60% of full sunlight irradiance (Smith, 1985). Apart from the effects of canopy cover, cloud conditions during afternoon periods reduce incident sunlight an estimated 40% over the day (Young and Smith, 1983).

Sunfleck activity varied considerably among microsites in red fir forests of California (Ustin *et al.*, 1984). The frequency of readings at relatively low PFD ($< 75\ \mu mol\ m^{-2}\ s^{-1}$) did not vary significantly among sites, but the frequency of readings above $1025\ \mu mol\ m^{-2}\ s^{-1}$ varied 3·5-fold. Along a line transect, PFD of sunflecks varied from a minimum of 31 to a maximum of $624\ \mu mol\ m^{-2}\ s^{-1}$. Among microsites, sunfleck size and frequency varied greatly, although sunfleck size tended to be inversely related to frequency.

Sunfleck measurements in the understorey of a redwood forest in California were made in three sites differing in total canopy cover and exposure (Powles and Björkman, 1981). In a deep shade site with little sunfleck activity, daily PFD was $0{\cdot}73\ mol\ m^{-2}\ d^{-1}$; the maximum 10 min average PFD measured was $153\ \mu mol\ m^{-2}\ s^{-1}$. At the edge of a gap, $2{\cdot}3\ mol\ m^{-2}\ d^{-1}$ was received, with as much as 71% contributed by two sunflecks. The third site, on the south-facing edge of a large clearing received $8{\cdot}14\ mol\ m^{-2}\ d^{-1}$; a major sunfleck at noon contributed as much as 83% of this total.

D. Tropical Evergreen Forests

Because of their equatorial proximity and tall, dense canopies, understories of tropical evergreen forests receive extremely low levels of diffuse solar radiation on a year-round basis (see review by Chazdon and Fetcher, 1984b). Since the pioneering studies of Evans (1939, 1956), sunflecks have been recognized as an important feature of the light environment within tropical forest understorey habitats. Using a Weston photoelectric cell and galvanometer, Evans (1939) made many observations of sunflecks in mature Nigerian rainforest and a nearby 14-yr-old secondary forest. In the mature forest understorey, sunflecks were generally of low irradiance compared with full sunlight, and were confined to a period of 4–5 h in the middle of the day. Over a 10-h period, fewer than 0·1% of the observations exceeded irradiances more than five times the mean shade irradiance, whereas 5·2% were greater than 20% above the mean shade irradiance (Evans, 1939). An estimated 10% of the total light energy was attributed to sunflecks (Evans, 1966; Table 4). Light conditions were very similar in the 14-yr-old secondary forest, but the incidence of high-irradiance sunflecks was lower than in the primary forest. Diffuse light irradiance was also lower in the secondary forest (Evans, 1939).

In another Nigerian forest, Evans (1956) calculated that sunflecks contributed about 70% of the total light energy reaching the forest floor between January and March (Table 4). Using the area-survey technique, he observed that 20–25% of the area of the forest floor was occupied by sunflecks at midday. On average, sunflecks were visible in a particular spot for about one hour a day. As in the other forest, sunflecks were rare during early morning and late afternoon, when the solar angle was below 30°. Photometer readings in a Brazilian rainforest understorey showed a similar pattern of sunfleck incidence, with sunflecks restricted to a 6 h period in the middle of the day (Ashton, 1958).

Extrapolations from similar studies in a tropical rainforest in Singapore showed that, during an entire year, approximately 50% of the total light energy received by understorey plants came from sunflecks (Whitmore and Wong, 1959; Table 4). Considerable variation was observed in the distribution of sunflecks and diffuse light in two forest plots (Evans *et al.*, 1960). Bright sunflecks, of equivalent irradiance to full sunlight, were rare (Whitmore and Wong, 1959; Evans *et al.*, 1960). In a lowland rainforest in Ecuador, Grubb and Whitmore (1967) found that sunflecks contributed 60% of the light energy on a sunny day (Table 4). Less sunfleck light was received

Table 4

The percentage of total radiant energy contributed by sunflecks in the understorey of temperate and tropical forests. Definitions of sunfleck activity and the number of days measured are specific to each study.

Forest type/site	Percentage total energy	Reference
Temperate deciduous forest		
Tennessee, USA (summer)	50	Hutchison and Matt (1976, 1977)
Coniferous forest		
Connecticut, USA (summer)	50	Reifsnyder *et al.* (1971)
Lowland tropical evergreen forest		
Nigeria (1 day)	10	Evans (1939, 1966)
Nigeria (Jan–Mar)	70	Evans (1956)
Singapore (entire year)	50	Whitmore and Wong (1959)
Ecuador (1 day)	60	Grubb and Whitmore (1967)

in a montane forest in Ecuador, although diffuse light readings were higher during midday compared to the lowland forest (Grubb and Whitmore, 1967).

Patterns of light distribution in the understorey under cloudy conditions were stable and reproducible, according to studies by Evans *et al.* (1960) in a Singapore rainforest. No correspondence was observed, however, between mean irradiance at the same microsite measured under sunny and cloudy conditions on different days. Evans *et al.* (1960) therefore concluded that measurements of light distribution on cloudy days could not be used to predict light distribution during sunny conditions. In contrast, Sasaki and Mori (1981a) found that the frequency and illuminance of sunflecks was strongly correlated with the illuminance of diffuse light in a Malaysian Dipterocarp forest. In the darkest understorey microsites, sunflecks were infrequent and of low illuminance, whereas microsites characterized by higher levels of diffuse light received more sunflecks of higher illuminance.

Based on measurements of photon flux in a deeply-shaded rainforest microsite in Queensland, Australia, Björkman and Ludlow (1972) found that sunflecks contributed 62% of daily photon flux on a clear day (Table 5). Photon flux density during sunflecks reached a maximum of 350 μmol $m^{-2} s^{-1}$, equivalent to 20% of PFD above the canopy. Most sunflecks were less than 2 min long. In a considerably brighter understorey site in Queensland, sunflecks contributed 38% of daily PFD (Pearcy, 1987). In the understorey of a Hawaiian rainforest, over 60% of the sunflecks received during the summer were less than 30 s long, and few were over 5 min long

Table 5

The percentage of total photon flux contributed by sunflecks in the understorey of temperate and tropical forests. Definitions of sunfleck activity and the number of days measured are specific to each study.

Forest type/site	Percentage total photon flux	Reference
Temperate deciduous forest		
Michigan, USA (summer)	45–55	Weber *et al.* (1985)
Lowland tropical evergreen forest		
Queensland, Australia (1 day)	62	Björkman and Ludlow (1972)
" (1 day)	12–65	Pearcy (1988a)
Hawaii, USA (5 weeks)	40	Pearcy (1983)
Costa Rica (3 days)	10–78	Chazdon (1986)
Mexico (1 day)	16–44	R. L. Chazdon, C. B. Field and R. W. Pearcy (unpublished data)

(Pearcy, 1983). These brief sunflecks largely reflect the windy conditions characteristic of this forest. The median maximum PFD during sunflecks was 250 μmol m^{-2} s^{-1}, with only a small proportion of sunflecks reaching PFD equivalent to full sunlight above the canopy. Arbitrarily defining sunflecks as PFD observations above 150 μmol m^{-2} s^{-1}, the average minutes of sunflecks per day during the summer months were 10·6 and 21 min, for two successive years. On relatively clear days, sunflecks contributed as much as 80% of daily PFD; over a 5-week measurement period, the estimated contribution of sunflecks decreased to 40% of daily PFD (Table 5). Potential sunfleck incidence, based on hemispherical photographs, ranged from an average of 5 min in the winter months to 61 min during the summer months (Pearcy, 1983).

Measurements of photon flux in the understorey of a Costa Rican rainforest indicate that sunflecks may contribute from 10 to 78% of daily PFD on clear days (Chazdon and Fetcher, 1984a; Chazdon, 1986; Table 5). The relative proportion of daily PFD contributed by sunflecks increased as the PFD of background diffuse radiation decreased (Chazdon, 1986). During sunflecks, PFD rarely exceeded 500 μmol m^{-2} s^{-1}. In this forest, sunflecks tend to be more frequent in the morning, because of afternoon cloud-cover and rain (R. L. Chazdon, unpublished data).

Sunfleck activity can vary greatly over a small spatial scale. In a Mexican rainforest, mean sunfleck incidence (observations above 50 μmol m^{-2} s^{-1}) for a single day ranged from 0 to 42 min among 16 leaves within the same understorey plant (Chazdon *et al.*, 1988). Mean minutes of sunflecks per day in four understorey plants from the same site ranged from 4 to 22 min per day. Average potential minutes of sunflecks per day for these same plants, based on hemispherical photographs, ranged from 11 to 64 min. Measurements of sunfleck activity along a 2·1 m line transect showed that minutes of sunflecks per day varied from 7 to 33 min among 16 sensors spaced 15 cm apart (R. L. Chazdon and C. B. Field, unpublished data). Daily PFD varied among sensors by a factor of 5. Sunfleck activity was so localized that daily patterns of PFD among sensors at distances greater than 0·6 m were not significantly correlated (Chazdon *et al.*, 1988).

Detailed sunfleck measurements in a closed-canopy site in this same forest using 16 sensors placed in a two-dimensional array showed that sunflecks were brief, of low intensity relative to full sun, and extremely variable on a spatial scale of 0·25 m^2 or less. On average, 56% of the sunflecks received were less than 4 s long, and over 90% were less than 32 s long (R. L. Chazdon, C. B. Field, and R. W. Pearcy, unpublished data). Mean sunfleck length was 13 s. Nearly 64% of the sunflecks had a maximum PFD below 100 μmol m^{-2} s^{-1}, whereas less than 2% exceeded 500 μmol m^{-2} s^{-1}. Furthermore, sunflecks were clumped in their temporal distribution. Among the 16 sensors, which were located within 0·5 m of each other, total sunfleck duration ranged from 10·6 to 29·2 min, and the contribution of sunflecks to daily PFD ranged from 16 to 44% (Table 5). Total sunfleck duration was positively correlated with daily PFD ($P < 0{\cdot}001$), whereas mean sunfleck length was negatively correlated with the total number of sunflecks received ($P < 0{\cdot}01$, R. L. Chazdon, C. B. Field, and R. W. Pearcy, unpublished data).

In the understorey of a Queensland rainforest, sites only 0·5 m apart can show two-fold variation in daily PFD and the number of sunflecks received (Pearcy, 1988a). Within a 5-m radius, daily PFD ranged from 0·47 to 1·5 mol $m^{-2}\,d^{-1}$, and the percentage contributed by sunflecks ranged from 12 to 65% (Table 5). Maximum PFD during sunflecks was almost always below full-sun levels; only 1% of the sunflecks measured exceeded 1200 $\mu mol\ m^{-2}\,s^{-1}$.

Within sapling crowns of two canopy tree species in a Costa Rican lowland forest, total minutes of sunfleck activity per day ranged from 2 to 106 for *Dipteryx panamensis* and from 1 to 94 for *Lecythis ampla* (Oberbauer *et al.*, 1988). For both species, total sunfleck exposure per day averaged 18–20 minutes.

E. Summary: Generalizations About Sunfleck Activity

The studies described above clearly illustrate the tremendous variation in sunfleck activity within and among forest types. Despite this variation, estimates of the percent of total light energy contributed by sunflecks are remarkably similar in temperate and tropical forests (Table 4). Sunflecks contribute 50% of the total light energy received during the summer in both coniferous and deciduous temperate forests, and 50–70% in tropical evergreen forests (Table 4).

The percentage of total PFD contributed by sunflecks is also quite similar in temperate deciduous forests during summer and tropical forests (Table 5). The two long-term estimates, 45–55% for a temperate deciduous forest during the summer (Weber *et al.*, 1985), and 40% for a five-week period in a Hawaiian subtropical forest (Pearcy, 1983), are consistent with the range of variation measured among sensors within a single day in three different tropical forests (Table 5). No yearly estimates of the contribution of sunflecks to total PFD are available for comparison. These studies, carried out independently by different researchers using a variety of techniques and assumptions, clearly illustrate the importance of sunflecks as sources of radiant energy and photosynthetically active radiation in all forest types.

To the extent that sunflecks are a consequence of forest canopy structure, some generalizations can be made with regard to the frequency, duration, and peak intensities of sunflecks in different forest types during sunny periods. Tall, multi-layered forest canopies have many openings that are smaller than the diameter of the solar disk from the perspective of the forest floor. Thus, penumbral effects often dominate in the understorey. Sunflecks, when they occur, are brief (usually less than 1 min long), extremely localized (less than 0·5 m in length or width), and tend to have peak PFD well below full-sunlight irradiance. Sunflecks tend to be clustered in distribution, and are rare during early morning and late afternoon.

In more open forest types and leafless forests (temperate or tropical deciduous forests), sunflecks are longer in duration (often up to 10 min) and occupy a larger area on the forest floor (up to several m in length or width). During sunfleck periods, peak PFD will frequently reach full-sun irradiance. Moreover, "sunpatches" will also tend to have higher diffuse PFD between sunfleck periods than more shaded microsites.

Within a particular understorey habitat, the number of sunflecks received per day tends to be negatively correlated with the mean length of sunflecks. Daily PFD is positively correlated with the total minutes of sunflecks received. Further generalizations can be made by incorporating known weather and atmospheric conditions within each forest. To the extent that overcast skies, rain, and wind are predictable, microsite variation in the frequency and duration of sunflecks of different peak intensity can be estimated for any particular spatial scale. It is important to note, however, that descriptions of sunfleck activity are specific to the spatial and temporal scale at which they are measured.

V. PHOTOSYNTHETIC RESPONSES TO SUNFLECKS

Sunflecks are a common and important feature of the light environment in all forest understorey habitats. In light-limiting habitats, the extent to which sunflecks influence plant growth, reproduction, and microsite distribution ultimately depends on their importance for leaf carbon gain. In this section, I review what is known about carbon gain during sunflecks in natural habitats and constraints on photosynthetic responses to sunflecks. These studies provide a basis for deeper understanding of the ecological significance of sunflecks for forest understorey plants.

A. Sunflecks and Carbon Gain in Understorey Habitats

Two general approaches have been used to evaluate the relative proportion of total carbon gain attributed to sunflecks in forest understorey plants: field studies and computer simulations. Ideally, photosynthetic measurements should be made on plants growing under natural patterns of light variation. Field studies, however detailed, are frequently limited by the inability to generalize results to other species or microsites. Simulations using modelled photosynthetic responses, when combined with field data, provide a means to extrapolate from an initially limited set of results. The most powerful approach involves the collection of detailed field measurements, modelling of leaf responses, and subsequent computer simulations using field-collected light and photosynthesis data.

1. Field Studies

To determine the importance of sunflecks for daily carbon gain in forest understorey plants, continuous measurements of PFD and CO_2 exchange are needed. Accurate instantaneous measurements of gas exchange during naturally fluctuating light conditions further require that instrument responses are faster than the physiological responses of leaves. In addition, large quantities of data must be recorded quickly and stored for subsequent analysis. Because of these technical difficulties, few detailed measurements of daily patterns of CO_2 exchange in forest understorey plants have been made (Table 6). The first measurements of gas exchange under natural sunfleck conditions in a Queensland rainforest understorey showed that sunflecks contributed substantially to daily carbon gain (Björkman *et al.*, 1972b). Nevertheless, leaves of the understorey herbs *Alocasia macrorrhiza* and *Cordyline rubra* were capable of achieving positive daily carbon balance in the absence of sunflecks. Both CO_2 assimilation and stomatal conductance increased rapidly in response to sunflecks. Although some sunflecks did exceed the light saturation point for both species (about 100 μmol m^{-2} s^{-1}), the efficiency of light use during both clear and overcast days was high (Björkman *et al.*, 1972b). Based on data presented by Björkman *et al.*

Table 6

Percentage of daily carbon gain contributed by sunflecks in natural habitats based on field measurements. Weather conditions and number of days measured varied among sites and species.

Forest type/species	Percentage daily carbon gain	Reference
Temperate deciduous forest		
Acer saccharum (summer)	35	Weber *et al.* (1985)
Tropical rainforest		
Euphorbia forbesii (1 day)	60	Pearcy and Calkin (1983)
Claoxylon sandwicense (1 day)	40	Pearcy and Calkin (1983)
Argyrodendron peralatum (1 day)	32	Pearcy (1987)

(1972b), Weber *et al.* (1985) calculated that, in the absence of sunflecks, total carbon gain for *Alocasia* would be reduced by about 10%.

In the subalpine understorey species *Arnica cordifolia*, photosynthetic characteristics differed between plants growing in relatively shaded and sunlit areas (Young and Smith, 1980): in shaded sites, photosynthesis during sunflecks was twice that during shaded periods. Carbon gain in understorey species is not always positively correlated with daily PFD. Photosynthetic rates of *Arnica latifolia* in a mixed spruce-fir forest were often light-saturated during sunflecks (Young and Smith, 1983). On cloudy days, net carbon gain of *A. latifolia* was 37% greater than on clear days, despite a 30% decrease in daily PFD. Diffuse light levels on cloudy days were often higher than on sunny days. Therefore, photosynthetic rates approached light saturation on cloudy days, partially accounting for the observed increase in daily carbon gain.

Pearcy and Calkin (1983) studied field gas exchange of two tree species in the understorey of a Hawaiian forest. In the absence of sunflecks, both *Euphorbia forbesii* and *Claoxylon sandwicense* were able to maintain positive rates of CO_2 assimilation over virtually the entire day because of low light compensation points. Photosynthesis during sunflecks, however, accounted for a substantial fraction of daily carbon gain. Photosynthetic responses to brief sunflecks were very rapid, and were effectively integrated by the gas-exchange apparatus. Sunflecks contributed an estimated 60% of the carbon gain in *Claoxylon* on a relatively clear day, and 40% of the carbon gain in *Euphorbia* on a less clear day (Table 6). At light saturation, net photosynthesis of *Euphorbia* was 50–60% higher than that of *Claoxylon*, and, consequently, *Euphorbia* appeared to utilize longer sunflecks more efficiently than *Claoxylon* (Robichaux and Pearcy, 1980; Pearcy and Calkin, 1983). On the other hand *Claoxylon* had higher rates of CO_2 assimilation under diffuse light in the understorey, such that growth rates of the two species were similar (Pearcy and Calkin, 1983; Pearcy, 1983).

In a Puerto Rican forest understorey, sunflecks increased photosynthetic rates of the shrub *Piper treleaseanum* by a factor of 5·8 over rates measured under diffuse light conditions (Lawrence, 1984). Photosynthesis was barely above the light compensation point under diffuse illumination, emphasizing the extreme importance of sunflecks for daily carbon gain in this species (Lawrence, 1984).

Field studies of gas exchange in seedlings of *Acer saccharum* in the understorey of a mixed hardwood forest in Michigan showed that, during summer, approximately 35% of daily carbon gain occurred during sunflecks exceeding 50 $\mu mol\ m^{-2}\ s^{-1}$ (Weber *et al.*, 1985; Table 6). Fluxes of diffuse and direct radiation in this forest were an order of magnitude greater than those in the Hawaiian forest studied by Pearcy (1983) and Pearcy and Calkin (1983). During the day, CO_2 assimilation closely tracked variation in PFD. Weber *et al.* (1985) calculated that daily carbon gain of *Acer* seedlings would be reduced by about 5% in the absence of sunflecks.

Pearcy (1987) measured the diurnal pattern of CO_2 assimilation of seedlings of *Argyrodendron peralatum* in the understorey of a Queensland rainforest (Fig. 3). The daily photon flux was about 3% of that received by leaves in the canopy, whereas daily carbon gain was nearly 10% of that of canopy leaves. Photosynthesis during sunflecks contributed 32% of the daily carbon gain (Table 6).

2. *Computer Simulations*

When it has not been feasible to measure diurnal variation in CO_2 assimilation in natural understorey habitats, the importance of sunflecks for daily carbon gain has been investigated using computer simulations based on actual or modelled photosynthetic responses, together with data on light variation. Gross (1982) used this approach to estimate the importance of sunflecks for carbon gain in *Fragaria virginiana.* Based on studies of dynamic photosynthetic responses to step-changes in PFD (Gross and Chabot, 1979; see below), he showed that sunflecks often made a significant contribution to carbon gain. When the canopy was in full leaf, appoximately 50% of daily carbon gain was attributed to sunflecks. The closer the mean low light level was to the light compensation point, the more important sunflecks became to daily carbon gain. Single sunflecks of similar duration to those observed in temperate deciduous forest were found to contribute a rather small percentage of daily carbon uptake if they occurred infrequently, even if the photon flux due to the sunfleck was quite high (Gross, 1982).

Chazdon (1986) used computer simulations to estimate the significance of light variation to total carbon gain in three species of rainforest understorey palms. These simulations were based on steady-state CO_2 assimilation rates measured in the field and daily patterns of PFD measured in a wide range of understorey and gap microsites in a Costa Rican lowland rainforest. Daily carbon gain was positive under most understorey conditions, even in the absence of sunflecks. When midday diffuse PFD was below 5 $\mu mol\ m^{-2}\ s^{-1}$, however, sunflecks provided the light energy needed to maintain positive carbon balance. Although daily carbon gain was linearly related to daily PFD when sunflecks were absent or infrequent, daily carbon gain was not a simple function of daily PFD when sunflecks contributed over 50% of daily photon flux. In the latter case, daily carbon gain was 33–35% lower than when the same daily PFD was achieved through higher flux densities of diffuse light. On days with sunny periods, simulations indicated that sun-

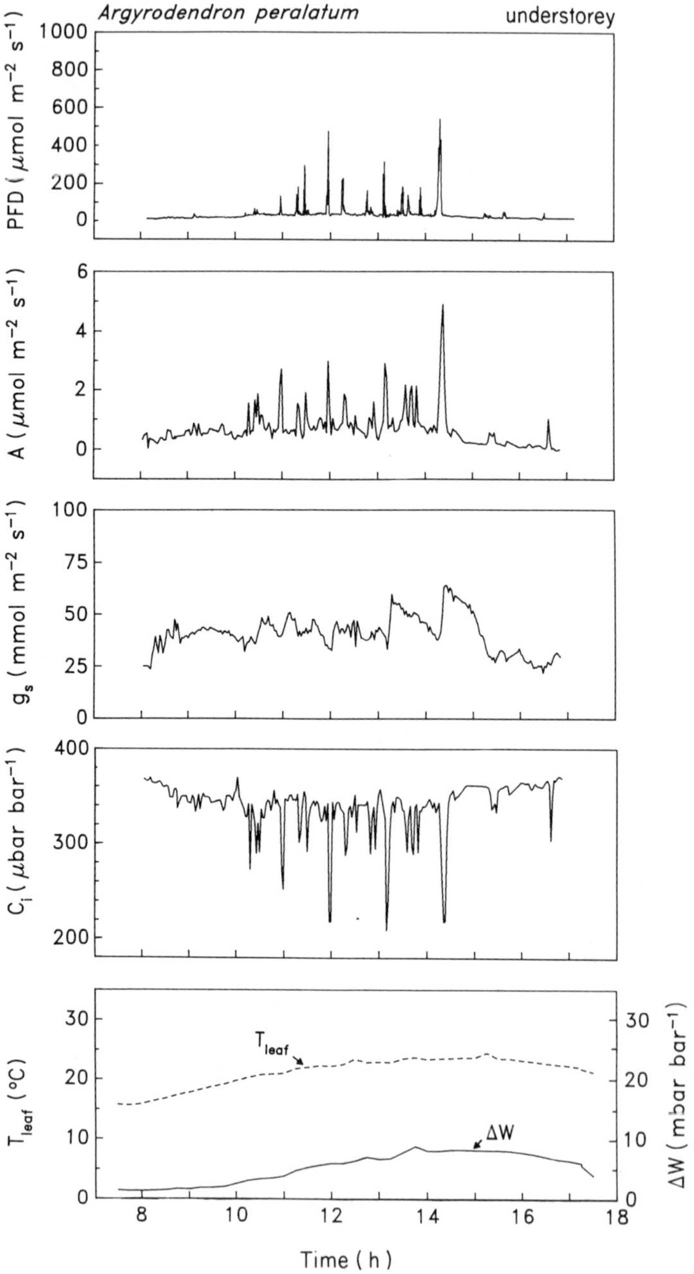

Fig. 3. Daily course of photon flux density (PFD; μmol m^{-2} s^{-1}), CO_2 assimilation, stomatal conductance, internal CO_2 pressure, and leaf temperature for a seedling of *Argyrodendron peralatum* in a Queensland rainforest. From Pearcy (1987), with permission of the publisher.

flecks accounted for 15–60% of daily carbon gain. In accordance with the findings of Gross (1982), the percentage of daily carbon gain contributed by sunflecks increased as midday diffuse PFD decreased (Chazdon, 1986).

B. Determinants of Sunfleck Utilization

Sunflecks are a critical resource for forest understorey plants. Although understorey plants are usually able to maintain a positive carbon balance in the absence of sunflecks, light is the major environmental factor limiting growth and reproduction in deeply-shaded understorey environments. We might therefore expect understorey plants to utilize sunflecks efficiently. Analyses of steady-state light responses can offer limited insight into patterns of sunfleck utilization. Based on determinations of a leaf's light compensation point and light saturation point, it is possible to predict photosynthetic responses to known intensities of diffuse and direct radiation (Harbinson and Woodward, 1984). Estimates of photosynthetic responses during sunflecks based on steady-state rates can be grossly misleading, however, particularly when averaged light measurements are used (Gross, 1982; 1984; but see McCree and Loomis, 1984). Steady-state photosynthetic responses simply do not apply when light is fluctuating rapidly. Ultimately, studies of transient photosynthetic responses are needed to determine how sunflecks of different frequency, duration, intensity, and temporal distribution are utilized by leaves (see review by Pearcy, 1988a).

For the present discussion, it is useful to distinguish between two types of transient photosynthetic responses; induction reponses and photosynthetic dynamics. Induction responses involve relatively slow (from several minutes to over an hour) increase in CO_2 assimilation in leaves equilibrated in darkness or low light following a sudden step-wise increase in light (Rabinowitch, 1956; Marks and Taylor, 1978; Pearcy *et al.*, 1985; Chazdon and Pearcy, 1986a; Kirschbaum and Pearcy, 1988). These induction responses affect the "readiness" of a leaf to respond to short-term light fluctuations (Chazdon and Pearcy, 1986a,b). Photosynthetic dynamics, on the other hand, involve short-term photosynthetic responses (seconds long) to light fluctuations, such as sunflecks (Gross, 1986). This distinction does not imply that dynamic responses are independent of the state of induction of a leaf; on the contrary, dynamic responses are highly dependent on whether or not a leaf has undergone induction (Chazdon and Pearcy, 1986a,b and see below). Rather, induction and photosynthetic dynamics are both transient responses that occur on different time-scales. Transient photosynthetic responses are best understood as dynamic responses superimposed on the background of leaf induction state.

1. Photosynthetic Induction

A typical induction response is shown for *Alocasia macrorrhiza* in Fig. 4(a). Following a long period at 10 μmol m^{-2} s^{-1}, PFD was suddenly increased to 400 μmol m^{-2} s^{-1}. In this example, it took 35 min for the leaf to reach the steady-state rate at saturating PFD (Chazdon and Pearcy, 1986a). Induction in rainforest understorey species may take from 20 to over 60 min (Pearcy *et al.*, 1985; Chazdon and Pearcy, 1986a). Although the time period for induction may vary somewhat between species, the induction requirement for maximum photosynthetic rates is an intrinsic feature of photosynthesis in all plants (Rabinowitch, 1956; Walker, 1981).

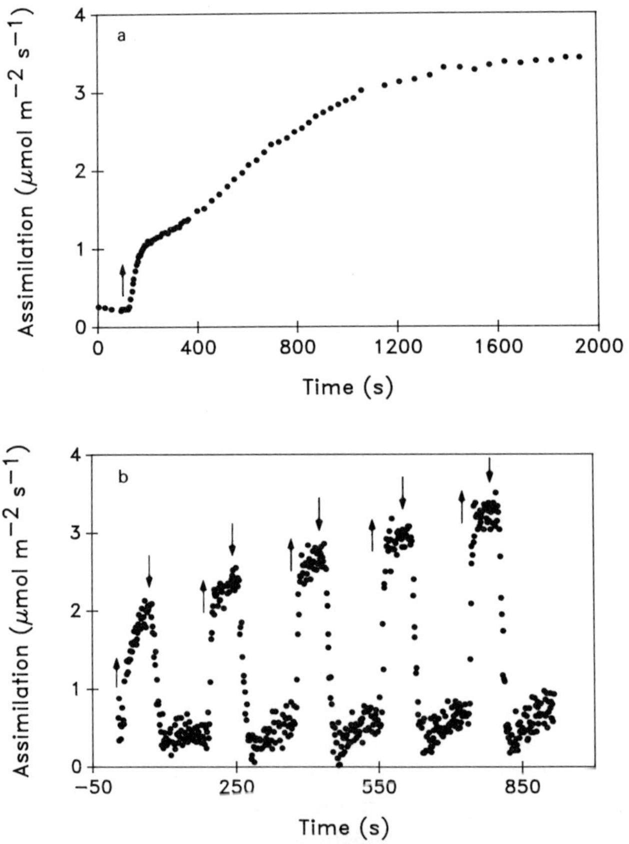

Fig. 4. The time course of photosynthetic induction in *Alocasia macrorrhiza* grown in low light. (a) Induction during a step-change in photon flux density from 10 to 400 μmol m^{-2} s^{-1} (b) Induction during 60-s lightflecks (400 μmol m^{-2} s^{-1}) separated by 2 min of low light (10 μmol m^{-2} s^{-1}). From Chazdon and Pearcy (1986a), with permission of the publisher.

Recent studies of photosynthetic induction in the rainforest understorey species *Alocasia macrorrhiza* (Chazdon and Pearcy, 1986a) indicate that, during the first 5–10 min following the light increase, increases in CO_2 assimilation are primarily limited by biochemical factors, such as the activity of the carboxylating enzyme ribulose-1,5-bisphosphate carboxylase (RuBP-Case). Generally, photosynthetic limitations by CO_2 diffusion into the mesophyll become more important during the later phases of induction. Studies of induction in other rainforest understorey species confirm the importance of biochemical limitations during the early phases (Pearcy *et al.*, 1985; Chazdon and Pearcy, 1986a). The nature and extent of stomatal limitation during induction may vary among species and in different environmental conditions. In the C_4 species *Euphorbia forbesii*, internal CO_2 pressures decreased from 300 to a minimum of 60–80 μbar following the light increase, suggesting that stomatal limitation may play a greater role in the induction response of this species (Pearcy *et al.*, 1985).

The degree to which stomatal conductance limits CO_2 assimilation immediately following the light increase may also vary according to initial stomatal conductance. Kirschbaum and Pearcy (1988) found that, in low-light grown plants of *Alocasia*, when initial conductances were low (resulting in internal CO_2 pressures below 100 μbar), increases in internal CO_2 pressure were largely responsible for increases in CO_2 assimilation during the first 10 min following the light increase. When stomatal conductances are low, a failure to partition transpiration between stomatal and cuticular paths can result in a significant overestimation of internal CO_2 pressure (Kirschbaum and Pearcy, 1988).

Field studies initially suggested, and laboratory studies later confirmed, that constant high light is not required to effect induction. Studies of the Hawaiian understorey species *Euphorbia forbesii* and *Claoxylon sandwicense* showed that, during a sequence of artificial sunflecks (lightflecks) 1 min in length, maximum CO_2 assimilation rates increased with successive lightflecks (Pearcy *et al.*, 1985). In the Australian rainforest species *Alocasia macrorrhiza* and *Toona australis*, leaf induction state increased 2- to 3-fold during a sequence of five 30 or 60 s lightfleck separated by 2 min of low light (Chazdon and Pearcy, 1986a; Fig. 4(b)). For *Alocasia*, low PFD and sunfleck PFD were 10 and 500 μmol m^{-2} s^{-1}, respectively, whereas for *Toona*, low PFD and sunfleck PFD were 15 and 1200 μmol m^{-2} s^{-1}. The rate of induction during the 60 s lightfleck sequence was not substantially different from the rate observed during constant illumination at the same high-light level (Chazdon and Pearcy, 1986a). Thus, sunflecks that occur early in a series can increase leaf "readiness" to respond to subsequent sunflecks.

Once leaves have undergone induction, and are returned to low light, they do not remain indefinitely in a state of photosynthetic "readiness". In low-light grown leaves of *Alocasia*, the loss of induction in low light followed a negative exponential function with a half-time of approximately 25 min (Chazdon and Pearcy, 1986a). Complete induction loss required over 60 min of exposure to constant low light. The rate of induction loss in high-light grown leaves of *Toona australis* was considerably faster (Chazdon and Pearcy, 1986a).

2. *Photosynthetic Dynamics and Carbon Gain During Sunflecks*

Few studies have examined photosynthetic dynamics of forest understorey species. Because of the difficulties in making accurate, rapid measurements and in carefully controlling and replicating experimental conditions, most of these studies have been carried out under laboratory conditions. Several studies on photosynthetic dynamics during high frequencies of flashing light have shown that photosynthetic rates are often higher than rates predicted from steady-state responses (Rabinowitch, 1956; McCree and Loomis, 1969; Pollard, 1970; Kriedemann *et al.*, 1973). These studies, however, are insufficient to assess photosynthetic responses to sunflecks in understorey plants, because they were not conducted using shade-grown understorey species, and they did not consider photosynthetic responses to measured frequencies of light variation in natural habitats.

Photosynthetic responses to sudden increases and decreases in PFD similar to those during naturally occurring sunflecks were studied in *Fragaria virginiana*, the common wild strawberry of the eastern USA (Gross and Chabot, 1979). Leaves grown at low- and high-light levels were subjected to step-increases and decreases in PFD. The sudden change in PFD was always followed by a time-lag before a change in CO_2 assimilation was first observed. Time-lags of about 10 s were observed over all the PFD increases and decreases, but were somewhat longer when PFD was initially very low. Following the time-lag, leaves responded rapidly to decreases in PFD and more slowly to increases in PFD. For light increases, the time constants (time required to reach 65% of the increment to the new steady-state rate) were less then 60 s, whereas time constants for light decreases were from 1 to 5 s.

Pearcy and Calkin (1983) investigated photosynthetic dynamics during a step-change from shade light to 700 μmol m^{-2} s^{-1} in *Euphorbia forbesii* and *Claoxylon sandwicense*. Increases in CO_2 uptake briefly lagged behind the light change and then increased rapidly to the new steady-state rate. Responses were virtually complete within 60 s. When light decreased to initial levels the rate of response was similar, except for a post-illumination CO_2 release in *Claoxylon*, which is characteristic of C_3 species.

Laboratory studies on photosynthetic dynamics of *Euphorbia* and *Claoxylon* by Pearcy *et al.* (1985) showed that, following induction, carbon gain during lightflecks depended strongly on lightfleck length. When lightflecks were less than 40 s long, carbon gain was 20–80% higher than that estimated from steady-state photosynthetic rates during the high- and low-light periods. Photosynthetic responses to 5 s intervals of high- and low-light also showed substantial enhancement when light-saturating PFD was used. Enhancement of carbon gain during brief lightflecks and flashing light was attributed to post-illumination CO_2 fixation, which contributed a large proportion of total carbon gain during brief sunflecks, but only a small proportion during long sunflecks. Pearcy *et al.* (1985) further observed that photosynthetic responses to lightflecks were strongly influenced by whether or not leaves had been previously exposed to high light. Following a 2 h exposure to low light, the carbon gain achieved during a 1 min lightfleck was only 44.5% and 47.3% of expected carbon gain for *Euphorbia* and *Claoxylon*, respectively. They attributed the lower carbon gain to inactivation of the photosynthetic apparatus (induction loss) during the long exposure to low light.

Studies by Chazdon and Pearcy (1986b) and Pearcy *et al.* (1987a) also showed that in low-light grown *Alocasia*, carbon gain and photosynthetic efficiency during lightflecks were greatly affected by leaf induction state, lightfleck length, and lightfleck PFD. Net carbon gain during lightflecks at saturating PFD (530 μmol m^{-2} s^{-1}) increased with the steady-state PFD applied before the lightfleck sequence (Fig. 5(a)). Increases in lightfleck PFD from 25 to 120 μmol m^{-2} s^{-1} also led to higher carbon gain during lightflecks presented following induction (Fig. 5(b)). Net carbon gain attributed to the lightfleck increased with lightfleck length, but the efficiency of light use decreased with lightfleck length (Fig. 5). An index of light utilization efficiency during lightflecks was calculated by comparing integrated carbon gain during a lightfleck with predicted carbon gain based on steady-state rates at the low- and high-light levels (Chazdon and Pearcy, 1986b). Efficiency following a 2 h period at low light ranged from 110% for 5-s lightflecks to 60% for 40-s lightflecks. Following induction, these efficiencies increased to 160% and 100%, respectively (Pearcy *et al.*, 1987a). Regardless of leaf induction state, however, photosynthetic efficiency decreased with lightfleck length. Similar responses to lightflecks were observed in leaves of *Toona australis* and *Alocasia* grown in high light, but efficiency was almost always below 100% (Chazdon and Pearcy, 1986b; Pearcy *et al.*, 1987a).

Enhancement of carbon gain during sunflecks does not require that leaves receive light-saturating PFD, especially when leaves have not undergone photosynthetic induction. Photosynthetic efficiency during 5-s lightflecks before induction exceeded 100% for shade-grown *Alocasia*, even when sunfleck PFD was below 100 μmol m^{-2} s^{-1} (Chazdon and Pearcy, 1986b). Following induction, however, the degree of enhancement during 5-s lightflecks increased as the PFD of lightflecks increased. Recent studies by

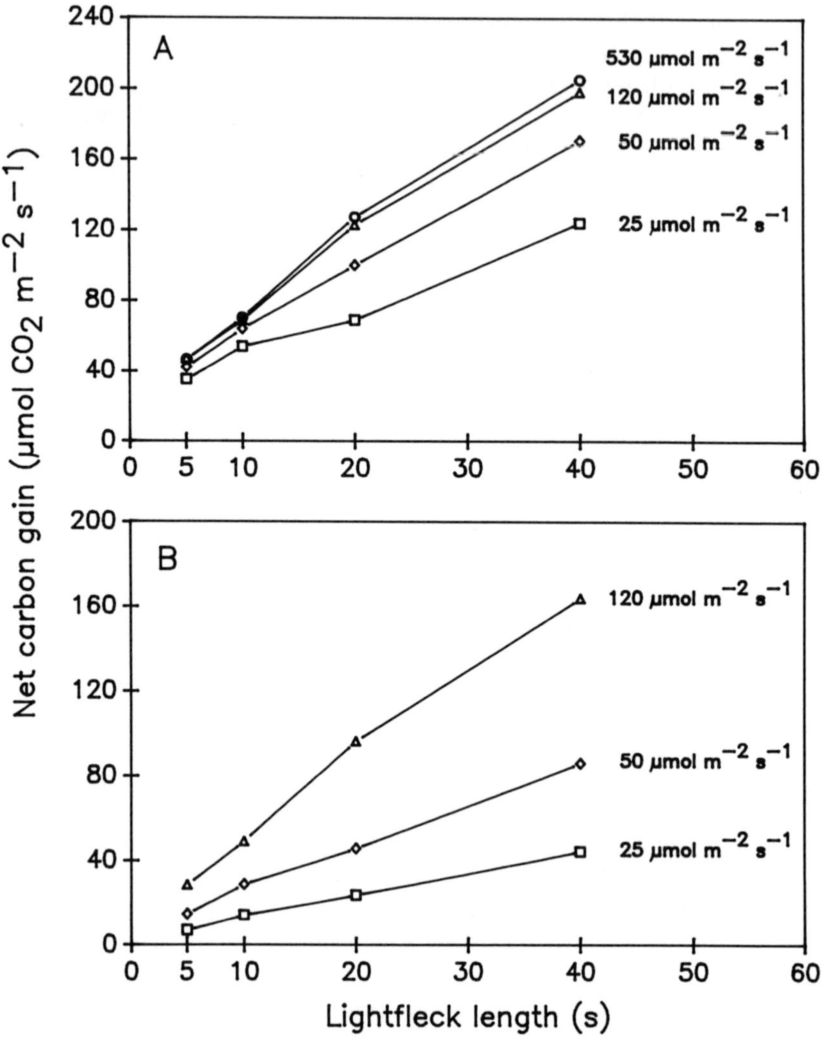

Fig. 5. Net carbon gain (μmol CO_2 m^{-2} s^{-1}) as a function of lightfleck length for low-light grown leaves of *Alocasia macrorrhiza*. (a) Responses to lightflecks at 530 μmol m^{-2} s^{-1} presented following equilibration of leaves at each of four different light levels. (b) Responses to lightflecks of different PFD in leaves following induction. In all cases, low-light PFD was 10 μmol m^{-2} s^{-1}. Data are from Chazdon and Pearcy (1986b).

Sharkey *et al.* (1986) indicate that the observed build-up of pools of triose-phosphates during 5-s lightflecks following induction in *Alocasia* was sufficient to account for the enhancement of carbon gain due to post-illumination CO_2 fixation. They hypothesize that extensive grana stacking, large intrathylakoid space, and high levels of chlorophyll in low-light grown plants enable significant post-illumination ATP synthesis, which is required to produce ribulose-1,5-bisphosphate (RuBP) from accumulated triose-phosphates. This hypothesis is consistent with observed differences between predicted and observed photosynthesis during brief sunflecks at different PFD (Chazdon and Pearcy, 1986b; Pearcy *et al.*, 1987a).

3. Stomatal Responses to Sunflecks

Patterns of stomatal opening and closure during fluctuating light conditions can be affected by endogenous rhythms (Gregory and Pearse, 1937), leaf water status (Davies and Kozlowski, 1975), the length of the previous dark period (Brun, 1972), PFD during high-light periods (Woods and Turner, 1971), and leaf induction state (Chazdon and Pearcy, 1986a). In general, stomatal responses tend to lag behind changes in irradiance and CO_2 uptake (Pearcy *et al.*, 1985). Woods and Turner (1971) studied the time required to reach equilibrium stomatal conductance following a light change in four tree species. Stomatal opening was always faster than closure, regardless of the magnitude of the light change. Stomatal opening took from 3 to 20 min, whereas closure required from 12 to 36 min. In three of the four species, stomatal opening and closing was faster when the magnitude of the light change was greater. The most shade-tolerant species, *Fagus grandifolia*, had the fastest rates of stomatal opening and closure. In a comparative study of seedlings of six hardwood species, Davies and Kozlowski (1975) also found that the three most shade-tolerant species had the fastest stomatal responses to increases in irradiance. These studies suggest that relatively rapid stomatal responses to light fluctuations served to maximize photosynthesis during sunfleck periods.

Field measurements of diurnal variation in stomatal conductance of forest understorey species show that, over long time periods, stomatal conductance follows changes in PFD (Björkman *et al.*, 1972b; Young and Smith, 1979, 1983; Elias, 1983; Pearcy and Calkin, 1983; Masarovicova and Elias, 1986; Pearcy, 1987). Stomatal responses to sunflecks, however, are often considerably slower than responses in CO_2 assimilation (Knapp and Smith, 1987; Weber *et al.*, 1985). In seedlings of *Argyrodendron* in a Queensland forest understorey, Pearcy (1987) noted that peak stomatal conductances were not reached until after a sunfleck had passed. During sunflecks, internal CO_2 pressures decreased from 240 to 200 μbar. In the subalpine understorey species *Arnica cordifolia*, decreases in stomatal conductance during simulated cloud cover lagged behind photosynthetic decreases by about 4 min (Knapp and Smith, 1987). Not all understorey species exhibit changes in stomatal conductance during sunflecks, however. In a northern hardwood forest, stomatal conductances of leaves of *Viola blanda* exposed to 10 min of saturating PFD did not differ significantly from those of shaded leaves (Curtis and Kincaid, 1984).

Although stomatal conductance often fluctuates with PFD during the day, in many understorey species stomatal conductance tends to remain high under low-light conditions (Mooney *et al.*, 1983). During periods of diffuse light, stomatal conductance in *Argyrodendron* remained above 25 mmol m^{-2} s^{-1} (Pearcy, 1987; Fig. 3). Measurements of stomatal conductance in other forest understorey plants confirm that, even under extremely low diffuse PFD, stomata typically remain open (Björkman *et al.*, 1972b; Mooney *et al.*, 1983; Pearcy and Calkin, 1983; Chazdon, 1984). Because of the apparent inability of stomata to respond as rapidly as do CO_2 assimilation pathways to light fluctuations, photosynthetic utilization of sunflecks may be impeded if stomata are closed during low-light conditions. Therefore, a high stomatal conductance relative to CO_2 assimilation rate under shaded conditions may serve to ensure that photosynthesis during sunflecks is not limited by internal CO_2 pressures (Mooney *et al.*, 1983; Pearcy, 1983; Fig. 3). A high ratio of internal CO_2 pressure to ambient CO_2 pressure could also be advantageous because of the resulting increased quantum yield for CO_2 uptake (Pearcy, 1987). Pearcy (1987) calculated that, in a C_3 plant, maintaining internal CO_2 pressures at 320 rather than 220 μbar at a leaf temperature of 25°C should result in a 14% increase in quantum yield. An interesting exception to this case is the Hawaiian C_4 species *Euphorbia forbesii*. In this species, stomatal conductance was very low under diffuse PFD, but increased rapidly in reponse to sunflecks. Nevertheless, stomatal conductance was found to limit photosynthetic responses during at least some sunflecks (Pearcy and Calkin, 1983).

C. Constraints on Sunfleck Utilization in Understorey Habitats

Photosynthetic utilization of sunflecks may be constrained by a variety of factors including loss of induction during low-light periods, restricted stomatal opening to conserve water, photoinhibition, wilting, and high leaf temperatures during prolonged high-light periods. Under field conditions, photosynthesis is influenced by a combination of environmental factors, of which PFD is only one. We are far from understanding how these different environmental factors affect carbon gain during sunflecks and during the intervening low-light periods. Although we have learned a great deal about constraints on carbon gain in natural understorey habitats (Pearcy *et al.*,

1987b), far less is known about how these constraints operate under fluctuating environmental conditions, such as during sunflecks. In this section, I examine what is known about these constraints and their relative importance for carbon gain during sunflecks in plants from different understorey habitats.

1. Leaf Induction State

Laboratory studies suggest that, when sunflecks are few and far between, the efficiency of light utilization during some sunflecks may be partially constrained by leaf induction state (Chazdon and Pearcy, 1986a,b). Field studies by Pearcy (1987), however, show little evidence that induction limited photosynthesis of *Argyrodendron* seedlings during sunflecks in a Queensland understorey. In this case, sunflecks were distributed throughout the day, such that leaves maintained a high induction state (Fig. 3). Daily PFD measurements for eight sensors in this site showed that 70% of the sunflecks occurred within one minute of the preceding sunfleck, and only 5% of the sunflecks were preceded by low-light periods of an hour or longer (Pearcy, 1988a). The temporal patterning of sunflecks in any particular microsite will influence the degree to which leaves remain in a state of "readiness" to respond to sunflecks. Sunflecks are often clumped in their distribution (Pearcy, 1983, 1988a), creating the potential for rapid leaf induction during a series of closely spaced sunflecks. On days when sunshine is interrupted by overcast or cloudy skies for several hours, the first sunflecks hitting a leaf may not yield as much carbon gain as similar sunflecks occurring later in the sequence. Although shorter sunflecks are utilized with greater efficiency, longer sunflecks are more effective for photosynthetic induction. After a series of five 30-s lightflecks separated by 2 min of low light, leaf induction state of *Alocasia* was only 48% of that measured for fully-induced leaves, whereas during a similar sequence using 60-s lightflecks, relative induction state reached 75% (Chazdon and Pearcy, 1986a).

2. Water-use Efficiency

In forest understorey conditions, light is usually a more important limiting factor than water stress. As discussed above, stomatal limitations to CO_2 uptake are rarely observed in field studies of forest understorey species. Under drought conditions, or during prolonged sunfleck exposures, however, regulation of stomatal opening to conserve water may impose limitations on CO_2 uptake during sunflecks. Stomatal responses to light increases in the sun-loving species *Pelargonium* slowed when plants were subjected to water stress, and responses to light decreases became faster (Willis and Balasubramaniam, 1968). Similar changes in stomatal responses during water stress were observed in seedlings of six hardwood tree species (Davies and Kozlowski, 1975). During a summer drought in a temperate deciduous forest, rates of CO_2 assimilation decreased in *Impatiens parviflora* and *Aegopodium podagraria* (Masarovicova and Elias, 1986).

Studies of stomatal responses to light fluctuations simulating natural cloud cover show that the stomata of the subalpine understorey species *Arnica cordifolia* responded rapidly to light increases and decreases in a manner similar to CO_2 assimilation (Knapp and Smith, 1987). This species routinely undergoes severe wilting during extended periods of full sunlight (Young and Smith, 1979). Coupling of stomatal opening and CO_2 assimilation resulted in a 30% increase in water-use efficiency compared to values calculated with a constant high-light rate of stomatal conductance (Knapp and Smith, 1987). Thus, although water-use efficiency of this species does not remain constant under fluctuating light conditions, coupled stomatal and photosynthetic responses ensure that water-use efficiency is high during sunfleck exposures.

Water-use efficiency of *Arnica latifolia* increased seven-fold on cloudy days compared to clear days (Young and Smith, 1983). Furthermore, daily carbon gain was 37% higher on cloudy days. On clear days, long sunflecks (sunpatches) led to increased leaf temperatures and leaf-to-air water vapor differences, higher transpiration rates, and lower water-use efficiency. Although the physiological basis for the decreased carbon gain on clear days is not known, lower xylem pressure potentials on clear days may cause decreases in photosynthesis in *A. latifolia*. Studies of *A. cordifolia* by Young and Smith (1979) showed that increased water loss may indirectly reduce photosynthesis during prolonged sunflecks through a decrease in xylem pressure potential followed by decreased stomatal conductance.

Studies of seasonal variation in stomatal responses to light in Douglas fir saplings show that, during autumn, winter and early spring, stomatal conductance was weakly related to PFD (Meinzer, 1982). During these periods of plentiful soil moisture, maximization of carbon gain may be more important than regulation of water-use efficiency. In the summer, when water conservation was most important, stomatal opening was tightly coupled with PFD, even under low-light conditions. Under field conditions, sudden changes in both PFD and vapor-pressure deficit (VPD) may occur simultaneously during a sunfleck. Dynamic stomatal responses of Douglas fir saplings to step-changes in VPD were in the opposite direction to that predicted: step-increases in VPD resulted in increases in stomatal conductance. Although these VPD responses resulted in increased rates of water loss, they served to enhance the speed of stomatal opening during brief sunflecks (Meinzer, 1982).

In the Mexican understorey species *Piper hispidum*, stomatal responses to humidity were very strong, whereas responses to PFD were weak (Mooney *et al.*, 1983). When relative humidity decreased from 95 to 85%, stomatal conductance decreased by over 50%. This response serves to reduce water loss when evaporative demand is high, and to maximize photosynthesis when evaporative demand is low. Thus, it is unlikely that stomatal conductance restricts CO_2 assimilation during sunflecks under humid conditions. During the dry season, however, relative humidity often drops below 90% in the tropical rainforest understorey (Fetcher *et al.*, 1985). Under these conditions, decreased stomatal conductances in this species may impose a strong limitation on carbon gain during sunflecks.

3. *Photoinhibition*

During sunflecks, PFD may suddenly increase 200-fold over diffuse light levels. If light intensities during sunflecks exceed light saturation for long periods, photoinhibition may occur. Shade-grown plants are highly susceptible to photoinhibition because they have low light-saturated photosynthetic capacities (Boardman, 1977; Björkman, 1981). In the shade fern *Pteris cretica*, exposure of fronds to PFD greater than 300 μmol m^{-2} s^{-1} caused a continuous decrease in CO_2 assimilation with time (Hariri and Prioul, 1978).

Despite a number of laboratory studies on photoinhibition in shade plants (Kozlowski, 1957; Björkman *et al.*, 1972a; Powles and Thorne, 1981), few studies have documented photoinhibition during sunflecks under field conditions. The understorey herb *Oxalis oregana*, common in the deeply shaded redwood forests of northern California, exhibits leaflet movements in response to sunflecks when PFD exceeds 300–400 μmol m^{-2} s^{-1} (Björkman and Powles, 1981). Leaflet movement was sensitive only to wavelengths between 375 and 490 nm, which is characteristic of blue-light-induced phototropic movements. These leaflet movements significantly reduced PFD incident on the leaflet surface; PFD decreased from an initial value of 1590 to 295 μmol m^{-2} s^{-1} for two leaflets and 492 μmol m^{-2} s^{-1} for the third leaflet, based on calculations of leaf angle and azimuth (Powles and Björkman, 1981). When leaflet movement was restrained during an 18-min sunfleck having an average PFD of 1500 μmol m^{-2} s^{-1}, CO_2 assimilation in diffuse light following the sunfleck was reduced by 30%. Decreases in photosynthesis following the sunfleck could not be attributed to stomatal conductance, because both stomatal conductance and internal CO_2 pressure were higher than before the sunfleck exposure. Furthermore, variable fluorescence significantly declined when *Oxalis* leaves were exposed to intense PFD similar to natural sunflecks. This decrease was associated with photoinhibitory inactivation of the photosystem II reaction centers. In the same understorey site, Powles and Björkman (1981) observed that leaves of *Trillium ovatum* growing next to *Oxalis* also suffered a 40% reduction in variable fluorescence following exposure to an intense sunfleck. Leaves of *Trillium* are incapable of the protective movements exhibited by *Oxalis*.

Measurements of CO_2 assimilation of *Oxalis* showed that the decrease in PFD resulting from leaflet movements did not result in lower photosynthetic rates during the sunfleck. Leaves of *Oxalis* reached light saturation at relatively low PFD; at 100 μmol m^{-2} s^{-1} leaves attained 90% of their light saturated rate. Leaflet movements caused by sunflecks of intermediate PFD appeared to adjust leaf angles so that PFD incident on leaflets was sufficiently high to permit light saturation, but sufficiently low to avoid photoinhibition (Powles and Björkman, 1981). Species that do not possess the protective mechanisms against photoinhibition shown by *Oxalis* may suffer reductions in CO_2 assimilation following intense sunflecks. In the absence of more field observation, however, the role of photoinhibition during sunflecks in restricting subsequent utilization and direct PFD remains unknown.

4. Leaf Temperature and Water Relations

Leaf temperatures during sunflecks may affect photosynthesis directly, through temperature dependence of enzymatic reactions, or indirectly, through effects on stomatal conductance and water relations. During an intense sunfleck, leaf temperature may rise as much as 18°C above air temperature (Ellenberg, 1963; Rackham, 1975; Young and Smith, 1979). In some cases, these temperatures may cause permanent heat damage and leaf necrosis, as has been observed in *Mercurialis perennis* in a deciduous woodland near Cambridge, UK (Rackham, 1975). Normally, leaf temperatures on the forest floor are slightly below air temperature and are unlikely to exceed air temperature, even during small sunflecks (Rackham, 1975; Chiariello, 1984). In the subalpine understorey species *Heracleum lanatum*, temperatures of shaded leaves were 2–5°C below air temperature, whereas sunlit leaves were 3–5°C above air temperature (Young, 1985). Although leaf temperatures may increase during sunflecks, thermal damage resulting from an unusually intense sunfleck is a rare phenomenon in the understorey.

In a Hawaiian understorey, leaf temperatures of *Euphorbia forbesii* and *Claoxylon sandwicense* may increase 5°C during the first 50 s of typical sunflecks, and may reach 30°C during long sunflecks (Robichaux and Pearcy, 1980). Leaf temperatures during sunflecks are closer to the photosynthetic temperature optimum of *Euphorbia*, however, conferring a carbon gain advantage over *Claoxylon* during sunflecks.

Changes in leaf temperature, water relations, and shoot extension were followed in *Circaea lutetiana* during a 7-min sunfleck in a *Fagus sylvatica* woodland in England (Woodward, 1981). Leaf temperature and transpiration rose rapidly during the first minute of exposure, but transpiration subsequently declined following stomatal closure. Convective (sensible) heat transfer increased and remained high during the entire 7-min period. Because of stomatal closure, radiation could not be dissipated by latent heat transfer through transpiration. Leaf water potential and pressure potential both declined rapidly for the first 2 min of the sunfleck and then remained constant for the remainder of the sunfleck. Consequently, shoot extension rates declined significantly during the sunfleck. Observations of 24 sunflecks over a 2-d period showed that in 75% of the cases, the post-sunfleck rate of shoot extension was less than the pre-sunfleck rate. Changes in photosynthetic rates during the sunfleck were then calculated based on modelled responses of CO_2 uptake to flux resistances to CO_2 transfer, assuming an instantaneous change in photosynthetic rate following the light increase. Predicted photosynthesis during the sunfleck reached 90% of the maximum light-saturated rate, a substantial increase over rates in diffuse light.

In the understorey species *Arnica cordifolia* and *A. latifolia*, exposures to long-term sunflecks resulted in elevated leaf temperatures and transpiration rates, which often led to various degrees of wilting (Young and Smith, 1979). Microhabitats occupied by *A. cordifolia* received more frequent, longer, more intense sunflecks than those occupied by *A. latifolia*. Leaf temperatures of *A. latifolia* were generally well below air temperatures, even during sunflecks, and xylem water potentials for *A. latifolia* remained much higher during most of the day than for *A. cordifolia*. Consequently, midday wilting occurred more frequently in *A. cordifolia*. *A. cordifolia*, however, maintained turgor following sunfleck exposures of up to 165 min, whereas *A. latifolia* permanently wilted after 90 min exposure (Young and Smith, 1979). Photosynthesis of *A. cordifolia* remained positive, even after plants had wilted (Young and Smith, 1980).

Similar responses were observed for six other understorey species from the same subalpine habitat (Smith, 1981). During sunflecks, decreases in xylem water potential led to midday wilting for four of the seven species studied. No stomatal closure was observed during sunflecks in any of the species, however. The increase in stomatal conductance and transpiration during sunflecks may be advantageous for two reasons. Higher rates of photosynthesis were permitted, and excessively high leaf temperatures were avoided. Following sunfleck periods, plants rapidly regained turgor and xylem water potential returned to pre-sunlit levels.

D. Sunfleck Regimes and Light Acclimation

Plants living in exposed habitats exhibit different photosynthetic properties than exhibited by those living in shaded conditions (Böhning and Burnside, 1956; Boardman, 1977; Björkman, 1981). Moreover, many plants have the capacity to shift photosynthetic responses following a change in growth conditions. These acclimatory responses are generally in a direction that improves growth under the new environmental conditions. Unlike short-term fluctuations in CO_2 assimilation during sunflecks, acclimatory changes occur on a time-scale from days to weeks (Gross, 1986; Table 1). Leaves at different stages of expansion may show differential abilities to acclimate to a change in light conditions (Pearce and Lee, 1969; Jurik *et al.*, 1979). In the understorey species *Fragaria virginiana*, acclimation potential was greatest during early stages of leaf expansion, and decreased as expansion was completed (Jurik *et al.*, 1979).

Typically, light acclimation of photosynthesis is measured by comparing steady-state light responses of plants grown under different light conditions (Björkman and Holmgren, 1963). Light conditions during growth are carefully controlled, and are usually maintained constant over the entire photoperiod. Some studies, however, have investigated acclimation of morphological and photosynthetic characteristics under conditions where peak light levels and photoperiods were varied to yield different integrated as well as instantaneous PFD (Nobel, 1976; Chabot *et al.*, 1979; Nobel and Hartsock, 1981). These studies elegantly demonstrated that photosynthesis and leaf structure were determined by integrated PFD, rather than by peak PFD.

Within the forest understorey, daily PFD is positively correlated with the total minutes of sunflecks received and with the relative contribution of sunflecks to daily PFD (R. L. Chazdon, C. B. Field, and R. W. Pearcy, unpublished data). The potential therefore exists for light acclimation in response to consistent microsite variation in sunfleck activity. As discussed previously, variation in sunfleck activity occurs on both temporal and spatial scales. Understorey microsites that receive few minutes of sunflecks during one month may receive significantly longer sunfleck exposures several months later. Similarly, some microsites within the forest understorey will tend to receive more PFD from sunflecks than others (Chazdon, 1986). To what extent does light acclimation occur *within* the understorey in relation to spatial or temporal variation in sunfleck activity?

1. Spatial Variation

To answer this question, comparisons of photosynthesis and leaf structure must be made within forest microsites that have demonstrably different sunfleck regimes. Such a study was done by Young and Smith (1980) on the subalpine understorey species *Arnica cordifolia*. This species exhibits considerable phenotypic plasticity in photosynthetic characteristics, leaf structure, and water relations. Sun and shade plants occur in relatively open and densely shaded areas, respectively. The sun plants received nearly twice as much energy and photon flux during sunfleck periods, and had photosynthetic capacities 2·5 times greater than shade plants. Sun plants also had greater stomatal conductances, higher light saturation points, higher photosynthetic temperature optima, greater water-use efficiency, smaller leaf area, thicker leaves, higher specific leaf mass, and less chlorophyll per dry mass compared with shade plants (Young and Smith, 1980).

Plants of *Aster acuminatus* grown under canopy gap (approximately 25% of full sun) and understorey conditions (approximately 3% of full sun) differed significantly in maximum photosynthetic rates measured during May, but not in July, when photosynthetic rates of both groups of plants declined (Pitelka and Curtis, 1986). Photosynthetic differences were even more pronounced when plants were grown under low- and high-light conditions in growth chambers.

An increasing number of laboratory studies indicate that, compared to plants from relatively open habitats, forest understorey plants exhibit less potential for light acclimation (Björkman, 1981; Bazzaz and Carlson, 1982; Langenheim *et al.*, 1984). In a comparison of six rainforest species in the genus *Piper*, Chazdon and Field (1987a) found that photosynthetic capacity showed little variation among leaves of understorey plants, despite high variation in light availability among leaf microsites. Plants in a nearby clearing exhibited considerable variation in photosynthetic capacity among leaves in relation to leaf light environment. Other studies, however, have shown that acclimation potential is not always clearly correlated with successional status or ecological conditions (Osmond, 1983; Fetcher *et al.*, 1987; Walters and Field, 1987).

In a study of light acclimation of tropical tree seedlings, Kwesiga *et al.* (1986) found that seedlings grown under light with a reduced red:far-red ratio had higher maximum rates of photosynthesis and higher quantum efficiency than seedlings grown under high red:far-red ratios. Integrated PFD was maintained constant. The low red:far-red ratio used in the experiment was similar to values measured in diffuse light in a Costa Rican rainforest understorey, although daily PFD was considerably higher (Chazdon and Fetcher, 1984b; Lee, 1987). In contrast, Corré (1983) found no significant difference in photosynthetic characteristics of herbaceous species grown under different red:far-red ratio.

Light acclimation has traditionally been measured by changes in steady-state photosynthetic responses. No field studies have addressed the issue of acclimatory responses in photosynthetic dynamics, such as photosynthetic efficiency during sunflecks. In laboratory investigations, high-light grown leaves of *Alocasia macrorrhiza* and *Toona australis* exhibited lower photosynthetic efficiency during sunflecks compared to low-light grown leaves of *Alocasia* (Chazdon and Pearcy, 1986b; Pearcy *et al.*, 1987a). Moreover, high- and low-light grown leaves of *Phaseolus* exhibited different capacities for post-illumination CO_2 fixation during sunflecks (Sharkey *et al.*, 1986). Until further study, we do not know whether microsite variation in sunfleck activity can effect dynamic photosynthetic responses.

2. Seasonal Variation

In deciduous forest understorey habitats, the availability of diffuse and direct radiation varies greatly over the year (see Subsection IV.B). Deciduous herbaceous species in these forests show a variety of growth patterns in relation to seasonal variation in light availability (Salisbury, 1916; Sparling, 1967). Photosynthetic light responses of these species strongly reflect the light conditions prevailing during leaf development (Sparling, 1967; Taylor and Pearcy, 1976; Kawano *et al.*, 1978; Hicks and Chabot, 1985; Masarovicova and Elias, 1986). Spring ephemerals develop their leaves under conditions of high light availability, before leaf expansion in the canopy. In April, leaves of the spring ephemeral *Erythronium americanum* exhibited photosynthetic light responses similar to herbs found in open habitats (Sparling, 1967; Taylor and Pearcy, 1976). As summer began, and the forest canopy closed, light-saturated photosynthetic rates and light saturation points of *Erythronium* declined to less than 50% of early spring values. Rates of dark respiration, however, remained high (Taylor and Pearcy, 1976). Decreases in CO_2 assimilation were correlated with decreased RuBP carboxylase activity (Taylor and Pearcy, 1976). Similar seasonal changes in photosynthetic responses and leaf biochemistry were found in the spring-active *Anemone raddeana* in a Japanese deciduous forest (Yoshie and Yoshida, 1987).

In another group of deciduous herbs, leaf expansion occurs during or after canopy closure, in early May. A representative of this group. *Trillium grandiflorum*, exhibited photosynthetic rates more typical of shade species. These rates also declined throughout the summer, and reached a minimum in July (Taylor and Pearcy, 1976). Species such as *Parthenocissus quinquefolia* and *Solidago flexicaulis*, in which leaf expansion occurred during midsummer, had the lowest light-saturated photosynthetic rates and RuBP carboxylase activity of the herb species studied (Taylor and Pearcy, 1976).

Seasonal variations in photosynthetic activity observed within the species studied were parallel to variations among species (Sparling, 1967; Taylor and Pearcy, 1976; Hicks and Chabot, 1985). Rates of dark respiration also decreased among species from spring to summer, enabling positive carbon balance to be maintained as light availability decreased (Taylor and Pearcy, 1976). These patterns suggest that deciduous understorey herbs possess a generalized acclimatory response to *decreasing* light availability from spring to fall. Leaves of *Fragaria virginiana* developed in low light were not able to acclimate completely to high light, whereas high-light grown leaves were able to acclimate completely to low light (Jurik *et al.*, 1979). These results suggest that light acclimation potential in all but spring ephemerals was restricted only when light availability increased dramatically (Fonteno and McWilliams, 1978). Based on measurements of seasonal variation in sunfleck activity and diffuse light penetration in deciduous forests, it is unlikely that daily PFD will increase throughout the summer growth season within the understorey, except in the event of a tree or branch fall.

Evergreen woodland herbs also exhibit variation in photosynthetic light responses, which parallel seasonal changes in light availability (Kawano *et al.*, 1983; Yoshie and Kawano, 1986). In *Pachysandra terminalis*, one-year-old leaves rapidly increased photosynthetic capacity following snow melt, and reached a yearly maximum in late April. Photosynthetic capacity then decreased to a minimum in July, when light availability in the understorey was lowest. In mid-August, photosynthetic capacity increased again, reaching a second peak in early October, when canopy leaves were senescing. Subsequently, photosynthetic capacity declined through the winter months. Current-year leaves appeared in early June, and photosynthetic capacity increased throughout the rest of the summer to a maximum in late September. Stomatal conductance varied in parallel throughout the year for all leaves (Yoshie and Kawano, 1986).

During early spring, over-wintering leaves of *Pachysandra* exhibited acclimation to high light availability, and did not show any evidence of photoinhibition (Yoshie and Kawano, 1986). These acclimatory responses, however, occur over a relatively long period of gradually increasing light availability from mid-March to early April. In contrast, laboratory studies of light acclimation expose plants to sudden, dramatic changes in irradiance. Results from these laboratory studies may therefore not apply to the gradual seasonal changes in PFD that occur in deciduous forest understories.

The semi-evergreen herb *Hepatica acutiloba* underwent major changes in chlorophyll and carotenoid content, and photosynthetic unit size in an oak forest understorey from April through October (Harvey, 1980). These changes were presumably associated with efficient light utilization as light availability decreased. In contrast, spring ephemerals in the same forest did not exhibit the same degree of plasticity in allocation to light-absorbing pigments when light availability decreased (Harvey, 1980).

E. Photosynthesis in Understorey Plants Revisited

Forest understorey plants exhibit a variety of photosynthetic characteristics that enable them to maintain positive carbon balance under extremely low PFD (Boardman, 1977; Björkman, 1981). Among these are low rates of dark respiration and high quantum efficiency under low light. Leaf anatomy, biochemistry, and chloroplast structure of understorey plants have also been interpreted as adaptations for maximizing the efficiency of light utilization at low irradiances under steady-state conditions (Björkman, 1968; Goodchild *et al.*, 1972; Nobel, 1976; Caemmerer and Farquhar, 1981).

As we begin to accumulate information on sunfleck activity and its importance for leaf carbon gain in understorey habitats, these photosynthetic characteristics may need to be interpreted more broadly. Recent studies suggest that growth in low light may effect changes in the regulation of pool sizes of Calvin cycle intermediates that ultimately control the efficiency of light use during transient sunflecks (Sharkey *et al.*, 1986). Extensive grana stacking in chloroplasts of low-light grown leaves may create a "capacitance" in the photosynthetic system that allows for transient build-up of a proton gradient for ATP formation following a sunfleck (Sharkey *et al.*, 1986; Pearcy *et al.*, 1987a). Both factors lead to substantial post-illumination CO_2 fixation, and enhancement of photosynthetic efficiency during brief sunflecks (Pearcy *et al.*, 1987a).

In many understorey habitats, the great majority of sunflecks are very brief, and of fairly low PFD (Figs 2, 3; see Section IV). It is during these brief sunflecks that light is most efficiently utilized. Furthermore, if the time interval between sunflecks is not too long, leaves of forest understorey plants will remain at fairly high induction states. There is no evidence that the photosynthetic characteristics responsible for efficient utilization of sunflecks impose any constraint on efficient utilization of low intensities of diffuse light. Some evidence does indicate, however, that photosynthetic adaptation to high irradiance imposes constraints on photosynthetic efficiency during sunflecks (Chazdon and Pearcy, 1986b; Sharkey *et al.*, 1986).

An accurate picture of the photosynthetic characteristics of forest understorey plants must incorporate transient photosynthetic responses. It is true that extremely low levels of diffuse light predominate over 75% of the time in forest understorey habitats. Sunflecks, despite their relatively low frequency, often contribute over 50% of the daily photon flux (Table 5). The ability to take advantage of these sunflecks, however brief and unpredictable, may prove to be at least as important for long-term carbon gain as maintaining positive carbon balance under diffuse light conditions.

VI. SEED GERMINATION, ESTABLISHMENT AND GROWTH IN RELATION TO SUNFLECK ACTIVITY

All phases of a plant's life-cycle may be influenced by light variation in understorey habitats. Sunfleck activity affects both the quantity and quality of light available within a microsite (Holmes and Smith, 1977a; Chazdon and Fetcher, 1984b; Lee, 1987). Light quality has well-characterized morphogenetic effects apart from the effects of light quantity on growth processes (Holmes and Smith, 1977b; Morgan and Smith, 1978; Corré, 1983). To the extent that growth is light-limited, relatively small differences in light availability may have significant effects on plant growth (Shirley, 1929; Blackman and Rutter, 1946; Hughes, 1966).

Relatively open understorey microsites will, on average, receive more direct PFD from sunflecks as well as higher levels of diffuse PFD (Young and Smith, 1979; Sasaki and Mori, 1981; Chazdon, 1986). When the incidence of diffuse and direct PFD are correlated, it is difficult, if not impossible, to determine whether microsite variation in sunfleck activity can account for differences in seedling establishment, growth, and distribution of plants in understorey microsites. Atkins *et al.* (1937) concluded that sunflecks were of relatively little concern for plant growth and distribution in deciduous forests because they were thought to contribute a relatively small percentage of total radiation at any particular site. Recent studies, however, indicate that, at least in some forest types, sunfleck activity is an excellent predictor of plant growth (Pearcy, 1983).

Responses to irradiance at the whole-plant level often do not correspond with predicted responses based on light responses of individual leaves. These relationships are complicated because of changes in leaf size, structure, and duration in response to changes in irradiance (Hughes, 1959; Blackman and Wilson, 1951, 1954). In some cases, increased production of leaf area may compensate for lower unit leaf rate in low-light conditions, such that relative growth rates remain constant or decrease only slightly (Blackman and Wilson, 1954; Hughes, 1966). Moreover, plant growth rates are also affected by self-shading and by plant size in ways that may be only indirectly related to light conditions. The light compensation point for an individual leaf, for example, may be significantly lower than that for an entire plant.

In this section, I review studies of the influence of sunfleck activity on seed germination, early establishment, and growth of forest understorey plants. Although these studies describe plant responses in relation to microsite variation in light availability in natural habitats, not all of them consider how these responses are specifically affected by differences in sunfleck activity among microsites. Despite a long history of research on light relations of forest understorey plants, the influence of sunflecks on plant establishment and growth in natural forest understorey sites remains a relatively unexplored area of research.

A. Seed Germination and Establishment in Understorey Habitats

Differences in light conditions between understorey and gap environments are known to affect seed germination in several tropical pioneer species (see review by Vázquez-Yánes and Orozco-Segovia, 1984). Far less is known about the extent to which smaller-scale light differences, such as those created by sunfleck activity, can affect seed germination of species that establish in the forest understorey. Comparisons of seed germination and subsequent establishment in a Costa Rican wet forest showed no significant difference in seed germination between high- and low-cover plots (Marquis *et al.*, 1986). In this study, understorey vegetation cover was removed in half of the plots, which decreased total cover from 90 to 85%.

Sunfleck activity in a Mexican rainforest was found to affect seed germination in the photoblastic pioneer species *Piper auritum* and *P. umbellatum* (Orozco-Segovia, 1986). Among seeds placed in three understorey microsites, percentage germination was significantly higher in the microsite that received longer sunflecks. In the understorey species *P. aequale*, percentage germination after one month was also significantly higher in the microsite with longer sunflecks, but after six months no significant differences were found among the three microsites (Orozco-Segovia, 1986). Not all species of *Piper* require red light for germination; some forest species exhibit high germination rates in darkness (Vázquez-Yánes, 1976). Many temperate forest species produce seeds that are capable of germination under low-light conditions (Angevine and Chabot, 1979).

B. Growth of Understorey Plants

1. Tree Seedlings and Saplings

Seedling growth of three dipterocarp species has been studied in relation to microsite variation in light availability. Within the study forest, Sasaki and Mori (1981) found that the frequency and intensity of sunflecks was correlated with measurements of diffuse irradiance. Growth of seedlings was closely correlated with diffuse light levels in a range below 20% of full sun. Within a given level of steady-state diffuse light, however, dipterocarp seedlings showed uniform growth, even though sunfleck incidence may have varied (Sasaki *et al.*, 1981). Although sunfleck activity was not measured in these microsites, these results suggest that growth was more dependent on diffuse light levels than on sunfleck activity. Similar results were obtained in a study of seedlings of the dipterocarp *Hopea pedicellata* (Gong, 1981). Survival and height growth under green mesh or under natural sunflecks did not differ significantly, although leaf length of shaded seedlings was significantly less than seedlings grown under sunflecks.

A striking relationship between sunfleck activity and growth was described by Pearcy (1983) for seedlings of *Euphorbia forbesii* and *Claoxylon sandwi-*

cense in a Hawaiian evergreen forest understorey (Fig. 6). Based on hemispherical photographs, the potential minutes of sunfleck activity for an entire year were estimated for each of 15 plants. The mean potential minutes of sunflecks per day was closely correlated with the relative growth rate of plants over the year. In contrast, the diffuse site factor estimated from the same photographs was not significantly correlated with growth. Growth rates of the two species were similar under similar sunfleck regimes. This study provides the strongest evidence available that sunfleck activity directly affects growth of understorey plants over an entire season.

Oberbauer *et al.* (1988) investigated growth and crown light environments of saplings of *Dipteryx panamensis* and *Lecythis ampla*, two rainforest canopy tree species. Height growth of *Lecythis*, but not of *Dipteryx*, was significantly correlated with the proportion of daily PFD contributed by sunflecks (instantaneous PFD above 50 μmol m^{-2} s^{-1}). In both species, height growth over a year was correlated with measurements of weekly total

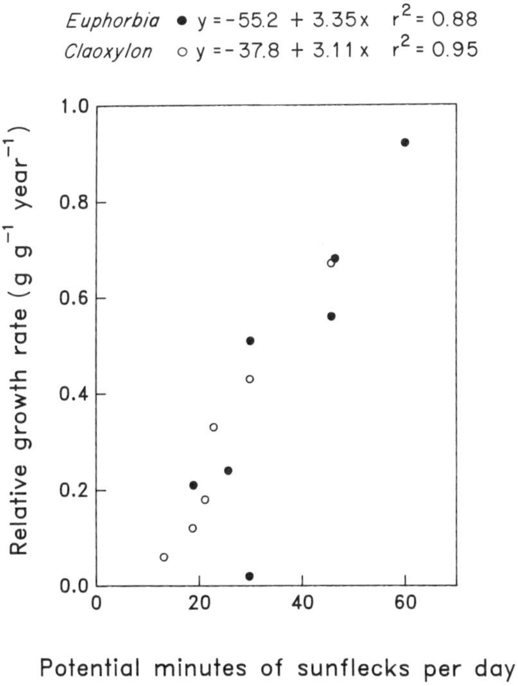

Fig. 6. Relative growth rate of *Euphorbia forbesii* and *Claoxylon sandwicense* as a function of average duration of potential sunflecks per day (in minutes), estimated from hemispherical photographs. From Pearcy (1983), with permission of the publisher.

PFD and the weekly percentage of full sun received, whereas diameter growth was only weakly correlated with light conditions.

2. Understorey Species

At light levels below 20% of full sun, irradiance usually limits growth of vegetation below forest canopies (Shirley, 1929). In temperate deciduous forests, growth and distribution of the understorey herb *Hyacinthoides (Scilla) non-scripta* was more dependent on the amount of light received during the high-light phase of spring than on that received during the low-light conditions following canopy closure (Blackman and Rutter, 1946). Thus, even if microsites varied significantly in sunfleck activity, the effects on plant growth would be relatively small compared to differences in growth during early spring. Growth studies of the forest annual *Impatiens parviflora* showed that when the diffuse site factor was constant, large increases in direct sunlight produced only a small increase in unit leaf rate, and very little change in leaf weight ratio (Coombe, 1966). Specific leaf area, however, showed a relatively large decrease.

The influence of light availability on growth of *Aster acuminatus*, a rhizomatous perennial of eastern deciduous forests in the USA, was also affected by seasonal distribution (Pitelka *et al.*, 1985). When light levels were initially high, as in the temperate forest in early spring, ramet growth increased more than when high-light levels occurred later in the growing season. Early season exposures to high light also appeared to enhance phenological development. Regardless of timing, high-light periods led to increased ramet height, weight, and rhizome production. According to Pitelka *et al.* (1985), it is likely that in at least some *Aster* patches there can be substantial seasonal variation in light availability because of different temporal patterns in sunfleck activity.

Long-term studies of *Aster acuminatus* have shown that light levels within understorey microsites were significantly correlated with average plant size within patches, and with the locations of patches (Pitelka *et al.*, 1980; Fig. 7). Although this species is often found in relatively open sites, such as treefalls, it exhibits extensive phenotypic plasticity and can occupy a wide range of microsites with different degrees of light availability (Pitelka *et al.*, 1980; Ashmun *et al.*, 1980). Microsite variation in light availability did not significantly affect patterns of biomass allocation to vegetative parts (Pitelka *et al.*, 1980). Transplant experiments in deciduous forest sites showed that mean ramet size increased with light level over a three-year period (Ashmun and Pitelka, 1984). Measurements of PFD in eight transplant gardens were positively correlated with survival of ramets, the number of new ramets produced, and the total number of ramets at the end of the experiment.

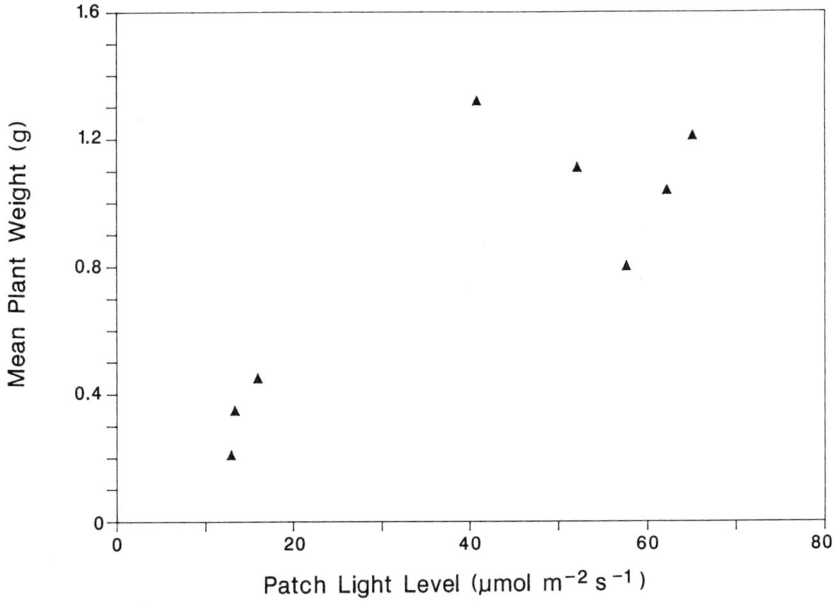

Fig. 7. Mean plant biomass (g) of *Aster acuminatus* as a function of patch light level (PFD; µmol m^{-2} s^{-1}). From Pitelka *et al.*, (1980), with permission of the publisher.

Moreover, ramets were correlated in size from one year to the next. In a different field site, similar relationships were observed between ramet size and light availability (Ashmun *et al.*, 1985). In four different growth seasons, multiple regression analysis using direct and diffuse site factors as independent variables showed a highly significant dependence of mean ramet weight, ramet density, and standing crop on light availability (Ashmun *et al.*, 1985).

Plants growing in the understorey of tropical evergreen forests are subjected to low daily PFD on a year-round basis, unless they are located near canopy gaps. Along the edge of gaps, seasonal variation in sunfleck activity was evident (Chazdon, 1986). Mean daily PFD at the northern edge of a gap was more than double that measured at the southern edge of the same gap in February, but was similar in March, when the sun was almost directly overhead (Chazdon, 1986). Further research is needed to determine whether differences in sunfleck activity associated with gap location significantly affect plant growth. Studies of understorey herbs in a seasonal tropical forest in Panama indicate that, for many species, growth and establishment are highly dependent on canopy gaps (Smith, 1987).

VII. THE INFLUENCE OF SUNFLECKS ON REPRODUCTIVE BEHAVIOR AND DISTRIBUTIONS OF UNDERSTOREY SPECIES

Although it is often assumed that reproduction and distribution of understorey species are limited by light availability, relatively little quantitative data has been gathered to support or refute this claim. Resource allocation to reproduction is an elusive quantity to measure (Bazzaz and Reekie, 1985; Bazzaz *et al.*, 1987). Attributing reproductive effort to measured light conditions during flowering or fruiting is also problematic. In some species, reproductive buds may be initiated at least a year before actual flowering, when light conditions may have been quite different. For insect-pollinated plants, seed set may be pollinator-limited, whereas resource allocation to flowering structures and fruit maturation may or may not be light-limited (Bierzychudek, 1981). Furthermore, the costs of reproduction may impose substantial constraints on vegetative growth in light-limited understorey species (Clark and Clark, 1987). Reduced vegetative growth may then lead to periods of little or no reproduction, despite relatively unchanged light conditions. Clearly, long-term studies are required to gain insights into the relationships between reproductive effort and light availability in understorey species.

Light availability is only one of the many possible causal factors in plant distribution; others include dispersal, herbivory, pathogens, disturbance, and both local and regional history (Augspurger, 1984; Augspurger and Kelly, 1984). Many understorey species are capable of vegetative reproduction, and individual clones may persist in a site for decades or longer. Therefore, what may, at first glance, appear as patches of current regeneration, may in fact represent the remnants of a persistent, long-lived clone that at some previous time had proliferated in response to increased light availability (Gómez-Pompa and Vázquez-Yánes, 1985; Smith, 1987). Studies of the distributions of long-lived perennials in relation to microenvironmental conditions must recognize this historical dimension. In this section, I review studies of the reproductive behavior and distribution of understorey species in relation to the patchiness of light availability in understorey habitats. Although the patchy nature of light availability is almost certainly correlated with microsite variation in sunfleck activity, more research is needed to determine the extent to which patterns of reproduction and distribution are affected by sunfleck activity rather than by other environmental factors.

A. Light Availability, Size Variation and Reproductive Behavior

Many demographic studies of forest understorey species have shown that plant size is often correlated with the frequency and amount of reproduction (Sohn and Policansky, 1977; Solbrig, 1981; Pitelka *et al.*, 1980; Sarukhán *et al.*, 1984). In the understorey palm *Astrocaryum mexicanum*, patterns of reproduction were associated with differences in leaf number and crown size (Piñero and Sarukhán, 1982). Moreover, plants with above- and below-average flowering frequency were clumped in their distributions. Individuals growing in gaps produced move leaves and fruits during a 2 yr period than plants growing in other areas (Piñero and Sarukhán, 1982). Inflorescence size and number were significantly higher in reproductive individuals of two Costa Rican understorey palm species growing in gap-edge plots compared to closed-canopy understorey (Chazdon, 1984). The frequency of reproduction in the understorey cycad *Zamia skinneri* was correlated with both plant size and an index of light availability (Clark and Clark, 1987). These studies provide anecdotal evidence that patterns of reproduction in understorey species are related to heterogeneity in light availability. To the extent that plant size is correlated with sexual expression in dioecious understorey species, differences in light availability may also be associated with changes in sex-ratio within plant populations (Bierzychudek, 1982).

Flowering of *Aster acuminatus* was highly dependent on both individual plant size and light availability (Pitelka *et al.*, 1980; Ashmun *et al.*, 1985). In all patches containing flowering plants, non-flowering plants were always smaller than flowering plants. In contrast, the proportion of total biomass allocated to vegetative reproduction (production of new rhizomes) remained constant (Pitelka *et al.*, 1980). Further studies of transplanted ramets showed that the percentage of ramets flowering increased with garden light level in each of three successive years (Ashmun and Pitelka, 1984). Variation in light availability alone was the principal factor that explained the large differences in reproduction observed among gardens. Plant size, however, was not the only determinant of flowering behavior. Reduction in light levels after deciduous canopy closure strongly affected sexual reproductive behavior, whereas ramet size was not significantly affected (Pitelka *et al.*, 1985). In this case, phases of vegetative growth and sexual reproduction became uncoupled, allowing plants to respond to seasonal changes in the availability of limiting resources.

In a study of understorey herbs of a seasonally dry tropical forest in Panama, Smith (1987) found that only a few species reproduce regularly in closed forest. Most species remain in a suppressed, vegetative state in the understorey until light availability increases following the creation of a tree-fall gap in the vicinity.

B. Vegetative and Sexual Reproductive Effort

Many, if not most, forest understorey species reproduce vegetatively as well as sexually. Allocation of resources (carbon or biomass) to both vegetative and sexual reproductive functions may vary according to light availability within the understorey. In *Aster acuminatus*, vegetative reproductive effort remained constant over a wide range of light levels, whereas sexual reproductive effort increased with greater light availability (Pitelka *et al.*, 1980). Plants transplanted to deeply-shaded sites did not flower, and few produced clonal offspring (Ashmun and Pitelka, 1984). In sites with higher irradiance, however, *Aster acuminatus* was capable of rapid clonal spread and a high incidence of sexual reproduction. When biomass allocation to the perennating rhizome was separated from allocation to clonal growth, Ashmun *et al.* (1985) found that vegetative reproductive allocation increased with light availability.

Other studies of reproduction in forest understorey plants indicate that sexual reproduction is not always more sensitive to light availability than is vegetative (clonal) reproduction, although both may be affected. Two strawberry species, *Fragaria virginiana* and *F. vesca*, both showed decreasing sexual reproductive effort in relatively shadier environments (Jurik, 1983, 1985). In this case, however, vegetative reproductive effort decreased more than sexual reproductive effort under shadier conditions, so that allocation was shifted in favor of sexual reproduction.

These studies show that the relationship between vegetative reproduction, sexual reproduction, and light availability is complex, and that patterns often differ among species and among forest types. Most studies agree, however, that seedling establishment of long-lived perennials is very rare in shaded forest understorey. Therefore, the ability of individuals to persist, may ultimately be determined by patterns of clonal growth and vegetative spread.

C. Sunflecks, Canopy Gaps and Species Distributions

Forest understorey species exhibit many different patterns of spatial distribution, ranging from random to highly clumped patterns. Relationships between distributional patterns and sunfleck activity have been studied in detail for only a few species. Seedings of red fir (*Abies magnifica*) show patchy distributions on south-facing slopes over much of their range. In a study by Ustin *et al.* (1984), low levels of sunfleck activity were the environmental factor that best accounted for the clumped distribution of these seedlings. Areas with low seedling density received long sunflecks at midday, with PFD at full-sun intensity. In contrast, areas with high seedling density had smaller canopy openings with shorter, less intense sunflecks. These areas closely resembled understorey sites on north-facing slopes, where seedlings were not patchily distributed. Ustin *et al.* (1984) suggest that aggregations of seedlings are the result of differential germination and seedling survival, which may be susceptible to water and thermal stresses during prolonged sunfleck exposures.

The small-scale distributions of *Arnica cordifolia* and *A. latifolia* in subalpine coniferous forests may also be strongly linked to the influence of sunfleck patterns on water relations (Young and Smith, 1979). Computer simulations of carbon gain and water-use efficiency of *A. cordifolia* indicate that water-use efficiency may be more important in microsite distributions than carbon gain (Young and Smith, 1982). Frequent cloud-cover during the summer growth season may mitigate the effects of sunflecks on water-use efficiency, however (Knapp and Smith, 1987).

Studies of the distribution of understorey herbs in a Panamanian forest showed that patterns of abundance and distribution can, to a large extent, be explained by temporal and spatial variation in canopy gaps (Smith, 1987). Most understorey herbs are capable of rapid growth and recruitment in gaps, but are also able to persist in closed-canopy forest, usually in a vegetative state. For many species, clumped spatial patterns are closely linked to the previous occurrence of a canopy gap in that location. Similar relationships between species distributions and canopy gaps were described for two understorey species in a wet lowland tropical forest in Costa Rica (Richards and Williamson, 1975).

Preliminary studies of the distributional patterns of understorey shrubs in the genus *Piper* within the understorey of primary and secondary rainforest suggest that some species are distributed differentially with regard to light availability and sunfleck activity (C. B. Field and R. L. Chazdon, unpublished data). *Piper hispidum* occurs in both early successional and primary forest habitats; two-thirds of the plants sampled received from 70 to 90 potential minutes of sunflecks per day (yearly average, based on hemispherical photographs). In contrast, all of the individuals of the understorey species *P. aequale* and *P. amalago* received less than 70 min of potential sunflecks per day.

D. Vertical Distribution of Understorey Species

A simple theoretical model of the three-dimensional distribution of light within forests shows that, immediately below the canopy, light levels exhibit extremely high variance. At this level, a point is either directly beneath a crown or directly in a gap (Terborgh, 1985). At greater depths below the canopy, however, a higher fraction of the space along a horizontal plane receives at least some direct radiation. The horizontal variance decreases, because the expanding cones of light beneath alternate canopy openings eventually intersect. Based on this "sunfleck" model of vertical canopy structure, Terborgh (1985) hypothesized that understorey (midlayer) trees with their crowns in this intersection plane will maximize photosynthetic production and reproductive output because of improved light conditions. Understorey trees, such as *Cornus florida*, should therefore grow in height up to, but not exceeding this level.

The height of the intersection plane can be predicted based on measured angular distributions of canopy openings. Terborgh (1985) compared heights of the understorey tree stratum with the predicted height of the intersection plane, and found close agreement. An implication of this model is that the vertical distribution of sunflecks is an important controller of the vertical distribution of plant species. The sunfleck model is also useful for predicting the extent of stratification within forests at different latitudes. In boreal forests, an intersection plane is not predicted above ground level; no woody substratum is observed in these forests. In contrast, tropical forests are predicted to have at least one additional canopy layer because light is able to penetrate the canopy at relatively shallow angles (Terborgh, 1985).

VIII. CONCLUSIONS

A. The Importance of Sunflecks: Scaling Up From Leaves to Whole Plants

Responses of understorey plants to sunflecks can be found at many different levels (Tables 1 and 2). Leaves show increased photosynthetic rates and stomatal conductance, plants often gain more biomass and produce more propagules, and some plant populations become restricted in their distribution within the forest. Although sunfleck activity may be highly correlated with these biological processes, these correlations do not necessarily imply that sunfleck activity plays a causal role. Despite greater logistical difficulties, the causal effects of direct radiation during sunflecks are far easier to interpret for individual leaves than for whole plants and populations.

Because of the non-linearity of photosynthetic responses, different patterns of light variation can yield different photosynthetic outcomes even when total PFD remains constant (Chazdon, 1986). High PFD during sunflecks may have detrimental effects on photosynthesis and water-use that could not be predicted from a knowledge of daily total PFD (Young and Smith, 1979; Powles and Björkman, 1981). For other biological processes, however, it appears that the incidence of direct radiation during sunflecks may be important in a strictly quantitative sense. Changes in photosynthetic capacity in leaves grown in different regimes appear to depend on integrated PFD rather than on instantaneous values (Chabot *et al.*, 1979). Recent studies have shown that relative growth rate may be linearly related to sunfleck activity (Pearcy, 1983), and that sexual reproductive allocation is linearly related to patch light level (Pitelka *et al.*, 1980). Thus, even though photosynthetic responses to light are non-linear, scaling up from leaves to whole plants may effectively linearize many of these relationships.

Whatever the basis, the apparent linearization and integration of organismal responses to changing light conditions make it exceedingly difficult to quantify the influence of sunflecks as opposed to other components of the light environment (diffuse irradiance, light quality) on plant responses. In this regard, computer simulations of whole-plant carbon balance and growth may be the most useful technique for elucidating the mechanisms by which whole plants respond dynamically to spatial and temporal light fluctuations.

B. Directions for Future Research

As this review amply demonstrates, many gaps remain in our understanding of sunfleck utilization by leaves, whole plants, and populations. Below, I discuss the subject areas that, in my view, are most in need of investigation.

We still know relatively little about sunfleck frequency, duration, and intensity within coniferous and deciduous temperate forests and tropical wet and dry forests. In particular, no sunfleck data have been published on tropical dry forests. Moreover, few studies have addressed the extent of seasonal variation in sunfleck activity in these different forest types. Light measurements need to be sufficiently frequent to account for light variation during even the shortest sunflecks. These studies should also incorporate analyses of the spatial scale of sunfleck activity.

Except for investigations of leaf movements in *Oxalis* (Powles and Björkman, 1981), morphological responses of leaves and whole plants to sunflecks have not been investigated in natural populations. These responses range from modifications of leaf structure to changes in leaf orientation and canopy structure. These studies should also take into account changes in spectral quality associated with sunfleck activity. The effect of plant canopy structure on variation in sunfleck activity among leaf microsites is yet another relatively unexplored area.

Studies of light acclimation in forest understorey plants have traditionally been concerned with steady-state photosynthetic responses and with plants grown under steady-state conditions. Acclimatory responses to constant light conditions may well affect photosynthetic dynamics, as suggested by laboratory studies (Chazdon and Pearcy, 1986b; Sharkey *et al.*, 1986). Moreover, we do not know whether growth under fluctuating light conditions may lead to changes in dynamic or steady-state photosynthetic responses.

Relatively little is known about constraints on sunfleck utilization in natural populations. These constraints may operate seasonally, such as water stress during the dry season in tropical forests, or on a shorter time-scale, such as photoinhibition and induction loss. Studies of daily courses of light, photosynthesis, stomatal conductance, and leaf temperature within a particular forest are needed during different times of the year as well as under different weather conditions. Extensive field studies in conjunction with laboratory investigations will provide a detailed understanding of potential and actual constraints on sunfleck utilization.

Although many studies have shown that increases in light availability are often associated with increases in plant growth and reproductive output, most of these studies have not specifically focused on the role of sunflecks. Long-term field studies are needed to assess the ecological significance of sunflecks as opposed to other components of the environment. Complicating circumstances such as storage, time-lags, and interactions with other environmental factors may require long-term studies, experimental approaches, and computer simulations in these investigations.

In this era of unprecedented deforestation in the tropics, studies of regeneration of secondary and primary forest trees are greatly needed. During tropical forest succession, rapidly growing trees, such as *Ochroma* and *Cecropia*, quickly form a thin canopy. Within a few years, however, a relatively tall, dense canopy is formed, producing a heavily-shaded understorey. It is under these heavily-shaded conditions that longer-lived trees of secondary and primary forests initially become established. There is a great need for comparative studies of sunfleck activity in successional forests of different age and composition and of physiological responses of regenerating tree seedlings to sunflecks. Studies in this area would greatly contribute to our understanding of tropical forest regeneration, forest management, and reforestation efforts.

ACKNOWLEDGEMENTS

I thank A. H. Fitter and an anonymous reviewer for their helpful comments on the manuscript.

REFERENCES

Anderson, M. C. (1964a). Studies of the woodland light climate. I. The photographic computation of light conditions. *J. Ecol.* **52,** 27–41.

Anderson, M. C. (1964b). Studies of the woodland climate. II. Seasonal variation in the seasonal light climate. *J. Ecol.* **52,** 643–663.

Anderson, M. C. (1966). Some problems of simple characterization of the light climate in plant communities. In: *Light as an Ecological Factor* (Ed. by R. Bainbridge, G. C. Evans and O. Rackham), pp. 77–90. Oxford: Blackwell.

Anderson, M. C. and Miller, E. E. (1974). Forest cover as a solar camera: penumbral effects in plant canopies. *J. appl. Ecol.* **11,** 691–697.

Angevine, M. W. and Chabot, B. F. (1979). Seed germination syndromes in higher plants. In: *Topics in Plant Population Biology* (Ed. by O. T. Solbrig, S. Jain, G. B. Johnson and P. H. Raven), pp. 188–206. New York: Columbia Univ. Press.

Ashmun, J. W. and Pitelka, L. F. (1984). Light-induced variation in the growth and dynamics of transplanted ramets of the understory herb, *Aster acuminatus*. *Oecologia (Berlin)* **64,** 255–262.

Ashmun, J. W., Brown, R. L. and Pitelka, L. F. (1985). Biomass allocation in *Aster acuminatus*: variation within and among populations over 5 years. *Can. J. Bot.* **63,** 2035–2043.

Ashton, P. S. (1958). Light intensity measurements in a rain forest near Santarem, Brazil. *J. Ecol.* **46,** 65–70.

Atkins, W. R. G. and Poole, H. H. (1926). Photo-electric measurements of illumination in relation to plant distribution. Part I. *Sci. Proc. Roy. Dublin Soc. N.S.* **18,** 277–298.

Atkins, W. R. G., Poole, H. H. and Stanbury, F. A. (1937). The measurement of the intensity and the colour of the light in woods by means of emission and rectifer photoelectric cells. *Proc. Roy. Soc. B.* **121,** 427–450.

Augspurger, C. K. (1984). Seedling survival of tropical tree species: Interactions of dispersal distance, light-gaps and pathogens. *Ecology* **65,** 1705–1712.

Augspurger, C. K. and Kelly, C. K. (1984). Pathogen mortality of tropical tree seedlings: Experimental studies of the effects of dispersal distance, seedling density, and light conditions. *Oecologia (Berlin)* **61,** 211–217.

Baldocchi, D. D., Hutchison, A., Matt, D. R. and McMillen, R. T. (1984). Seasonal variations in the radiation regime within an oak–hickory forest. *Agric. For. Meteorol.* **33,** 177–191.

Bazzaz, F. A. and Carlson, R. W. (1982). Photosynthetic acclimation to variability in the light environment of early and late successional plants. *Oecologia (Berlin)* **54,** 313–316.

Bazzaz, F. A. and Reekie, E. G. (1985). The meaning and measurement of reproductive effort in plants. In: *Studies on Plant Demography* (Ed. by J. White), pp. 373–387. London: Academic Press.

Bazzaz, F. A., Chiariello, N. R., Coley, P. D. and Pitelka, L. F. (1987). Allocating resources to reproduction and defense. *Bioscience* **37,** 58–67.

Bierzychudek, P. (1981). Pollinator limitation of plant reproductive effort. *Am. Naturalist* **117,** 838–840.

Bierzychudek, P. (1982). The demography of Jack-in-the-pulpit, a forest perennial that changes sex. *Ecol. Monogr.* **52,** 335–351.

Biggs, W. W., Edison, A. R., Easton, J. D., Brown, K. W., Maranville, J. W. and Clegg, M. C. (1971). Photosynthesis light sensor and meter. *Ecology* **52,** 125–131.

Björkman, O. (1968). Carboxydismutase activity in shade-adapted and sun-adapted species of higher plants. *Carnegie Inst. Wash. Yearbook* **67,** 487–488.

Björkman, O. (1981). Responses to different quantum flux densities. In: *Physiological Plant Ecology I. Encyclopedia of Plant Physiology* (Ed. by O. L. Lange, P. S. Nobel, C. B. Osmond, H. Ziegler), New Series, Vol. 12A, pp. 57–107. New York: Springer-Verlag.

Björkman, O. and Holmgren, P. (1963). Adaptability of the photosynthetic apparatus to light intensity in ecotypes from exposed and shaded habitats. *Physiol. Plant.* **16,** 889–914.

Björkman, O. and Ludlow, M. M. (1972). Characterization of the light climate on the floor of a Queensland rainforest. *Carnegie Inst. Wash. Yearbook* **71,** 85–94.

Björkman, O., Boardman, N. K., Anderson, J. M. and Thorne, S. W. (1972a). Effect of light intensity during growth of *Atriplex triangularis* on the capacity of photosynthetic reactions, chloroplast components and structure. *Carnegie Inst. Wash. Yearbook* **71,** 115–135.

Björkman, O., Ludlow, M. M. and Morrow, P. S. (1972b). Photosynthetic performance of two rainforest species in their native habitat and analysis of their gas exchange. *Carnegie Inst. Wash. Yearbook* **71,** 94–102.

Björkman, O. and Powles, S. B. (1981). Leaf movement in the shade species *Oxalis oregana.* I. Response to light level and light quality. *Carnegie Inst. Wash. Yearbook* **80,** 59–62.

Blackman, G. E. and Rutter, A. J. (1946). Physiological and ecological studies in the analysis of plant environment. I. The light factor and the distribution of the bluebell (*Scilla non-scripta*) in woodland communities. *Ann. Bot. n.s.* **10,** 361–390.

Blackman, G. E. and Wilson, G. L. (1951). Physiological and ecological studies in the analysis of plant environment. VII. An analysis of the differential effects of light intensity on the net assimilation rate, leaf-area ratio, and relative growth rate of different species. *Ann. Bot. n.s.* **15,** 373–408.

Blackman, G. E. and Wilson, G. L. (1954). Physiological and ecological studies in the analysis of plant environment. IX. Adaptive changes in the vegetative growth and development of *Helianthus annuus* induced by an alteration in light level. *Ann. Bot. n.s.* **18,** 71–94.

Boardman, N. K. (1977). Comparative photosynthesis of sun and shade plants. *Ann. Rev. Plant Phys.* **28,** 355–377.

Böhning, R. H. and Burnside, C. A. (1956). The effect of light intensity on rate of apparent photosynthesis in leaves of sun and shade plants. *Am. J. Bot.* **43,** 557–561.

Brun, W. A. (1972). Rhythmic stomatal opening response in banana leaves. *Physiol. Plantarum* **15,** 623–630.

Caemmerer, S. von and Farquhar, G. D. (1981). Some relationships between the biochemistry of photosynthesis and the gas exchange of leaves. *Planta* **153,** 376–387.

Chabot, B. F., Jurik, T. W. and Chabot, J. F. (1979). Influence of instantaneous and integrated light flux density on leaf anatomy and photosynthesis. *Am. J. Bot.* **86,** 940–945.

Chan, S. S., McCreight, R. W., Walstad, J. D. and Spies, T. A. (1986). Evaluating forest vegetative cover with computerized analysis of fisheye photographs. *Forest Sci.* **32,** 1085–1091.

Chazdon, R. L. (1984). *Ecophysiology and architecture of three rain forest understory palm species*, Ph.D. Thesis, Cornell University, Ithaca, New York.

Chazdon, R. L. (1986). Light variation and carbon gain in rain forest understorey palms, *J. Ecol.* **74,** 995–1012.

Chazdon, R. L. (1987). Aspectos importantes para el estudio de los regímenes de luz en bosques tropicales. *Rev. Biol. Trop.* **35** (Suppl. 1), 191–196.

Chazdon, R. L. and Field, C. B. (1987a). Determinants of photosynthetic capacity in six rainforest *Piper* species. *Oecologia (Berlin)* **73,** 222–230.

Chazdon, R. L. and Field, C. B. (1987b). Photographic estimation of photosynthetically active radiation: Evaluation of a computerized technique. *Oecologia (Berlin)* **73,** 525–532.

Chazdon, R. L. and Fetcher, N. (1984a). Photosynthetic light environments in a lowland tropical rainforest in Costa Rica. *J. Ecol.* **72,** 553–564.

Chazdon, R. L. and Fetcher, N. (1984b). Light environments of tropical forests. In: *Physiological Ecology of Plants of the Wet Tropics* (Ed. by E. Medina, H. A. Mooney and C. Vázquez-Yánes), pp. 27–36. The Hague: Junk.

Chazdon, R. L. and Pearcy, R. W. (1986a). Photosynthetic responses to light variation in rain forest species. I. Induction under constant and fluctuating light conditions. *Oecologia (Berlin)* **69,** 517–523.

Chazdon, R. L. and Pearcy, R. W. (1986b). Photosynthetic responses to light variation in rainforest species. II. Carbon gain and photosynthetic efficiency during lightflecks. *Oecologia (Berlin)* **69,** 524–531.

Chazdon, R. L., Williams, K. and Field, C. B. (1988). Interactions between crown structure and light environment in five rainforest *Piper* species. *Am. J. Bot.*, in press.

Chiariello, N. (1984). Leaf energy balance in the wet lowland tropics. In: *Physiological Ecology of Plants of the Wet Tropics* (Ed. by E. Medina, H. A. Mooney and C. Vázquez-Yánes), pp. 85–98. The Hague: Junk.

Clark, D. B. and Clark, D. A. (1987). Leaf production and the cost of reproduction in a tropical rain forest cycad, *Zamia skinneri. J, Ecol.*, in press.

Coombe, D. E. (1957). The spectral composition of shade light in woodlands. *J. Ecol.* **45,** 823–830.

Coombe, D. E. (1966). The seasonal light climate and plant growth in a Cambridgeshire wood. In: *Light as an Ecological Factor* (Ed. by R. Bainbridge, G. C. Evans, and O. Rackman), pp. 148–166. Oxford: Blackwell.

Corré, W. J. (1983). Growth and morphogenesis of sun and shade plants. II. The influence of light quality. *Acta. Bot. Neerlandica* **32,** 185–202.

Curtis, W. F. and Kincaid, D. T. (1984). Leaf conductance responses of *Viola* species from sun and shade habitats. *Can. J. Bot.* **62,** 1268–1272.

Davies, W. J. and Kozlowski, T. T. (1975). Stomatal responses to changes in light intensity as influenced by plant water stress. *For. Sci.* **21,** 129–133.

Elias, P. (1983). Water relations pattern of understorey species influenced by sunflecks. *Biol. Plant.* **25,** 68–74.

Ellenberg, H. (1963). *Vegetation Mitteleuropas mit den Alpen.* Stuttgart: Eugen Ulmer. 989 pp.

Evans, G. C. (1939). Ecological studies on the rain forest of southern Nigeria II. The atmospheric environmental conditions. *J. Ecol.* **27,** 436–462.

Evans, G. C. (1956). An area survey method of investigating the distribution of light intensity in woodlands, with particular reference to sunflecks. *J. Ecol.* **44,** 391–428.

Evans, G. C. (1966). Model and measurement in the study of woodland light climates. In: *Light as an Ecological Factor* (Ed. by R. Bainbridge, G. C. Evans, and O. Rackman), pp. 53–76. Oxford: Blackwell.

Evans, G. C. and Coombe, D. E. (1959), Hemispherical and woodland canopy photography and the light climate. *J. Ecol.* **47,** 103–113.

Evans, G. C., Whitmore, T. C. and Wong, T. K. (1960). The distribution of light reaching the ground vegetation in a tropical rainforest. *J. Ecol.* **48,** 193–204.

Evans, G. C., Freeman, P. and Rackham, O. (1975). Developments in hemispherical photography. In: *Light as an Ecological Factor. II* (Ed. by G. C. Evans, R. Bainbridge and O. Rackham), pp. 549–556. Oxford: Blackwell.

Federer, C. A. and Tanner, C. B. (1966). Spectral distribution of light in the forest. *Ecology* **47,** 555–560.

Fetcher, N., Oberbauer, S. F. and Strain, B. R. (1985). Vegetation effects on microclimate in lowland tropical forest in Costa Rica. *Int. J. Biometeorol.* **29,** 145–155.

Fetcher, N., Oberbauer, S. F., Rojas, G. and Strain, B. R. (1987). Efectos del régimen de luz sobre la fotosíntesis y el crecimiento en plántulas de árboles de un bosque lluvioso tropical de Costa Rica. *Rev. Biol. Trop.* **35** (Suppl. 1), 97–110.

Fonteno, W. C. and McWilliams, E. L. (1978). Light compensation points and acclimatization of four tropical foliage plants. *J. Am. Soc. hort. Sci.* **103,** 52–56.

Gómez-Pompa, A. and Yázquez-Yánes (1985). Estudios sobre la regeneración de selvas en regiones cálido-húmedas de México. In: *Investigaciones sobre la regeneración de selvas altas en Veracruz, México. II* (Ed. by A. Gómez-Pompa and S. del Amo R.), pp. 1–25. México: Editorial Alhambra Mexicana.

Gong, W. K. (1981). Studies on the natural regeneration of a hill Dipterocarp species, *Hopea pedicellata. Malaysian Forester* **44,** 357–369.

Goodchild, D. J., Björkman, O. and Pyliotis, N. A. (1972). Chloroplast ultrastructure, leaf anatomy, and content of chlorophyll and soluble protein in rainforest species. *Carnegie Inst. Wash. Yearbook* **71,** 102–107.

Gregory, F. G. and Pearse, H. L. (1937). The effects on the behavior of stomata of alternative periods of light and darkness of short duration. *Ann. Bot. n.s.* **1,** 3–10.

Gross, L. J. (1982). Photosynthetic dynamics in varying light environments: a model and its application to whole leaf carbon gain. *Ecology* **63,** 84–93.

Gross, L. J. (1984). Reply to McCree and Loomis. *Ecology* **65,** 1018–1019.

Gross, L. J. (1986). Photosynthetic dynamics and plant adaptation to environmental variability. *Lectures Math. Life Sci.* **18,** 135–169.

Gross, L. J. and Chabot, B. F. (1979). Time course of photosynthetic response to changes in incident light energy. *Plant Physiol.* **63,** 1033–1038.

Grubb, P. J. and Whitmore, T. C. (1967). A comparison of montane and lowland forest in Ecuador. III. The light reaching the ground vegetation. *J. Ecol.* **55,** 33–57.

Gutschick, V. P., Barron, M. H., Waechter, D. A. and Wolf, M. A. (1985). Portable monitor for solar radiation that accumulates irradiance histograms for 32 leaf-mounted sensors. *Agric. For. Meteorol.* **33,** 281–290.

Harbinson, J. and Woodward, F. I. (1984). Field measurements of the gas exchange of woody plant species in simulated sunflecks. *Ann. Bot.* **53,** 841–851.

Hariri, M. and Prioul, J. L. (1978). Light-induced adaptive responses under greenhouse and controlled conditions in the fern *Pteris cretica* var *ouvardii*. II. Photosynthetic capacities. *Physiol. Plantarum* **42,** 97–102.

Harvey, G. W. (1980). Seasonal alteration of photosynthetic unit sizes in three herb layer components of a deciduous forest community. *Am. J. Bot.* **67,** 293–299.

Herbert, T. J. (1988). Area projections of fisheye photographic lenses. *Agric. For. Meteorol.*, in press.

Hicks, D. J. and Chabot, B. F. (1985). Deciduous forest. In: *Physiological Ecology of North America Plant Communities* (Ed. by B. F. Chabot and H. A. Mooney), pp. 257–277. London: Chapman and Hall.

Hill, R. (1924). A lens for whole sky photographs. *Q. J. Roy. met. Soc.* **50,** 227–235.

Holmes, M. G. and Smith, H. (1977a). The function of phytochrome in the natural environment. II. The influence of vegetation canopies on the spectral energy distribution of natural daylight. *Photochem. Photobiol.* **25,** 539–545.

Holmes, M. G. and Smith, H. (1977b). The function of phytochrome in the natural environment. IV. Light quality and plant development. *Photochem. Photobiol.* **25,** 551–557.

Horowitz, J. L. (1969). An easily constructed shadow-band for separating direct and diffuse solar radiation. *Solar Energy* **12,** 543–545.

Hughes, A. P. (1959). Effects of the environment on leaf development in *Impatiens parviflora* D.C. *J. Linn. Soc. (Bot.)* **56,** 161–165.

Hughes, A. P. (1966). The importance of light compared with other factors affecting plant growth. In: *Light as an Ecological Factor* (Ed. by R. Bainbridge, G. C. Evans and O. Rackham), pp. 121–147. Oxford: Blackwell.

Hutchison, B. A. and Matt, D. R. (1976). Beam enrichment of diffuse radiation in a deciduous forest. *Agric. Meteorol.* **17,** 93–110.

Hutchison, B. A. and Matt, D. R. (1977). The distribution of solar radiation within a deciduous forest. *Ecol. Monogr.* **47,** 185–207.

Jupp, D. L. B., Anderson, M. C., Adomeit, G. M. and Witts, S. J. (1980). Pisces—a computer program for analysing hemispherical canopy photographs. *CSIRO Technical Memorandum 80/23.* Canberra.

Jurik, T. W. (1983). Reproductive effort and CO_2 dynamics of wild strawberry populations. *Ecology* **64,** 1329–1342.

Jurik, T. W. (1985). Differential costs of sexual and vegetative reproduction in wild strawberry populations. *Oecologia (Berlin)* **66,** 394–403.

Jurik, T. W., Chabot, J. F. and Chabot, B. F. (1979). Ontogeny of photosynthetic performance in *Fragaria virginiana* under changing light regimes. *Plant Physiol.* **63,** 542–547.

Kawano, S., Takasu, H. and Nagai, Y. (1978). The productive and reproductive biology of flowering plants. IV. Assimilation behavior of some temperate woodland herbs. *J. Coll. Lib. Arts Toyama Univ. (Nat. Sci.)* **11,** 33–60.

Kawano, S., Masuda, J., Takasu, H. and Yoshie, F. (1983). The productive and reproductive biology of flowering plants. XI. Assimilation behavior of several evergreen temperate woodland plants and its evolutionary–ecological implications. *J. Coll. Lib. Arts Toyama Univ. (Nat. Sci.)* **16,** 31–65.

Kirschbaum, M. and Pearcy, R. W. (1988). Gas exchange analysis of the relative importance of stomatal and biochemical factors in photosynthetic induction in *Alocasia macrorrhiza. Plant Physiol.* **86,** 782–785.

Knapp, A. K. and Smith, W. K. (1987). Stomatal and photosynthetic responses during sun/shade transitions in subalpine plants: influence on water use efficiency. *Oecologia (Berlin)* **74,** 62–67.

Kozlowski, T. T. (1957). Effect of continuous high light intensity on photosynthesis of forest tree seedlings. *For. Sci.* **3,** 221–224.

Kriedemann, P., Torokfalvy, S. and Smart, R. E. (1973). Natural occurrence and photosynthetic utilization of sunflecks in grapevine leaves. *Photosynthetica* **7,** 18–27.

Langenheim, J. H., Osmond, C. B., Brooks, A. and Ferrar, P. J. (1984). Photosynthetic responses to light in seedlings of selected Amazonian and Australian rainforest tree species. *Oecologia (Berlin)* **63,** 215–224.

Lawrence, W. T. (1984). Photosynthetic response of a tropical understory species to naturally occurring sunflecks. *Plant Physiol.* (Suppl. 1) **5.**

Lee, D. W. (1987). The spectral distribution of radiation in two neotropical rainforests. *Biotropica* **19,** 161–166.

Lundegardh, H. (1922). Zür Physiologie und Okologie der Kohnlensaurassimilation. *Biol. Zbl.* **42,** 337–358.

McCree, K. J. and Loomis, R. S. (1969). Photosynthesis in fluctuating light. *Ecology* **50,** 422–428.
McCree, K. J. and Loomis, R. S. (1984). Photosynthetic dynamics—a comment. *Ecology* **65,** 1016–1018.
Marks, T. C. and Taylor, K. (1978). The carbon economy of *Rubus chamaemorus* L. I. Photosynthesis. *Ann. Bot.* **42,** 165–179.
Marquis, R. J., Young, H. J. and Braker, H. E. (1986). The influence of understory vegetation cover on germination and seedling establishment in a tropical lowland wet forest. *Biotropica* **18,** 273–278.
Masarovicova, E. and Elias, P. (1986). Photosynthetic rate and water relations in some forest herbs in spring and summer. *Photosynthetica* **20,** 187–195.
Meinzer, F. C. (1982). Models of steady-state and dynamic gas exchange responses to vapor pressure and light in Douglas Fir (*Psudotsuga menziesii*) saplings. *Oecologia (Berlin)* **55,** 403–408.
Miller, E. E. and Norman, J. M. (1971). A sunfleck theory for plant canopies. I. Lengths of sunlit segments along a transect. *Agron. J.* **63,** 735–738.
Mooney, H. A., Field, C. B., Vázquez-Yánes, C. and Chu, C. (1983). Environmental controls on stomatal conductance in a shrub of the humid tropics. *Proc. nat. Acad. Sci.* **80,** 1295–1297.
Morgan, D. C. and Smith, H. (1978). Simulated sunflecks have large, rapid effects on plant stem extension. *Nature* **273,** 534-536.
Nobel, P. S. (1976). Photosynthetic rates of shade versus shade leaves of *Hyptis emoryi* Torr. *Plant Physiol.* **58,** 218–223.
Nobel, P. S. and Hartsock, T. L. (1981). Development of leaf thickness for *Plectranthus parviflorus*—Influence of photosynthetically active radiation. *Physiol. Plant.* **51,** 163–166.
Norman, J. M., Miller, E. E. and Tanner, C. B. (1971). Light intensity and sunfleck-size distributions in plant canopies. *Agron. J.* **63,** 743–748.
Oberbauer, S. F., Clark, D. B., Clark, D. A. and Quesada, M. A. (1988). Crown light environments of saplings of two species of rain forest emergent trees. *Oecologia (Berlin)* **75,** 207–212.
Oker-Blom, P. (1984). Penumbral effects of within-plant shading on radiation distribution and leaf photosynthesis: a Monte-Carlo simulation. *Photosynthetica* **18,** 522–528.
Orozco-Segovia, A. D. L. (1986). *Fisiología ecología del photoblastismo en semillas de cuatro especies del género* Piper *L.* Ph.D. Thesis, Universidad Nacional Autonoma de México, México, 123 pp.
Osmond, C. B. (1983). Interactions between irradiance, nitrogen nutrition, and water stress in the sun–shade responses of *Solanum dulcamara. Oecclogia (Berlin)* **57,** 316–321.
Pearce, R. B. and Lee, D. R. (1969). Photosynthetic and morphological adaptation of alfalfa leaves to light intensity at different stages of maturity. *Crop Sci.* **9,** 791–794.
Pearcy, R. W. (1983). The light environment and growth of C_3 and C_4 species in the understory of a Hawaiian forest. *Oecologia (Berlin)* **58,** 26–32.
Pearcy, R. W. (1987a). Photosynthetic gas exchange responses of Australian tropical forest trees in canopy, gap and understory microenvironments. *Functional Ecology* **1,** 169–178.
Pearcy, R. W. (1988a). Photosynthetic utilization of lightflecks by understory species. *Austr. J. Plant Physiol.* **15,** in press.
Pearcy, R. W. (1988b). Radiation and light measurements. In: *Physiological Plant Ecology: Field Methods and Instrumentation.* (Ed. by R. W. Pearcy, J. R. Ehleringer, and P. W. Rundel). London: Chapman and Hall, in press.

Pearcy, R. W. and Calkin, H. (1983). Carbon dioxide exchange of C_3 and C_4 tree species in the understory of a Hawaiian forest. *Oecologia (Berlin)* **58,** 26–32.

Pearcy, R. W., Osteryoung, K. and Calkin, H. (1985). Photosynthetic responses to dynamic light environments by Hawaiian trees. *Plant Physiol.* **79,** 896–902.

Pearcy, R. W., Chazdon, R. L. and Kirschbaum, M. U. F. (1987a). Photosynthetic utilization of lightflecks by tropical forest plants. In: *Progress in Photosynthesis Research*, Vol. IV (Ed. by J. Biggens), pp. 257–260. Dordrecht: Martinus Nijhoff.

Pearcy, R. W., Björkman, O., Caldwell, M. M., Keeley, J. E., Monson, R. K. and Strain, B. R. (1987b). Carbon gain by plants in natural environments. *BioSci.* **37,** 21–29.

Piñero, D. and Sarukhán, J. (1982). Reproductive behavior and its individual variability in a tropical palm, *Astrocaryum mexicanum*. *J. Ecol.* **70,** 461–472.

Pitelka, L. F. and Curtis, W. F. (1986). Photosynthetic responses to light in an understory herb, *Aster acuminatus*. *Am. J. Bot.* **73,** 535–540.

Pitelka, L. F., Stanton, D. S. and Peckenham, D. O. (1980). Effects of light and density on resource allocation in a forest herb, *Aster acuminatus* (Compositae). *Am. J. Bot.* **67,** 942–948.

Pitelka, L. F., Ashmun, J. W. and Brown, R. L. (1985). The relationships between seasonal variation in light intensity, ramet size, and sexual reproduction in natural and experimental populations of *Aster acuminatus* (Compositae). *Am. J. Bot.* **72,** 311–319.

Pollard, D. F. W. (1970). The effect of rapidly changing light on the rate of photosynthesis in largetooth aspen (*Populus grandidentata*). *Can. J. Bot.* **48,** 823–829.

Powles, S. B. and Björkman, O. (1981). Leaf movement in the shade species *Oxalis oregana*. II. Role in protection against injury by intense light. *Carnegie Inst. Wash. Yearbook* **80,** 63–66.

Powles, S. B. and Thorne, S. W. (1981). Effect of high light treatments in inducing photoinhibition of photosynthesis in intact leaves of low-light grown *Phaseolus vulgaris* and *Lastreopsis microsora*. *Planta* **152,** 471–477.

Rabinowitch, E. I. (1956). *Photosynthesis and Related Processes*, Vol. II Part 2. New York: Interscience.

Rackham, O. (1975). Temperatures of plant communities as measured by pyrometric and other methods. In: *Light as an Ecological Factor. II* (Ed. by G. C. Evans, R. Bainbridge and O. Rackham), pp. 423–450. Oxford: Blackwell.

Reifsnyder, W. E., Furnival, G. M. and Horovitz, J. L. (1971). Spatial and temporal distribution of solar radiation beneath forest canopies. *Agric. Meteorol.* **9,** 21–37.

Rich, P. M., Clark, D. B., Clark, D. A. and Oberbauer, S. F. (1987). Canopy photography for assessment of local light environment of tropical forest trees and palms. *Bull. Ecol. Soc. Am.* **68,** 397.

Richards, P. and Williamson, G. B. (1975). Treefalls and patterns of understory species in a wet lowland tropical forest. *Ecology* **56,** 1226–1229.

Robichaux, R. H. and Pearcy, R. W. (1980). Photosynthetic responses of C_3 and C_4 species from cool shaded habitats in Hawaii. *Oecologia (Berlin)* **47,** 106–109.

Salisbury, E. J. (1916). The oak–hornbeam woods of Hertfordshire. Parts I and II. *J. Ecol.* **4,** 83–117.

Salminen, R., Nilson, T., Hari, P., Kaipiainen, L. and Ross, J. (1983). A comparison of different methods for measuring the canopy light regime. *J. appl. Ecol.* **20,** 897–904.

Sarukhán, J., Martinez-Ramos, M. and Piñero, D. (1984). The analysis of demographic variability at the individual level and its population consequences. In: *Perspectives on Plant Population Ecology* (Ed. by R. Dirzo and J. Sarukhan), pp. 83–106. Sinauer: Sunderland, MA.

Sasaki, S. and Mori, T. (1981). Growth responses of dipterocarp seedlings to light. *Malaysian Forester* **44,** 319–345.

Sasaki, S., Mori, T and Ng, F. S. P. (1981). Seedling growth under various light conditions in the tropical rain forest. In: *Proc. XVII IUFRO World Congress. Kyoto, Japan, 1981,* 79–85.

Sharkey, T. D., Seemann, J. R. and Pearcy, R. W. (1986). Contribution of metabolites of photosynthesis to postillumination CO_2 assimilation in response to lightflecks. *Plant Physiol.* **82,** 1063–1068.

Shirley, H. L. (1929). The influence of light intensity and light quality upon the growth of plants. *Am. J. Bot.* **16,** 354.

Smith, A. P. (1987). Respuestas de hierbas del sotobosque tropical a claros ocasionados por la caída de árboles. *Rev. Biol. Tropical* **35** (Suppl. 1), 111–119.

Smith, W. K. (1981). Temperature and water relation patterns in subalpine, understory plants. *Oecologia (Berlin)* **48,** 353–359.

Smith, W. K. (1985). Western montane forests. In: *Physiological Ecology of North American Plant Communities* (Ed. by B. F. Chabot and H. A. Mooney), pp. 95–126. London: Chapman and Hall.

Sohn, J. J. and Policansky, D. (1977). The costs of reproduction in the mayapple *Podophyllum peltatum* (Berberidaceae). *Ecology* **58,** 1366–1374.

Solbrig, O. T. (1981). Studies on the population biology of the genus *Viola*. II. The effect of plant size on fitness in *Viola sororia*. *Evolution* **35,** 1080–1093.

Sparling, J. H. (1967). Assimilation rates of some woodland herbs in Ontario. *Bot. Gaz.* **128,** 160–168.

Taylor, R. J. and Pearcy, R. W. (1976). Seasonal patterns of the CO_2 exchange characteristics of understory plants from a deciduous forest. *Can. J. Bot.* **54,** 1094–1103.

Terborgh, J. (1985). The vertical component of plant species diversity in temperate and tropical forests. *Am. Nat.* **126,** 760–776.

Ustin, S. L., Woodward, R. A., Barbour, M. G. and Hatfield, J. L. (1984). Relationships between sunfleck dynamics and red fir seedling distribution. *Ecology* **65,** 1420–1428.

Vázquez-Yánes, C. (1976). Estudios sobre ecofisiología de la germinación en una zona cálido-húmeda de México. In: *Regeneración de Selvas* (Ed. by A. Gómez-Pompa, S. del Amo and A. Butanda), pp. 279–387. México: Editorial Continental.

Vázquez-Yánes, C. and Orozco-Segovia, A. (1984). Ecophysiology of seed germination in the tropical humid forests of the world: a review. In: *Physiological Ecology of Plants of the Wet Tropics* (Ed. by E. Medina, H. A. Mooney and C. Vázquez-Yánes), pp. 37–50. The Hague: Junk.

Walker, D. A. (1981). Photosynthetic induction. In: *Proc 5th Int. Cong. Photosyn,* Vol. IV (Ed. by G. Akoyonoglou), pp. 189–202. Philadelphia: Balaban Int. Sci. Series.

Walters, M. B. and Field, C. B. (1987). Photosynthetic light acclimation in two rainforest *Piper* species with different ecological amplitudes. *Oecologia (Berlin)* **72,** 449–456.

Weber, J. A., Jurik, T. W., Tenhunen, J. D. and Gates, D. M. (1985). Analysis of gas exchange in seedlings of *Acer saccharum*: integration of field and laboratory studies. *Oecologia (Berlin)* **65,** 338–347.

Whitmore, T. C. and Wong, T. K. (1959). Patterns of sunfleck and shade in tropical rain forest. *Malayan Forester* **22,** 50–62.

Willis, A. J. and Balasubramaniam, S. (1968). Stomatal behavior in relation to rate of photosynthesis and transpiration in *Pelargonium. New Phytol.* **67,** 265–285.

Woods, D. R. and Turner, N. C. (1971). Stomatal responses to changing light by four tree species of varying shade tolerance. *New Phytol.* **70,** 77–84.

Woodward, F. I. (1981). Shoot extension and water relations of *Circaea lutetiana* in sunflecks. In: *Plants and Their Atmospheric Environment* (Ed. J. Grace), pp. 83–91. Oxford: Blackwell.

Woodward, F. I. and Yaqub, M. (1979). Integrator and sensors for measuring photosynthetically active radiation and temperature in the field. *J. appl. Ecol.* **16,** 545–552.

Yoshie, F. and Kawano, S. (1986). Seasonal changes in photosynthetic characteristics of *Pachysandra terminalis* (Buxaceae), an evergreen woodland chamaephyte, in the cool temperate regions of Japan. *Oecologia (Berlin)* **71,** 6–11.

Yoshie, F. and Yoshida, S. (1987). Seasonal changes in photosynthetic characteristics of *Anemone raddeana*, a spring-active geophyte in the temperate region of Japan. *Oecologia (Berlin)* **72,** 202–206.

Young, D. R. (1985). Microclimatic effects on water relations, leaf temperatures, and the distribution of *Heracleum lanatum* at high elevations. *Am. J. Bot.* **72,** 357–364.

Young, D. R. and Smith, W. K. (1979). Influence of sunflecks on the temperature and water relations of two subalpine understory congeners. *Oecologia (Berlin)* **43,** 195–205.

Young, D. R. and Smith, W. K. (1980). Influence of sunlight on photosynthesis, water relations, and leaf structure in the understory species *Arnica cordifolia. Ecology* **61,** 1380–1390.

Young, D. R. and Smith, W. K. (1982). Simulation studies on the influence of understory location on the water and photosynthetic relations of *Arnica cordifolia* Hook. *Ecology* **63,** 1761–1771.

Young, D. R. and Smith, W. K. (1983). Effect of cloudcover on photosynthesis and transpiration in the subalpine understory species *Arnica latifolia. Ecology* **64,** 681–687.

Geochemical Monitoring of Atmospheric Heavy Metal Pollution: Theory and Applications

ELIZABETH A. LIVETT

I. Summary 65
II. Introduction 67
III. Theoretical Considerations 69
A. The Accumulation of Heavy Metals 70
B. Chronology and Dating 89
C. Interpretation 99
IV. Practical Applications 106
A. Historical Perspectives 107
B. Present-day Sources of Atmospheric Heavy Metal Pollutants . 130
C. Present World-wide Occurrence of Atmospheric Heavy Metal Pollutants 143
V. Conclusions 154
Acknowledgements 157
References 157
Appendix 174

I. SUMMARY

The three types of natural deposit that have been used most extensively in monitoring the deposition of heavy metals from the atmosphere are peat, ice deposits and aquatic sediments. Their monitoring potential can be assessed from theoretical considerations and from proven applications.

On theoretical grounds it is known that the initial uptake of airborne metals is achieved most effectively by peat overlain by vegetation, and least effectively by ice and snow. The subsequent retention of metals by the three

ADVANCES IN ECOLOGICAL RESEARCH Vol. 18
ISBN 0-12-013918-9

types of deposit varies considerably. There are few problems with ice deposits, as heavy metal stability in them is governed primarily by temperature. In both peat and aquatic systems, however, the chemical and physical changes accompanying sediment accrual can, under some circumstances, bring about the loss of heavy metals from the system, or their remobilization within it. A knowledge of sediment accumulation and decomposition processes within peat and aquatic systems is therefore an essential prerequisite of the recognition of suitable deposits and circumstances for monitoring.

Peat, ice and aquatic sediments, with vastly differing accumulation characteristics and widely varying time-spans, provide pollution records with very different degrees of definition. The most detailed records are provided by rapidly accumulating ice deposits—particularly short-lived deposits in temperate regions. A wide range of dating techniques is applicable in geochemical monitoring. For ice deposits the choice is limited, but appropriate methods are subject to few inaccuracies. The choice is wider for peat deposits and aquatic sediments, but physical and chemical processes within them can reduce the reliability of the dating determinations. Especial difficulties are experienced with dating recently formed horizons, particularly in peat deposits.

It is easy to relate the heavy metal record in ice deposits to atmospheric heavy metal concentration or deposition rate, whereas in peat and aquatic sediments the relationship is more complicated and not always easy to quantify. Additionally, aquatic sediments can accumulate heavy metals by diverse pathways, and metals of specifically atmospheric pollutant origin are not always separable in the sediments.

Studies of the practical applications of geochemical monitoring have been carried out mainly in North-west Europe, North America and the remote polar regions, particularly in relation to the historical aspects of pollution. Records from peat deposits and lake sediments in Europe document the growth and development of industry there over the last 2000 years. Reconstructed historical deposition rates demonstrate the severity of pollution in urban areas of northern Britain in the eighteenth and nineteenth centuries, whilst the spread of pollution at this time is evidenced by dated profiles from many isolated regions in northern Britain and Scandinavia. Analyses of recent sediments reveal a variable pattern of present-day pollution, with some areas experiencing increasing levels of deposition and others decreasing ones.

In North America, deposition records reveal a much shorter history of atmospheric pollution, and document the rapid spread of heavy metals during the mid and late nineteenth century, both on a local and on a regional scale—though small early increases in heavy metal enrichments have proved difficult to recognize in deposits from some rural areas. As in Europe, the pattern of deposition in the twentieth century is patchy: in many areas heavy metal deposition is still increasing, and enrichments of some metals have occurred only recently.

The importance of different sources of pollution has been studied by a variety of geochemical monitoring approaches. In a few examples, the dispersal of metals from a single large stationary source has been assessed both qualitatively and quantitatively by analysis of surface deposits in the surrounding area. Much emphasis has been placed on the assessment of the relative importance of fossil fuel combustion and motor vehicles as sources of ubiquitous lead. The examination of detailed short-term trends, in both lead concentration and deposition, and of associated pollutants and stable lead isotopes, indicates that cars are now an increasingly important, if not a dominant, source of atmospheric lead in certain parts of Europe and over large areas of the United States.

A comparison of present-day deposition rates in Europe, North America and North Atlantic polar regions shows very wide variations in atmospheric heavy metal distribution in the northern hemisphere. Analyses of ice profiles and present-day ice and snow samples from Greenland suggest that pollutant lead is ubiquitous in the North Polar atmosphere, and there are some indications that the very low concentrations of lead in Antarctic snows may contain a significant anthropogenic fraction.

Ice, peat and aquatic sediments fulfil the basic criteria of monitoring media. The particular value of geochemical monitoring, in comparison with conventional and biological techniques, is that it portrays the global extent of atmospheric pollution in its full historical perspective.

II. INTRODUCTION

An awareness of the problem of air pollution is not a twentieth century phenomenon, but can be traced back as far as Ancient Rome and Palestine in the second century AD (Stern *et al.*, 1984; Mamane, 1987). Contemporary records reveal that the problem was also recognized in mediaeval England, and that there were repeated, but largely unsuccessful, attempts to combat it by legislation over the following centuries (Chambers, 1976). During this period, air pollution was seen as an essentially local phenomenon affecting only the larger industrial cities, and it was only in the mid-nineteenth century, when industrial expansion in Britain was at its zenith, that a more quantitative, analytical approach to air pollution was adopted, and that the concept of monitoring was born. This approach was pioneered in the period 1850–1870 by A. C. Smith, who distinguished the gaseous and particulate components of polluted air and rain, and attempted atmospheric sampling at a network of both rural and urban sites throughout Britain (e.g. Smith, 1872). Smith's work set the pattern for pollution monitoring over the next fifty years. By 1914 systematic atmospheric sampling had started at a number of sites in Britain (Press *et al.*, 1983), although for the most part these were restricted to measurements of rain acidity and sulphur concentration, and there was little attempt to distinguish the chemical properties of the "total particulate" fraction of polluted air and rain.

In 1956 and 1963 Clean Air Acts were implemented in Britain and the USA, and a co-ordinated programme of air pollution monitoring was established in both countries (Meetham, 1981). During the same period attention was focused on heavy metals, especially lead, in the atmosphere. Patterson (1965) recognized the significance of the historical aspects of lead in the environment and suggested that the atmospheric and human lead burden in the northern hemisphere had increased 1000-fold since prehistoric times, whilst Chow and his co-workers (Chow and Johnstone, 1965; Chow, 1970) established by means of isotopic measurements that automobile emissions were responsible for much of this increase. By the 1970s conventional monitoring networks for heavy metals in the atmosphere had been set up by AERE at Harwell in the UK and by the Department of Health, Education and Welfare in the USA. At the same time, however, an alternative means of monitoring heavy metals in the environment, using living and non-living components of ecosystems (biological and geochemical monitors), was developed; this offered considerable advantages, both in avoiding the need to detect trace concentrations of heavy metals in air and precipitation, and in providing a more realistic measurement of heavy metals entering natural ecosystems (Aaby *et al.*, 1979; Wagner and Müller, 1979). Bryophytes and lichens had been used for a number of years previously as indicators of air quality, but it was not until the work of Rühling and Tyler (1968, 1969, 1973) in Scandinavia that their full potential as quantitative monitors of atmospheric heavy metals was appreciated. This, and other aspects of biological monitoring, are reviewed by Martin and Coughtrey (1982) and MARC (1986). Airborne heavy metals are also incorporated into terrestrial and aquatic deposits, and early interest was focused especially on their accumulation by ice (Murozumi *et al.*, 1969), peat (Hvatum, 1971; Lee and Tallis, 1973) and lacustrine and marine sediments (Thomas, 1972; Aston *et al.*, 1973; Barnes and Schell, 1973; Chow *et al.*, 1973). This review explores the value of these deposits as monitors of atmospheric heavy metals, and evaluates their contribution to our understanding of the trends and sources of these pollutants in the environment.

During the last few decades the words "heavy metal" and "pollution" have acquired a rather imprecise meaning. As Nieboer and Richardson

(1981) have pointed out, the words "trace metal" and "heavy metal" are frequently used indiscriminantly to include both ferrous and non-ferrous metals, and even toxic non-metals and metalloids. They have proposed the following alternative classification of both toxic and non-toxic metals based on the kinetics of ligand formation: Class A Metals—those which preferentially bind to oxygen donor atoms in ligands (and includes the alkaline and alkaline-earth metals); Class B Metals—those which seek sulphur groups in ligand formation (and includes the precious and semi-precious metals, and lead and mercury); and Borderline Metals—those whose ligand-forming characteristics are intermediate between those of the other two groups (and includes cadmium, copper, iron, manganese, nickel and zinc). This classification relies on the fundamental chemical properties of the metals and embodies those characteristics which are responsible for their deleterious effects, both on organisms and ecosystems.

The term "pollution" is inherently less easy to define or replace than "heavy metal". The most important concepts have been discussed by, for example, Mellanby (1972), Chambers (1976), Hodges (1977) and Holdgate (1979): pollutants can be energy patterns, radiation levels, physical or chemical constituents or living organisms, which have been introduced into the environment as a consequence, direct or indirect, of man's activity, and in quantities which produce recognizably toxic effects. In a particularly broad definition, the deleterious effects produced by pollutants include the diminution of amenity and the quality of life (Holdgate, 1979).

For the sake of convenience, the terms "heavy metal" and "pollution" will be retained in the present paper. It should be noted that all the metals discussed belong either to Class B or to the Borderline Group of Nieboer and Richardson (1981), and that all have proven toxic properties at certain concentrations and in certain situations. It is primarily by virtue of their occurrence as a consequence of human activity, and not necessarily on the basis of their absolute concentrations in the environment, that these metals will be considered as pollutants.

III. THEORETICAL CONSIDERATIONS

The use of layered deposits to monitor atmospheric heavy metal pollution requires an understanding of the two main aspects of the development of the deposition record: the accumulation of heavy metals by the deposit as it is laid down (discussed in Section III.A), and the inclusion of intrinsic time-markers which establishes the chronological dimension of the heavy metal record (discussed in Section III.B). The interpretation of this record (discussed in Section III.C) must take into account the quantitative relationship between deposition and accumulation, and the broader catchment characteristics of the deposits.

A. The Accumulation of Heavy Metals

1. Deposition from the Atmosphere

Atmospheric heavy metals occur principally in particulate form (e.g. Corn, 1976); accordingly, the transfer of airborne particles to land or water surfaces by dry, wet and occult deposition, is the first stage in the development of the sedimentary record.

Dry deposition can be subdivided (Chamberlain, 1960) into four distinct processes: gravitational settling, impaction, turbulent transfer, and transfer by Brownian motion. The relative importance of these processes depends primarily on particle size, specifically MMED*. Gravitational settling is the most important means of deposition for particles of MMED > 20 μm, but with decreasing size, impaction and turbulent transfer become more effective, and for particles of MMED < 1 μm, transport by Brownian motion predominates. The net efficiency of these processes can be described by the deposition velocity V_g, which is defined as:

$$V_g = \frac{\text{flux of particles to the surface}}{\text{volumetric concentration of particles at a specified height above the surface}}$$

V_g has units of, for example, $gcm^{-2}s^{-1}/gcm^{-3} = cms^{-1}$, and is specific to particle type, surface type and meteorological conditions. Lowest values of V_g have been observed for particles in the size-range 0·1–1·0 μm (Chamberlain, 1986).

Anthropogenic aerosols show a broad size-distribution, from *c.* 0·001 μm to 50 μm (Fig. 1), with < 2 μm as the modal class (Corn, 1976). Within this range, size-distribution is strongly influenced by atmospheric transport processes (Parungo and Rhea, 1970; Newell, 1971); hence in rural regions anthropogenic particles > 5μm are absent (e.g. Ward *et al.*, 1975), and in polar regions, heavy metals are most frequently associated with small aggregated particles in the range 0·1–1·0 μm (Maenhaut *et al.*, 1979; Barrie, 1986). The principal mechanisms of dry deposition for heavy metals in rural and remote regions are therefore impaction and turbulent transfer, and the characteristics of the recipient surface then assume importance. Table 1 gives examples of the properties of some natural surfaces and their aerodynamic resistance to deposition (which is an inverse measure of the efficiency of

*MMED: mass median equivalent aerodynamic diameter, where equivalent aerodynamic diameter of a particle is the diameter of a sphere of density 1 g cm^{-3} which has the same falling velocity.

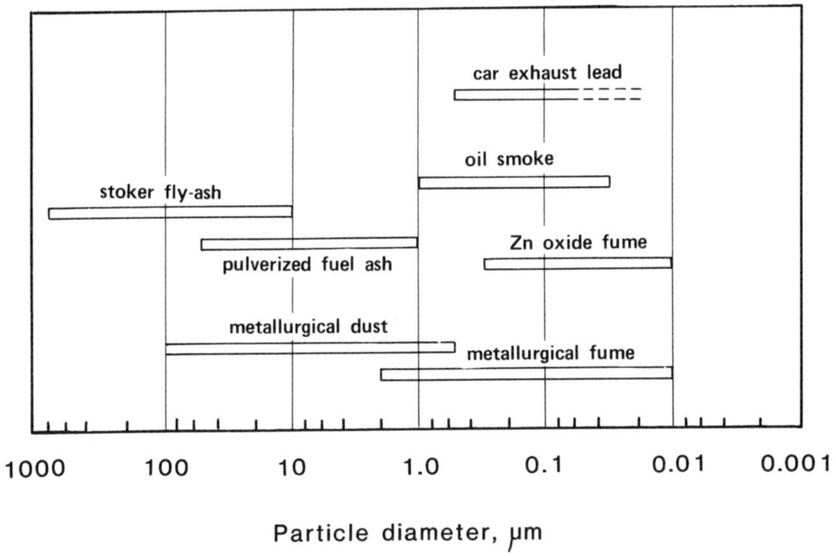

Fig. 1. Particle sizes of some atmospheric pollutants. Modified from Rupp (1956), with additional data from Lee *et al.* (1971).

Table 1

Microtopographic characteristics affecting dry deposition of particles to natural surfaces (data of Hosker and Lindberg, 1982).

Surface	Geochemical monitoring equivalent	Canopy height (m)	Roughness length (cm)[a]	Aerodynamic resistance (s cm^{-1})
Smooth ice	Ice deposits	—	0·003	0·46
Ocean	Aquatic sediments	—	0·005	0·44
Tilled soil	Bare peat/rough snow	—	0·1	0·32
Thick grass	Vegetated peat (e.g. *Eriophorum vaginatum*)	0·1	2·3	0·19
Dwarf shrubs	Vegetated peat (e.g. Ericaceae)	< 1·0	—	—
Shrubs	—	1·5	20·0	0·10

[a] Aerodynamic roughness.

deposition: Hosker and Lindberg, 1982; Chamberlain, 1986). These data indicate that dry deposition in rural and remote areas is most effective for densely vegetated surfaces and least effective for snow and ice.

As agencies of atmospheric cleansing, wet and occult deposition are complementary to, and at least as effective as, dry deposition. Wet deposition involves two processes (Junge, 1963; Wanta and Lowry, 1976): nucleation (whereby particles of MMED < 1 μm serve as foci for the condensation of water vapour droplets), and within- and below-cloud scavenging (whereby the wetted aerosol particles are collected by falling raindrops). In occult deposition, wetted aerosol particles (fog and mist) are deposited by impaction or turbulent transfer (Dollard *et al.*, 1983; Barrie and Schemenauer, 1986). The efficiency of wet (and occult) deposition can be described by a number of mathematical relationships (Hosker, 1986), the simplest being the washout factor, W, (Chamberlain, 1960), which is defined as:

$$W = \frac{\text{concentration of element in precipitation}}{\text{concentration of element in air}}$$

Representative values of W determined by Peirson *et al.* (1973) at Wraymires (Lancs.) range from 450 (Pb) to 1100 (Ni), and the order is approximately the same as the order of solubility of the metals and their compounds. Washout factor, or indeed any value which describes the efficiency of wet deposition, depends also on the absolute concentration of impurities in the aerosol; the efficiency of atmospheric cleansing is known to decrease logarithmically during the course of a precipitation event. Junge (1977) recognized that the extreme purity of the South Polar aerosol (approximating to "clean air" conditions) results in a very inefficient cleansing of the atmosphere and hence a close similarity between the chemical composition of precipitation and air. This relationship has been verified by Boutron and Lorius (1979), Peel and Wolff (1982) and Dick and Peel (1985), who reported ratios of between 0·4 and 5·9. In the relatively polluted North Polar region, however, the "clean air" condition of Junge (1977) is not fulfilled, and accordingly large discrepancies between heavy metal concentrations in snow and air have been reported (e.g. Rahn and McCaffrey, 1979).

The relative importance of dry and wet (and occult) deposition of heavy metals to aquatic sediments, peat and ice depends primarily on geographical location. The processes of nucleation and rainout are most productive when the aerosol is uniformly distributed to the rain-forming altitudes (Ter Haar *et al.*, 1967; Peirson *et al.*, 1973), so wet deposition would be expected to predominate in regions furthest from pollution sources. Published values for the contribution of wet deposition to total flux of metals at a number of "marine", "rural" and "urban" localities range from 20% to 90% (Peirson *et al.*, 1973; Galloway *et al.*, 1982), but show no obvious correspondence with metal or location of site. Wet deposition, and indeed total pollutant influx, would also be expected to vary according to precipitation regime within the same general locality, and this has been substantiated by observations in Scandinavia (Rühling and Tyler, 1969, 1973) and in Antarctica (Boutron *et al.*, 1972), as illustrated in Fig. 2. On a world-wide scale, ombrotrophic peat—which, by definition depends exclusively on precipitation for its mineral nutrients—is the only deposit which is associated consistently with a particular precipitation regime (that of cool-temperate and moist-temperate regions). Therefore, all other factors being equal, atmospheric pollutant influx would be expected to be higher to ombrotrophic peat than to either aquatic sediments or ice.

2. Cycling and Redistribution in Aquatic Systems

The cycling of heavy metals between particulate and aqueous phases in lakes and oceans comprises a number of distinct physical and chemical interactions with different components of the aquatic ecosystem; these have been described in a model by Imboden and Schwarzenbach (1985). For simplicity, the present study will consider the interactions under two headings: those which take place in the water column, and those which take place in the

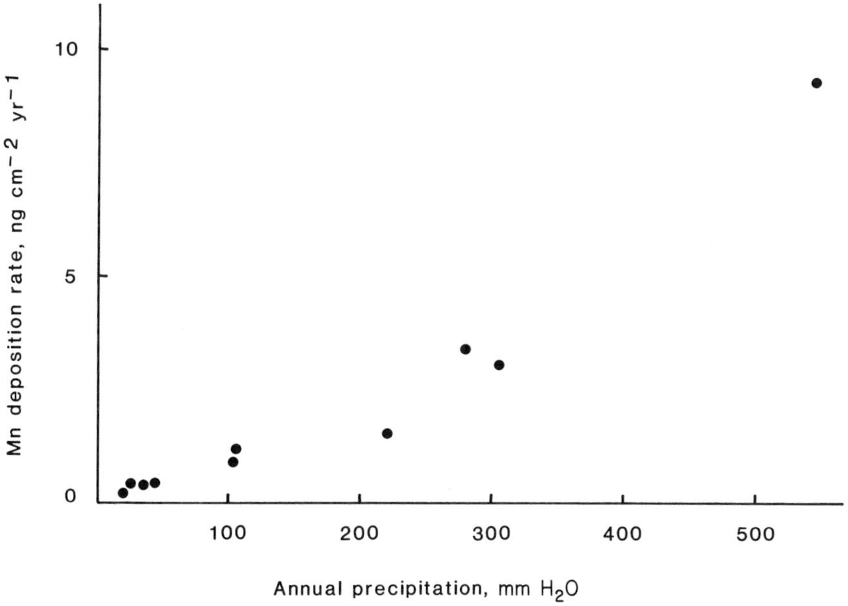

Fig. 2. Variation of deposition rate of manganese with annual precipitation along a transect from Mirny to Vostok, Antarctica. Data of Boutron *et al.* (1972).

unconsolidated and permanent sediments. Running water situations will not be considered. General sources for the discussion are Förstner and Wittman (1981), Förstner (1982) and Salomons and Förstner (1984).

(a) Interactions in the water column. A high proportion of atmospheric heavy metals entering aquatic systems is in small particulate ($< 1\ \mu m$) or soluble form, so transport within the system is influenced strongly by chemical interactions between the heavy metals and other particulate or non-particulate substances.

The first site of chemical interactions is the microlayer (a surface film of thickness *c.* 0·1–1·0 μm), where the often high concentrations of surface-active molecules act as a heavy-metal trap (Liss, 1974; Duce and Hoffman, 1976). High enrichments of heavy metals in the microlayer of lakes and oceans relative to bulk water have been reported: for example, 3- to 11-fold for cadmium, copper and zinc (Elzerman and Armstrong, 1979), and over 100-fold for lead (Pattenden *et al.*, 1981). On the other hand, the microlayer probably acts as a focus for heavy metal scavenging by micro-organisms (McIntyre, 1974), and relatively short residence times of, for example, 5–20 minutes for particles and *c.* 3 hours for soluble metal have been estimated (Pattenden *et al.*, 1981). It is unlikely, therefore, that the microlayer significantly delays heavy metal sedimentation (McIntyre, 1974).

In the main water body of lacustrine and marine systems, the following are the most important reactions in the transfer of metal from aqueous to particulate phases: precipitation and co-precipitation with hydroxides, sulphides or carbonates; cation exchange/sorption on clay minerals, iron/manganese oxides and sulphides, carbonates, phosphates and organic matter; and complexation with organic matter. Many of these reactions take place with coatings of the binding agent on unreactive minerals (Hart, 1982) and are therefore highly surface area-dependent, leading to a marked grain-size partitioning of heavy metal concentrations in the sediments (e.g. Cline and Chambers, 1977; Filipek and Owen, 1979) (Fig. 3). The relative importance of any particular reaction depends on a number of factors, notably the affinity of the heavy metal for a particular ligand and their respective concentrations, as well as independent variables such as redox potential and pH.

Widespread evidence suggests that organic matter in various forms is the most important scavenger of heavy metals in natural waters (Jackson, 1978; Balistrieri *et al.*, 1981; Hart, 1982; Nriagu *et al.*, 1982; Fischer *et al.*, 1986; Veron *et al.*, 1987), even in nutrient-poor or basic waters (e.g. Sigg, 1985; Baron *et al.*, 1986). Particularly high proportions of particulate heavy metal have been reported to be associated with phytoplankton in productive lakes and seas during bloom periods (Hamilton-Taylor, 1979; Hamilton-Taylor *et*

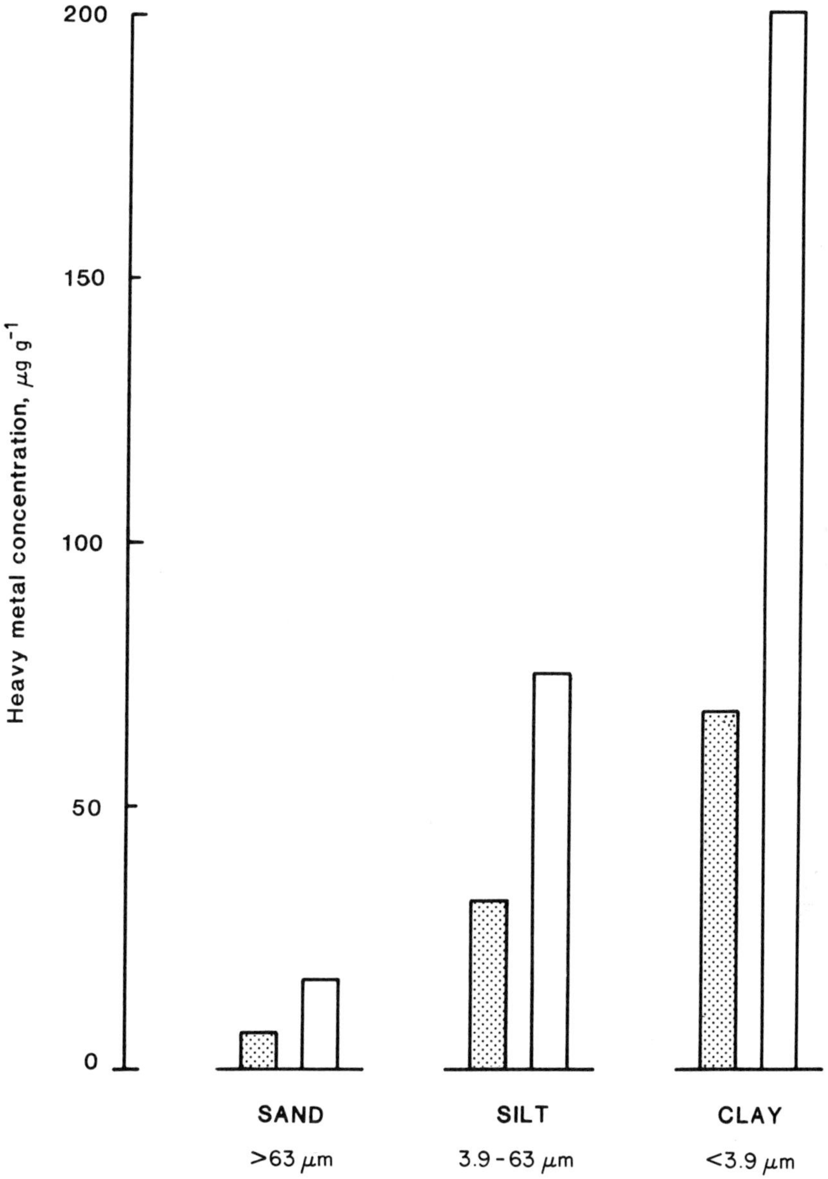

Fig. 3. Partitioning of non-lithogenous concentrations of copper (shaded bars) and zinc (unshaded bars) in sand, silt and clay fractions of sediments from Little Traverse Bay, Lake Michigan. Data of Filipek and Owen (1979).

al., 1984; Valenta *et al.*, 1986), and the apparently active accumulation of heavy metals by these organisms of the order of 100 000-fold for lead, 560 000-fold for zinc and 34 000-fold for copper has been observed (Denny and Welsh, 1979; Deniseger *et al.*, 1986). Consequently, when a diatom bloom sinks it can make a significant contribution to heavy metal sedimentation (Denny and Welsh, 1979; Sigg *et al.*, 1987). Deniseger *et al.* (1986) rightly pointed out that the poisoning of the diatom flora of a lake would have profound consequences for the cycling of heavy metals.

Heavy metal scavenging in the water column can be influenced by many independent factors but, in the context of pollution monitoring, probably the most important of these is pH. H^+ ions have a competitive advantage over heavy metals for exchange sites, and affect the equilibria between dissolved and particulate phases of many heavy metal compounds. Lowered pH therefore reduces the efficiency of many of the important scavenging mechanisms listed above. Increased concentrations of soluble metals—and an implied decrease in scavenging—have been widely observed in acidified lakes in Sudbury, Ontario (Nriagu *et al.*, 1982; Nriagu and Gaillard, 1984), Scandinavia (Henrikson and Wright, 1978; Dickson, 1980), and Switzerland (Sigg, 1985). Conclusive evidence to support this process was provided by Yan and Dillon (1984) in a study of lakes in the Sudbury district which had become acidified, were subsequently limed, and then became re-acidified.

The net efficiency of heavy metal scavenging, and of settling of the heavy-metal-rich particles, can be expressed as the residence time of the metal in the water column. Sigg (1985) has shown that the residence time of metal in a lake is limited at the lower extreme by the particulate settling rate, and at the upper extreme by the residence time of the water. Typical residence times for heavy metals in lakes and coastal waters range from 1–2 months for lead in the Southern California coastal zone (Bruland *et al.*, 1974), to 6–10 months for copper, nickel and zinc in lakes in the Sudbury area (Nriagu *et al.*, 1982), and 8·5 years for cadmium in Lake Michigan (Muhlbaier and Tisue (1981). Residence times of metals in oceans are several orders of magnitude higher, for example, 10^2 years for lead and copper (Goldberg and Arrhenius, 1958; Flegal and Patterson, 1983) and 10^4–10^5 years for cadmium, nickel and zinc (Goldberg and Arrhenius, 1958; Balistrieri *et al.*, 1981). Within any particular sedimentary system, the order of residence times is usually remarkably constant, namely Pb < Cu < Zn < Ni < V,Hg < Cd.

(b) The sediments. In the first stage of sedimentation, fine particulate matter, e.g. seston, forms a layer over the bottom sediments (epipelon) and on submerged rocks, plants and sedentary animals (epilithon, epiphyton and epizoon), before becoming part of this bottom ooze and mud deposit

(Verdouw *et al.*, 1987). This newly-settled material is particularly susceptible to physical and chemical remobilization.

The physical process of sediment redistribution—sediment focusing (Likens and Davis, 1975)—which is particularly relevant to lakes, consists of several distinct processes, of which the most significant are linked to the thermal cycle and seasonal turbulence (Hilton, 1985; Hilton *et al.*, 1986). The net chemical consequence of sediment focusing is a proportionately greater loading of heavy metal in the sediment at deeper zones in the lake (Evans and Rigler, 1980, 1985; Dillon and Evans, 1982). The intermediate stages of the horizontal redistribution of metal are, however, not well understood. Thomas (1972) related mercury focusing in Lake Huron to the higher productivity of localized areas of the basin, and corresponding enrichments of the sediments in these areas when heavy-metal-rich phytoplankton blooms sank, and Kahl *et al.* (1984) observed focusing of lead, but not of zinc, in the sediments of a New England lake. Some insight into metal cycling during this first stage of sedimentation was provided by Everard and Denny (1985a,b) in a study of Ullswater, Cumbria. They proposed that newly settled seston, which was repeatedly resuspended during turbulent periods, provided a means for the continual recycling of lead in the lake, both by contact with submerged plants and by redissolution of lead and uptake by phytoplankton. These processes could provide a mechanism for the chemical fractionation observed by Kahl *et al.* (1984).

When newly deposited material enters the primary sediment layer, heavy metal associations are subjected to diagenetic modification. The first of these is the degradation of organic matter, which takes place rapidly in aerobic conditions and slowly in anaerobic conditions. Metal release from associations with organic matter usually occurs in the top 1–3 cm of sediment (Garrett and Hornbrook, 1976; Fischer *et al.*, 1986) and metals are released in the order $Cd > Zn > Cu > Pb$ (Jackson, 1978). Thus, particularly high benthic fluxes of up to 90% of cadmium have been reported (Hamilton-Taylor *et al.*, 1984; Gobeil *et al.*, 1987), whereas lead displays more conservative behaviour.

The subsequent fate of the released metal is determined primarily by the redox potential of the sediments and the overlying water, which in turn is linked to fundamental lake parameters such as morphometry, thermal cycle and trophic level (Frevert, 1985). Lu and Chen (1977) have suggested that metals can be classified into three groups on the basis of their response to redox potential: those which are fixed in reducing conditions (Cd, Cu, Ni, Pb, Zn), those which are fixed in oxidizing conditions (Fe, Mn), and those which are unaffected by redox potential (Cr, Hg). In extremely reducing, organic-rich sediments (such as are found in deep, permanently anoxic basins), heavy metals can form highly stable complexes with transformed organic matter, and thus be retained in the pore waters (Presley *et al.*, 1972; Elderfield and Hepworth, 1975; van den Berg *et al.*, 1987). In seasonally anoxic, eutrophic sediments, however, sulphide has been shown to be an important sink for heavy metals, as the degradation of organic matter generates a plentiful supply of sulphide in the surficial sediments (Presley *et al.*, 1972; Thomas, 1972; Jackson, 1978; Gendron *et al.*, 1986). When the periodic oxygenation of the surface sediments occurs, oxidation of sulphide to sulphate results in the release of some heavy metal to the overlying water, as reported in the Saguenay Fjord, Quebec (Lu *et al.*, 1986), and Lake Kinneret, Israel (Frevert, 1987; Frevert and Sollmann, 1987). In such instances, the order of release of metals was reported to be Zn > Cd > Cu > Pb, whilst the absolute amounts released depended on the duration of oxidizing conditions and rate of sediment accrual.

In water bodies which experience more frequent oxygenation of the bottom waters and/or low productivity, co-precipitation with iron/manganese oxides is a more important mechanism of metal sedimentation than sulphide formation. The iron/manganese cycle is well known and is described, for example, by Davison (1985). Briefly, when the hypolimnion and/or sediments are anoxic, soluble Fe(II) and Mn(II) tend to diffuse upwards through the pore water and water column to the redox boundary where they are oxidized to Fe(III) and Mn(IV), and are precipitated back to the sediment surface. When the redox boundary occurs in the sediments, the accumulation of iron and manganese is below the sediment surface. These freshly formed iron and manganese oxides have been shown to be very efficient scavengers of heavy metals released by the degradation of organic matter (Loring, 1976; Sigg, 1985; Balistrieri and Murray, 1986); accordingly, surface or subsurface enrichments of Fe/Mn oxides in sediments are often associated with peaks in concentrations of lead, cadmium, zinc, nickel and vanadium (Mackereth, 1966; Ochsenbein *et al.*, 1983) (Fig. 4). Alternation of aerobic and anaerobic conditions at the sediment–water interface can, under some circumstances, give rise to a series of Fe/Mn peaks and associated enrichments of heavy metals in the sediment record (e.g. Ingri and Pontér, 1986). When there is an extended aerobic period in the hypolimnion, the prolonged deposition of iron and manganese at the sediment surface can lead to the formation of crusts, which can also act as a focus for heavy metal accumulation (Gorham and Swaine, 1965).

The acidification of lakes has recently received much attention, as a lowered pH may not only reduce heavy metal scavenging (as discussed above), but also remobilize metals from the surface sediment horizons. Decreases in the concentrations of heavy metals, especially zinc, in recent sediments have been widely reported and attributed to acidification (e.g.

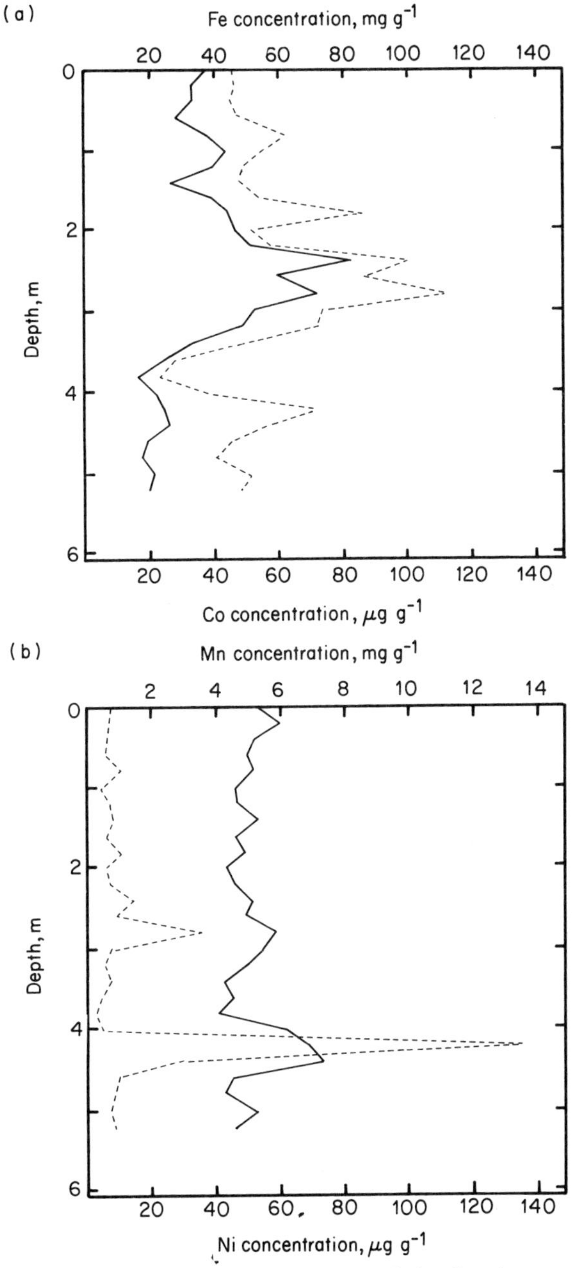

Fig. 4. Sediment profiles from Esthwaite Water, Cumbria, showing concentrations of: (a) cobalt (solid line) and iron (broken line); and (b) nickel (solid line) and manganese (broken line). Redrawn from Mackereth (1966).

Davis *et al.*, 1983; Norton, 1986; Nriagu and Rao, 1987). Experimental evidence confirms that zinc is more susceptible to acid-leaching than is lead (Kahl *et al.*, 1984), but the reported threshold of pH 3–4 for major solubilization of metals (Kahl *et al.*, 1984; Arafat and Nriagu, 1986) is too low to explain the depletion of metals in the surface sediments of, for example, the Adirondack Lakes, where the pH has not fallen below 5 (Davis *et al.*, 1982; Norton, 1986). An alternative hypothesis (Carignan and Tessier, 1985) proposes that subsurface peaks in heavy metals are generated as a result of the diffusion of soluble metals from the acidified bottom waters to the less acidic pore waters of the surficial sediments, where they are deposited as sulphides, as shown in Fig. 5.

Two other generally applicable, although less important, agents of metal remobilization are redistribution by aquatic plants and bioturbation. The possibility that rooted macrophytes may bring about metal remobilization from deep sediments is supported by observations: firstly, that some species growing in polluted sediments accumulate high heavy metal concentrations in their roots (Mayes *et al.*, 1977; McIntosh *et al.*, 1978), and secondly, that metals may be translated from roots to shoots, and thence released to the water (Welsh and Denny, 1980). The latter authors pointed out, however, that metal uptake would not necessarily involve depletion of the sediments at depth, but for lead, could involve metal recycling at the sediment–water interface. Clearly, if relocation does occur, it is restricted to the marginal zone of lakes, and is limited by metal toxicity to water plants (Harding and Whitton, 1978).

Bioturbation is a more widespread agent of metal redistribution, and is limited only by the extent of aerobic conditions in the sediments. The consequences are two-fold: firstly, the mechanical mixing of the surficial sediment layers; and secondly, the aeration of the sediments, which facilitates the chemical remobilization of metals. The phenomenon can be detected either visually, as a flattening of surface peaks in metal concentrations (e.g. Edgington and Robbins, 1976; Robbins, 1982; Veron *et al.*, 1987), or radioisotopically (Benninger *et al.*, 1979). A simple model to simulate the effect of bioturbation constructed by Johnson *et al.* (1986) showed that mixing to a depth of 3 cm could lead to an underestimation of surface concentrations of heavy metals of as much as 50%, whilst Christensen and Goetz (1987) have demonstrated recently that bioturbation can result in the pre-dating of the onset of pollutant deposition by as much as 15–30 years.

These conclusions suggest that the extent to which aquatic sediments constitute a conservative sink for heavy metals varies considerably. Mass balance calculations based on sediment analyses, and independent estimates of atmospheric flux and fluvial input and output, substantiate this conclusion. Retention rates of metals in lakes of between 70% and 90% (lead and

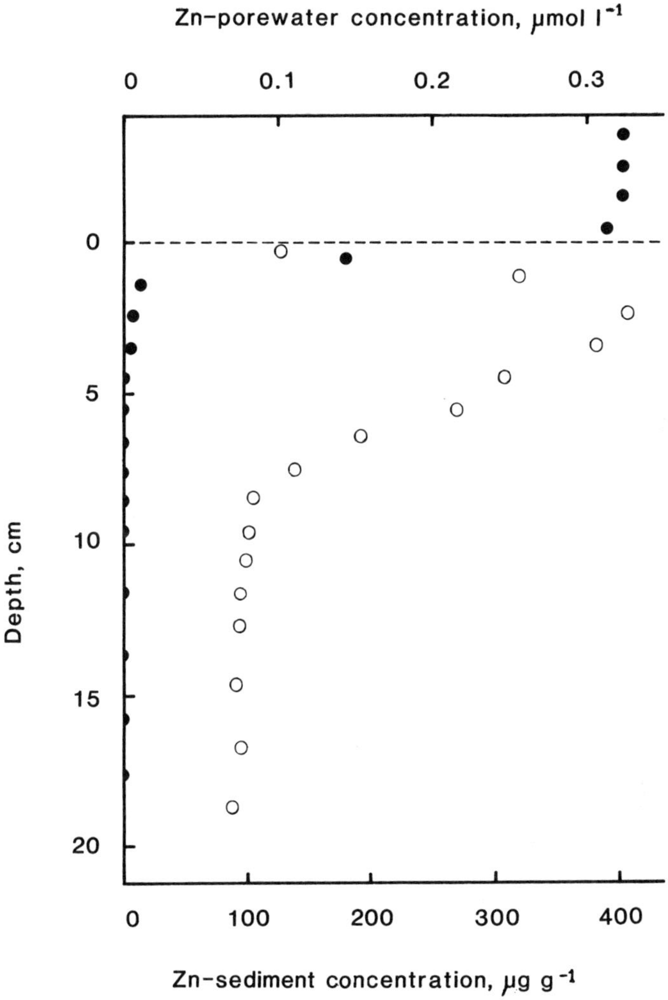

Fig. 5. Concentration of zinc in solid sediment (open circles) and porewater and overlying water (closed circles) in a profile through the lower part of the water column and the upper sediments from Clearwater Lake, Ontario. The broken line represents the sediment/water interface. Redrawn from Carignan and Tessier (1985).

zinc) and *c*. 60% (cadmium and copper) have been calculated by Baier and Healy (1977), Harding and Whitton (1978) and Sigg (1985). Losses are most often ascribed to the long residence times of seston-bound metals in the water column, combined with efficient flushing (Baier and Healy, 1977; Nriagu and Wong, 1986). Clearly, conditions for pollution monitoring are most favourable in non-acidified, productive lakes with a high sedimentation rate (Jackson, 1978; Sigg, 1985), and particularly in closed-basin, meromictic lakes (Evans and Dillon, 1982), where retention of metals may be virtually total.

3. Cycling and Redistribution in Peat Ecosystems

It is already well-established (e.g. Aaby *et al*., 1979; MARC, 1985) that the only types of peat deposits which are suitable for recording changes in atmospheric heavy metal deposition are those formed in ombrotrophic systems (raised mires and blanket mires), where the atmosphere is the sole source of minerals. Consideration is accordingly limited to these systems and minerotrophic peats and soils are not discussed.

(a) Heavy metal–peat associations. The initial associations of heavy metals in the peat system are determined by the nature of the surface on which the metals are deposited—whether bryophyte- or angiosperm-dominated, or bare. Heavy metals are incorporated particularly efficiently by bryophytes, where the effective absence of a vascular system necessitates a relatively unrestricted exchange of solutes between the atmosphere and the living plant tissue. In the aerial organs of higher plants, by contrast, there is poor penetration by atmospheric heavy metals, particularly when in larger particulate form (Hughes *et al*., 1980). Greater absorption may occur, however, after the death of a plant organ when the cuticle breaks down (Ault *et al*., 1970).

Once penetration has occurred, the principal site of accumulation of atmospheric heavy metals in both higher plants and cryptograms is the cell wall and intracellular membranes (Hughes *et al*., 1980). Knight *et al*. (1961) and Clymo (1963) have shown that the heavy metal binding ability (measured by cation exchange capacity) of a range of plant species is closely related to their polyuronic acid content, as shown in Table 2. The exceptional ability of *Sphagnum* spp. to accumulate heavy metals is thus largely attributable to their high (up to 30%) content of these acids (Clymo, 1963). Clymo (1963) has proposed that the marked differences in the heavy metal accumulative properties of different ombrotrophic mire species of *Sphagnum* may give rise to considerable vertical and horizontal heterogeneity in the distribution of ions in peat deposits.

Associations formed initially between heavy metals and bog plants

Table 2
Cation exchange capacity of bog plants, humic acids and peat.

Material	pH	CEC (mmol H^+ g^{-1})	Source
Flowering plants growing on peat	7·0	0·12–0·33	Knight *et al.* (1961)
Mosses growing on peat	7·0	0·30–1·0	Knight *et al.* (1961)
Sphagnum spp.	7·0	1·2	Clymo (1963)
Humic acid from bog peat	4·5	1·7	Kononova (1966)
Humic acid from bog peat	6·4	2·86	Kononova (1966)
Humic acid from bog peat	8·1	4·0	Kononova (1966)
Peat	—	0·2–2·0	Given and Dickinson (1975)
Raised bog peat	3·0–3·5	0·5–1·27	Gorham (1953a)
Blanket peat	5·2	0·78	Gorham (1953b)

undergo subsequent modification as a result of biomass cycling in the surface peat. In bryophyte-dominated communities, heavy metals are progressively concentrated in the ageing tissues as decay proceeds and more volatile constituents are lost as the lower portions of the shoots are converted into peat (Tamm, 1953; Rühling and Tyler, 1970; Tyler, 1972; Pakarinen, 1978). In vascular plant communities the boundaries between the different age-categories are more distinct. Passive heavy metal uptake increases sharply with death and the onset of decay, and continues to increase as decomposition proceeds (Nilsson, 1972; Tyler, 1972). In blanket mire communities the above-ground litter may act as a temporary heavy metal reservoir (Livett, 1982); the turnover rate of this litter, which can vary from one year in *Eriophorum vaginatum* (northern Pennines: Forrest, 1971) to as long as 20 years in *Calluna vulgaris* (Cumbria: Gore and Olson, 1967), determines the time-lapse before the heavy metals are incorporated into the main peat reservoir. If the metal has a long "residence time" in plant litter, sheet erosion may remove a considerable proportion before it can enter the stable peat matrix. This phenomenon is particularly prevalent in degraded blanket mires (Imeson, 1974; Tallis and Yalden, 1984), and is a major contributory factor to the loss of minerals (Crisp, 1966). Livett (1982) estimated that in a badly eroded area of blanket peat in the southern Pennines approximately 95% of all the lead deposited since the beginning of the Industrial Revolution had been lost as a result of sheet erosion. In such extreme circumstances, peat cannot be considered as even a semi-conservative reservoir for heavy metals, and is unsuitable for pollution monitoring.

As plant litter is converted into peat, its heavy metal content is increasingly influenced by the associations formed with newly-synthesized organic matter. The precise nature of the process whereby plant material is transformed into peat is still imperfectly understood, but it is acknowledged that the two complementary processes of decomposition and synthesis are involved. Micro-organisms break down the less resistant plant constituents and then reassemble them by condensation reactions into more resistant macromolecules of a sort which are not found in the original plants (Given and Dickinson, 1975; Schnitzer and Khan, 1978). Accordingly, peat of increasing age comprises a diminishing proportion of hemicelluloses, a nearly constant proportion of the more resistant lignins, and an increasing proportion of "humic substances"* (typically 1–2% in the litter horizon and 15–20% in deeper peat: Given and Dickinson, 1975).

Because humic substances are the major reactive constituents of peat, and also important agents of metal-binding (Schnitzer and Khan, 1978), it follows that the behaviour of heavy metals in peat is largely a function of humic acid-heavy metal associations. The associations between heavy metals and peat or humic acid are dominated by cation exchange/chelation processes (van Dijk, 1971; Livett *et al.*, 1979; Tummavuori and Aho, 1980), where the predominantly aromatic subunits of the humic acid macromolecules—phenolic, hydroxyl and carboxyl groups (Randhawa and Broadbent, 1965)—furnish ligands for the chelation of metal ions (Kononova, 1966; van Dijk, 1971). Complexation by porphyrin residues is thought to be of some importance in copper–peat associations (Goodman and Cheshire, 1976). Livett *et al.* (1979) reported that a significant proportion of zinc, but not of lead, in homogenized whole-peat samples could be removed by strong acid (0·05 M H_2SO_4), whereas over 60% of the copper resisted leaching by acids and EDTA.

The intrinsic ability of humic substances in peat to bind inorganic ions is often expressed as "cation-exchange capacity" (CEC), a loose term which carries no implications of the precise nature of the chemical binding involved. Values of cation-exchange capacity for humic acid isolated from peat are generally higher than those of whole-peat samples (Table 2). Clymo (1983) suggests that some of the variation between published values of CEC, however, may be attributable to the use of different analytical techniques. Nevertheless, different peat types certainly do exhibit different values of CEC, so that considerable variation in the concentration of inorganic ions may occur between, for example, blanket peat and raised mire peat (Gorham,

*The terms "humic substance" and "humic acid" denote a group of compounds which is structurally heterogenous, but which can be operationally defined by a simple chemical procedure (Clymo, 1983).

1953a; Walsh and Barry, 1958). It has been suggested that the CEC of "humic substance" increases progressively with time due to the increased polymerization of the constituent macromolecules (Hildebrand and Blum, 1975). Experimental evidence to support this theory, however, is conflicting and inconclusive (Matsuda and Ito, 1970; Stevenson, 1977; Aaby and Jacobsen, 1979).

In general usage, "cation-exchange capacity" is expressed as H^+ equivalents and defines the maximum ability of, for example, "humic substance" to bind cations. The ability of "humic substance" to bind heavy metals, however, is invariably lower than its CEC (H^+), and varies according to specific characteristics of the metal such as valency, ionic radius, and whether it is alone or in competition (van Dijk, 1971). Binding ability can be satisfactorily expressed by a stability constant, which measures the equilibrium reaction between the metal in solution and in the presence of humic acid. The relative affinities of "humic substances" for different heavy metals, described by these stability constants obtained experimentally, are in the order: Cu > Pb > Cd > Zn > Ni (van Dijk, 1971; Bunzl *et al.*, 1976; Stevenson, 1977). The markedly higher affinity of "humic material" for copper and lead in relation to other metals was also observed in experiments on whole-peat samples by Tummavuori and Aho (1980).

The most important independent factor to influence heavy metal binding by "humic material" is pH, as verified experimentally by, for example, Randhawa and Broadbent (1965) and Livett (1982). At high concentrations of H^+ ions, the ionization of carboxyl and hydroxyl groups is inhibited, and the number of sites which can participate in ion exchange is reduced (Clymo, 1983). Indeed, cation exchange is so strongly influenced by pH that Clymo (1983) proposed that the term "cation-exchange capacity" should be reserved for those specific instances where the complete ionization of the exchange groups can be demonstrated, and should be replaced by "cation-exchange ability" when the "humic material" is only partly ionized.

(b) The redistribution of heavy metals. The heavy metal associations with peat established during peat formation are further modified by the action of hydrological and biotic factors in the environment.

In ombrotrophic peat, the movement of water exerts a major influence on the redistribution of heavy metals—by seepage of precipitation water through the surface peat and vegetation, and by fluctuations of the water table in the subsurface peat horizons.

The permeability, and thus susceptibility to leaching, of the surface peat depends, firstly, on the extent to which the structure has collapsed (Clymo, 1983). In raised mire and blanket peat water flow is most rapid in hummocks, on account of their well-preserved structure and dryness (Pakarinen and

Tolonen, 1977a,c; Aaby *et al.*, 1979). The low density of hummock peat, and consequently low bulk concentration of exchange sites for the chelation of heavy metals, also increases the likelihood of redistribution of these metals (Pakarinen and Tolonen, 1976; Pakarinen, 1978). In saturated "hollow" peat, however, there is considerable resistance to passage of water; moreover, it has been shown that a high proportion of the total heavy metal content of such peat is strongly bound, and thus resistant to remobilization (Aaby *et al.*, 1979). The redistribution of metal in peat by this mechanism can be recognized by the presence of corresponding depletion and enrichment horizons in the peat profile (Tyler, 1972; Pakarinen and Tolonen, 1977c; Clymo, 1983; Pakarinen and Gorham, 1983), as shown in Fig. 6. Aaby and Jacobsen (1979) pointed out that the exact position of the enrichment horizon of leached metal, which may be mistakenly ascribed to an anthropogenic effect, often coincides with the junction of poorly- and well-humified peat, indicating an impedance of water flow at that horizon.

In subsurface peat, probably the most important agent of heavy metal remobilization is the fluctuation in redox potential associated with water-table movements on a seasonal basis, described by Clymo (1983). The redistribution of heavy metal by this process is particularly well illustrated by the study of metal distribution in Swedish raised mires by Damman (1978). Peaks in the concentrations of lead and zinc in adjacent hummock and hollow profiles corresponded with the position of mean water-table levels in these two profiles (Fig. 7); these peaks were accordingly attributed to the remobilization and accumulation of metals at the aerobic–anaerobic boundary in the peat. Aaby *et al.* (1979) later criticized Damman's conclusions on the grounds that the peat growth rate had not been taken into account, and that the heavy metal enrichments might be explained alternatively in terms of changing atmospheric inputs of metal with time.

In peats with a high degree of humification and low hydraulic conductivity even in the surface layers (e.g. the southern Pennine peats: Tallis, 1973), the concept of free water-table is practically meaningless, and bears little or no relation to the zone of sulphide formation. Accordingly, in studies of southern Pennine peats, Lee and Tallis (1973) and Livett *et al.* (1979) found no evidence to support the existence of large-scale redistribution of heavy metals connected with water-table movements.

As discussed above, a significant fraction of heavy metals in peat is in chelatable form (Livett *et al.*, 1979), and it is this fraction which is potentially accessible to uptake by vascular plants, and therefore susceptible to translocation. The extent to which inorganic constituents can be relocated by plants depends on the depth of root activity in the members of the mire community, which can vary considerably from, for example, 15 cm in ericaceous species of British heathland (Boggie *et al.*, 1958) to at least 150 cm in *Rubus*

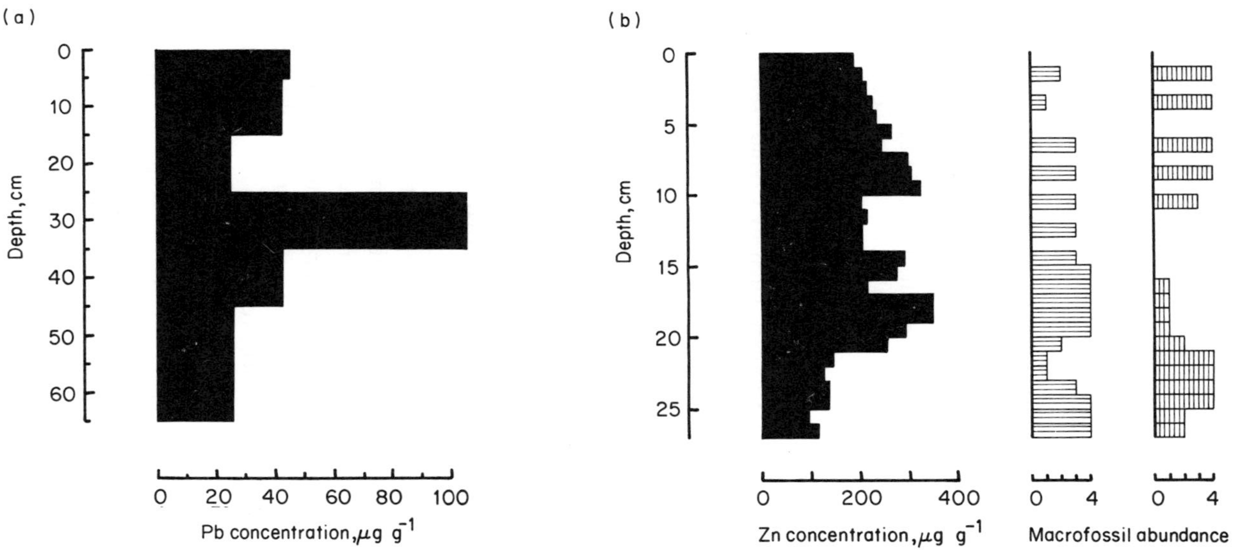

Fig. 6. Heavy metal distributions in peat profiles, showing: (a) lead concentration in a *Sphagnum fuscum* hummock from Myras Mire, Finland (redrawn from Pakarinen and Tolonen, 1977a); and (b) zinc concentration and abundance of macrofossils of *Sphagnum* sp. (horizontal hatching) and *Eriophorum vaginatum* (vertical hatching) in a hummock from Moor House, Cumbria (redrawn from (Livett *et al.*, 1979).

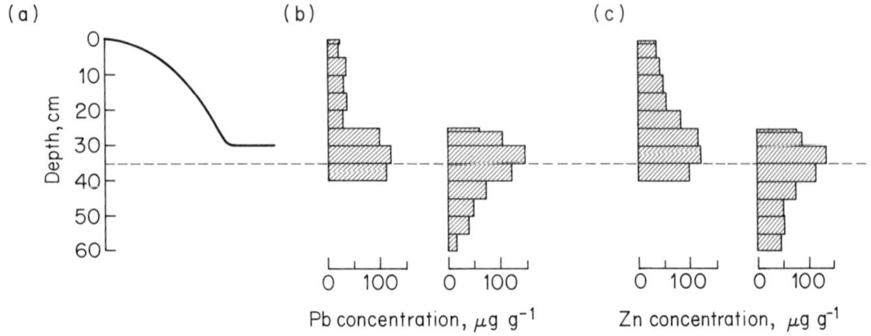

Fig. 7. Profile through a peat hummock and adjacent hollow on Store Mosse, Sweden, showing: (a) surface topography; (b) lead concentration; and (c) zinc concentration. Broken line shows the mean position of the water-table. The left-hand of each pair of concentration profiles represents the hummock, the right-hand the hollow. Redrawn from Damman (1978).

chamaemorus in northern Norway (Stavset, cited by Clymo, 1983), and on the diffusion coefficients of cations in peat (Clymo, 1983). The continuous upward recycling of heavy metals by plants has been widely cited as the mechanism of surface enrichment in Scandinavian peats (Hvatum, 1971; Sillanpää, 1972; Yliruokanen, 1976). More recent workers (e.g. Aaby *et al.*, 1979) have argued, however, that only some of the minor anomalies in heavy metal distribution, particularly in hummocks, are likely to be attributable to this mechanism, and that high enrichments at the surface are more likely to be of pollutant origin. Instances of small-scale redistribution of heavy metals by *Sphagnum* and by higher plants have been reported by Rühling and Tyler (1970), Pakarinen (1978), Pakarinen *et al.* (1980) and Livett *et al.* (1979).

The accumulation of heavy metals is, therefore, most effective in actively-growing peat with bryophyte-dominated vegetation, and subsequent retention is favoured in highly-humified peat or in waterlogged conditions. Hummocks provide unfavourable conditions for pollution monitoring, and the metals most likely to be remobilized are cadmium, nickel and zinc. Degenerative peat, which is very ineffective as a sink for heavy metals, is unsuitable for pollution monitoring.

4. Cycling and Redistribution in Ice Deposits

In ice and snow, in contrast to aquatic sediments and peat, heavy metals are held by physical bonds in the ice crystalline matrix, and virtually no chemical transformation occurs after deposition has taken place.

The stability of metals in ice deposits depends solely on meteorological factors. Newly-deposited snow layers containing contemporary inputs of heavy metals can be subject to mass redistribution by strong winds, notably in parts of Antarctica (Picciotto *et al.*, 1964; Boutron, 1982), but these effects are usually limited to the current year's or season's snow accumulation. The subsequent integrity of the stratified deposit, consisting of successive winter and summer layers, depends on the constant maintenance of a subzero temperature (Jaworowski *et al.*, 1975). Occasionally, episodes of melting may result in a slight merging of these annual layers, but evidence suggests that surface meltwater is usually re-frozen before it penetrates to any great depth (Murozumi *et al.*, 1969). Similarly, in non-permanent snow deposits, the distinct stratification produced by individual snow episodes attests to the integrity of the deposit (Elgmork *et al.*, 1973).

B. Chronology and Dating

In geochemical monitoring the age of deposit sets a limit on the time-span of the heavy metal record, whilst the rate of accrual determines the degree of definition of the record. The use of an appropriate dating technique enables these parameters to be defined.

1. Age and Accumulation Rate

Most deposits used for pollution monitoring date back much further than is required for the scale of the investigation (see, however, discussion in Section IV.C). Ocean sediments have the oldest origins—*c.* 160 million years (m.y.) BP—limited by the rate of sea-floor speading (Heirtzler, 1968). Inland water bodies exhibit a broad age-range, according to the diverse circumstances of their formation (Hutchinson, 1957). The oldest lakes are probably those formed during major tectonic activity in the Tertiary period (up to 65 m.y. BP), but glaciation is almost certainly the most important agent of formation of extant lakes on a world-wide scale. Most lakes in this category, for example the English Lakes, the North American Great Lakes, Lakes Constance and Geneva, therefore date back approximately 1 m.y. The youngest water bodies are those formed as a result of catastrophic volcanic activity and, often of particular interest in pollution monitoring, man-made reservoirs. The latter usually provide records dating back no further than 150–200 years.

Many of the world's major ice sheets were initiated before the onset of the Pleistocene glaciations, at *c.* 3 m.y. BP (Arctic: Andrews, 1979), 7 m.y. BP (Andes: Kennett, 1977), and before *c.* 22 m.y. BP (Antarctic: Mercer, 1984). Most of the world's peat deposits started forming no more than 9000 years ago (Rybniček, 1973), and are therefore young in a geological perspective. In particular, many ombrogenous peats, with which this review is primarily concerned (e.g. the blanket peats of the British Isles), were initiated considerably later, at *c.* 3000–7000 yr BP (Tallis, 1983).

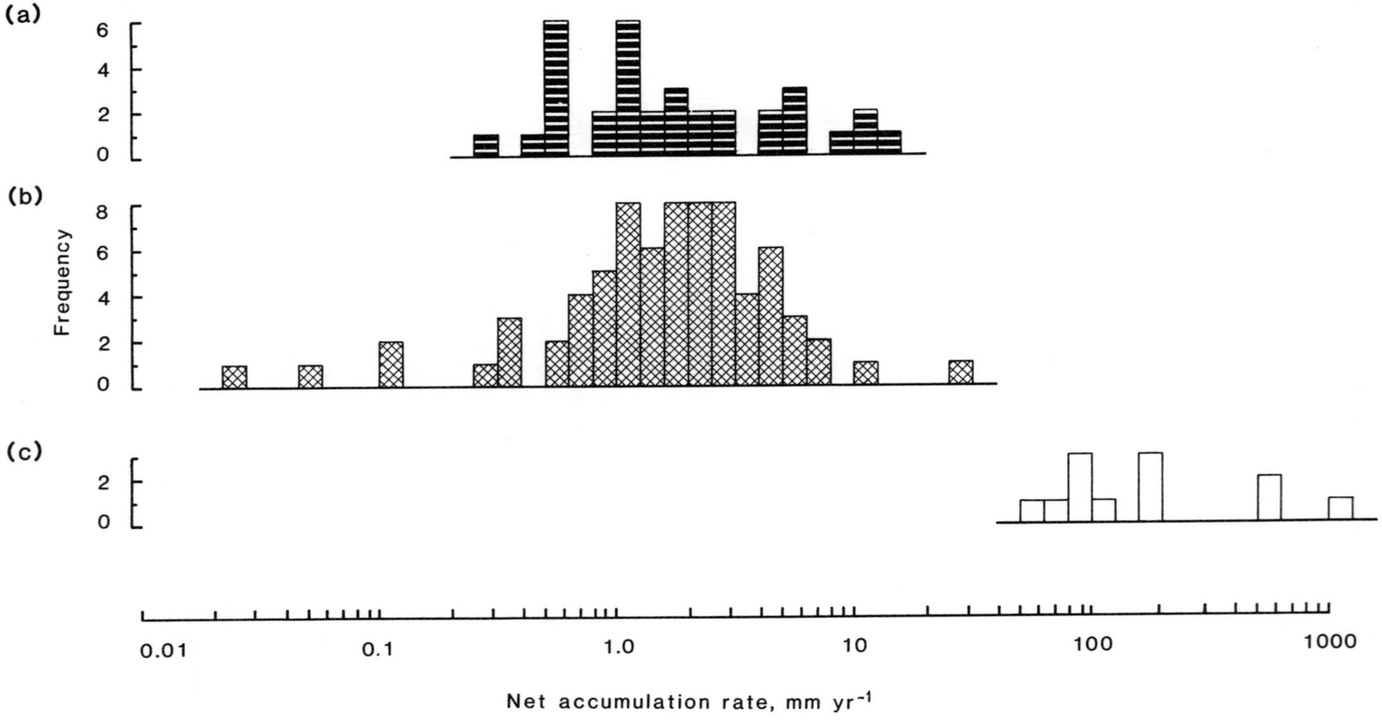

Fig. 8. Frequency distribution of net accumulation rates (log scale) of: (a) peat; (b) aquatic sediments; and (c) ice, snow and firn deposits. Data from various sources listed in the references.

There is a wide range, spanning five orders of magnitude, in the accumulation rates of deposits used in geochemical monitoring (Fig. 8). Ice and snow generally have high accumulation rates and hence reflect short-term trends in pollutant input with great accuracy, for example annual and seasonal fluctuations (Weiss *et al.*, 1971; Herron *et al.*, 1977). In temperate regions, the contributions made by individual snowfalls to ephemeral deposits can often be distinguished (e.g. Elgmork *et al.*, 1973). A major disadvantage of a high accumulation rate in a permanent deposit is that a considerable depth of firn needs to be excavated in order to observe pre-industrial to post-industrial trends in heavy metal deposition (Boutron, 1987).

Lake and continental shelf sediments accumulate at a much slower rate than do ice and snow, thus providing a more compact stratigraphic record, but one in which annual and seasonal layers are still occasionally recognizable. Sediments of the continental slopes and mid-oceans, in contrast, have accumulation rates at the extreme end of the range (of the order of a few mm per century), which creates considerable interpretative problems.

In peat deposits, the accumulation of material is a much more complex process than in ice or aquatic sediments, so, as Clymo (1984) explains, accumulation rate is a difficult concept. Peat formation consists of the addition of newly-dead material at the bog surface, partial decomposition of these remains in the acrotelm (the aerobic peat), and the continued, slower decay of material in the catotelm (the anaerobic peat). When the accumulation of dry matter at the surface is exactly equalled by the loss of matter by decay deeper down, the net accumulation of peat is zero. The terms accumulation rate and growth rate, therefore, are used here to describe the net, apparently linear, growth rate of a particular section of the peat profile (as described by Clymo, 1978).

The frequency of peat accumulation rates shown in Fig. 8 tends towards a bimodal distribution. This represents, on the one hand, the higher decay rates in the predominantly aerobic upper peat horizons of hummocks and retrogressive bog surfaces (e.g. southern Pennines), and on the other hand, the slow decay rates in the largely waterlogged peat below *Sphagnum* pools and lawns (e.g. Scandinavian *Sphagnum fuscum* peat). Although the higher accumulation rates might reflect small-scale changes in heavy metal deposition more accurately, the discussion in Section I.A suggests that the greater degree of heavy metal mobility in such peats may make them less suitable for pollution monitoring than slower-growing, well-humified peats.

2. Dating Techniques

The techniques which have been used for dating aquatic sediments, peat and ice in pollution monitoring can be divided into four groups: stratigraphic methods; biostratigraphic methods; radiometric methods; and geomagnetic methods. Not all of these provide the same type of information, so it is often useful to employ two or more complementary techniques. Most of the techniques are described in full in reviews by Wise (1980), Lowe and Walker (1984) and Bradley (1985). The following discussion will therefore briefly outline each method and its particular relevance in pollution monitoring. Table 3 summarizes the important features of the methods.

(a) Stratigraphic techniques. The commonest types of stratification in natural deposits are of climatic origin. In permanent ice deposits, the alternation of winter layers (produced by bulk deposition) with summer layers (consisting mainly of dry-deposited material) produces a stratification which can be characterized by measurements of the $^{18}O:^{16}O$ ratio (which is indicative of seasonal temperature fluctuation) and analysis of dust and sea salt concentrations (indicative of seasonal fluctuations in long-range transport of aerosols). Under ideal conditions, these layers are preserved for between 1000 and 7000 years (in the North Polar, and South Polar regions, respectively). Firn stratigraphy has been used by, for example, Murozumi *et al.* (1969).

Varves are the best known examples of seasonal stratification in aquatic deposits. These features occur in glaciolimnetic and glaciomarine sediments and comprise annual, paired laminations consisting of layers of coarse material eroded from the catchment during spring and summer, alternating with layers of finer clays which settled out during quiescent periods in the winter. Problems with dating by varve counts arise chiefly from variations in varve thickness between regions, and the absence or poor development of varves due to adverse or atypical climatic conditions. The need for accurate dating techniques for recent sediments in pollution studies has led to the utilization of a variety of varve-like formations which occur in sediments in temperate regions. The range of diversity of these features is described by O'Sullivan (1983) and Renberg (1984). Biogenic laminations—alternation of diatom-rich summer layers with predominantly minerogenic winter layers—are the commonest type of pseudo-varves, and have been used by Chow *et al.* (1973), Bruland *et al.* (1974) and Shirahata *et al.* (1980).

Peat profiles do not usually exhibit marked stratigraphic features of a seasonal origin. However, it was recognized by Overbeck and Happach (1957) and Tallis (1959) that seasonal growth periodicity in certain bryophytes is reflected by a cyclical zonation of branching patterns and pigmentation, and that these features can be used to estimate the growth rate of the plant. The moss growth increment method was developed from this basic concept as a dating technique for poorly-humified peats by Pakarinen and Tolonen (1977b) and later verified by ^{210}Pb dating by El Daoushy *et al.* (1982). This technique is useful in pollution studies when dating of the very recent peat horizons cannot be satisfactorily achieved by other methods (e.g.

Table 3
A summary of the principal characteristics of dating techniques used in geochemical monitoring.

Technique	Practicable time-span (years ago)	Type of chronology	Type of deposit[a]	Problems
Stratigraphic	0–1000	Continuous	a i	Geographical variation
Biostratigraphic:				
pollen	0–10 000	Discrete	a p	Geographical variation
diatoms	0–10 000	Floating	a	Geographical variation
Radiometric:				
carbon-14	50–20 000	Continuous	a i p	Distortion in polluted regions
lead-210	0–20	Continuous	a i p	Distortion in polluted regions
thorium-228	0–10	Continuous	a	—
caesium-137	0–(1954)	Discrete	a i p	Mobility and low concentrations in S. hemisphere
iron-55 plutonium-239 + 240 tritium	0–(1954)	Discrete	a i p	Mobility and low concentrations in S. hemisphere
Geomagnetic:				
palaeomagnetic	100–∞	Discrete	a	Coarse time-scale
small-scale	0–∞	Floating	a	Geographical variation

[a]Types of deposit are: a, aquatic sediments; i, ice and snow; and p, peat.

pollen analysis, see below). Its application is limited, however, to very poorly humified peats.

(b) Biostratigraphic techniques. Pollen analysis is the most widely used technique for peat and aquatic sediments, and the practical and theoretical aspects are well established (Moore and Webb, 1978; Birks and Birks, 1980).

As pollution studies are invariably concerned with recent deposits (the last *c.* 200 years), anthropogenic episodes in vegetation history figure prominently in the pollen diagram. For example, in North American studies of the Great Lakes, the sudden influx of *Ambrosia* pollen (marking the massive woodland clearance and cultivation by settlers from 1850 onwards) and the decline in *Castanea* pollen (due to the widespread blight in the 1930s) provide reference horizons for the extrapolation of sedimentation rates (e.g. Kemp *et al.*, 1974; Nriagu *et al.*, 1979). In contrast, Lee and Tallis (1973) and Livett *et al.* (1979) used a number of pollen features to build up a complete dating scheme, as illustrated in Fig. 9. This method has the advantage of providing a nearly continuous time–depth curve for the peat profile, although the authors point out that smaller scale fluctuations in peat growth rate, especially towards the peat surface, cannot be detected.

A source of error in pollen analysis which is peculiar to lake sediments arises from the inwash of pollen from the catchment. This phenomenon, and methods which have been developed to enable redeposited pollen to be distinguished from contemporaneous pollen, are discussed by Birks and Birks (1980). Recent work by Clymo and Mackay (1987) has suggested that redistribution of pollen in peat can also be a source of error. The authors showed that pollen grains and spores can be vertically displaced by at least 1–2 cm in very poorly humified peat so that, even allowing for compaction occurring in the catotelm, results from fine-resolution sampling (< 1 cm) would be meaningless.

A second biostratigraphic technique, diatom stratigraphy, is only applicable to dating lake deposits and is based on variations in the abundance of different diatom species in response to changing conditions, e.g. eutrophication. Although it may sometimes be possible to relate these changes to specific dates in history, the diatom sequence essentially constitutes a "floating" chronology. The construction and application of such a chronology is described by Battarbee (1978) and Rippey *et al.* (1982).

(c) Radiometric techniques. Radionuclides in natural deposits originate either from natural processes or from anthropogenic activities, and it is convenient to divide radiometric techniques into two corresponding groups. The techniques based on the occurrence of natural radionuclides usually provide continuous time-scales, whilst the techniques based on artificial radionuclides furnish single reference horizons.

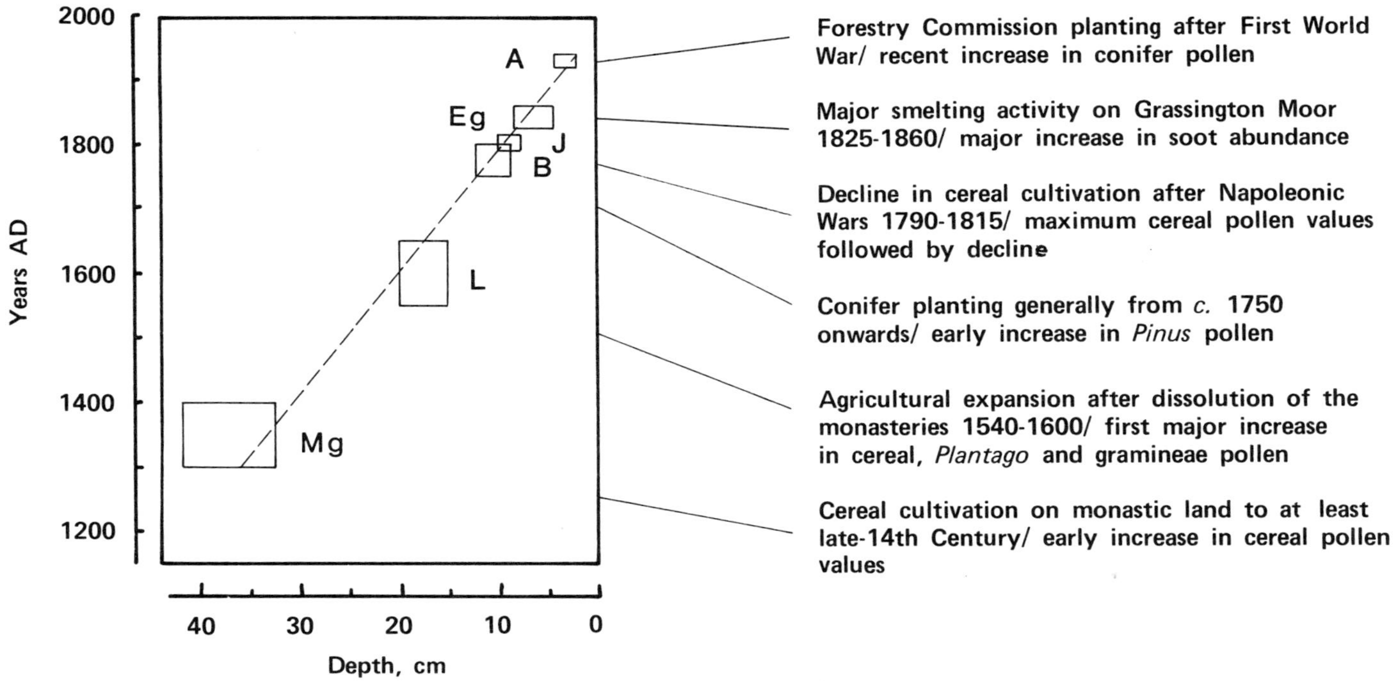

Fig. 9. Dating plot, based on pollen-analytical, macrofossil and documentary evidence, for a peat profile from Grassington Moor, UK. Letters refer to features on the pollen diagram, and rectangles represent their limits in time and space. A summary of the features and their interpretation is given at the right of the plot. Modified from Livett *et al.* (1979).

Radiocarbon dating was the first radiometric method to be developed and its practical and theoretical principles are well established. Although widely used for organic-rich deposits, such as peat (Lee and Tallis, 1973), the technique is also applicable to aquatic sediments and ice (e.g. Murozumi *et al.*, 1969; Boutron and Patterson, 1986).

The main disadvantage of radiocarbon dating in pollution studies is that the long half-life of ^{14}C ($5{\cdot}7 \times 10^3$ yr) means that its accuracy for recent deposits is low. Livett *et al.* (1979) concluded that this intrinsic error, amplified by the need to use a large volume of peat for the measurements, makes this method rather unsatisfactory for dating peat deposits of the last 100 years. A second drawback of this method when it is used for dating recent deposits is the error introduced by the "Suess effect", the dilution of natural ^{14}C in the atmosphere by ^{14}C from fossil fuel combustion, which has a low residual radioactivity (Suess, 1955). This effect is particularly important in heavily polluted deposits, for example, the Baltic Sea sediments (Erlenkeuser *et al.*, 1974). In a study of the South Wales uplands, Chambers *et al.* (1979) reported radiocarbon ages of *c.* 3000 and 3700 years BP for peat deposits which had formed after the start of the Industrial Revolution.

A second group of natural radionuclide dating methods are based on the decay products of ^{238}U, ^{235}U and ^{232}Th—the uranium series. Lead-210 dating is the most widely used method in this group for recent deposits. This method depends on the estimation of ombrogenic ^{210}Pb in natural deposits, and the relationship of its rate of decay to the age of the deposit (Robbins, 1978). The half-life of ^{210}Pb is only 22·3 years, so this technique is particularly useful for dating deposits of age 100–150 years, complementing the role of radiocarbon dating. The technique has been used widely in geochemical investigations for dating marine and lake sediments (Matsumoto and Wong, 1977; Edgington and Robbins, 1976), peat (Aaby *et al.*, 1979; Schell *et al.*, 1986), and ice (Weiss *et al.*, 1971).

A major problem with the use of lead-210 dating for aquatic sediments and peat is that ^{210}Pb, like its stable counterparts, may be susceptible to remobilization (see Section I.A). This phenomenon has been recognized in aquatic sediments by Koide *et al.* (1973) and Durham *et al.* (1980), and in peat by Oldfield *et al.* (1979). A further source of error in lead-210 dating arises from the enhancement of ombrogenic ^{210}Pb in lake sediments with terrigenic material, and from the enhancement of natural ^{210}Pb in all types of deposits with artificial ^{210}Pb from industrial emissions and atomic tests. In each instance, the contaminant material has a younger radiolead age than the ombrogenic material, producing anomalies in the curve of declining ^{210}Pb concentration with increasing age of the deposit (Jaworowski *et al.*, 1975; Oldfield *et al.*, 1978a). Lastly, Imboden and Stiller (1982) have recently suggested that the diffusion of ^{222}Rn (a precursor of ^{210}Pb) in soils may lead to error in the estimation of ombrogenic ^{210}Pb, and the underestimation of sedimentation rate by as much as 30–50%.

Another method based on the uranium series concerns the distribution of ^{228}Th, a granddaughter of ^{232}Th, in marine and fresh waters. The half-life of ^{228}Th is only 1·91 years, which makes this radionuclide very useful for dating sediments which have accumulated over the last decade or so, and for confirming that the sediment–water interface of a core is intact (Koide *et al.*, 1973). The isotopes of thorium are very stable in a fluctuating redox environment.

The second main group of radiometric dating methods are all based on the correlation of the distribution of fission-product radionuclides in natural deposits with particular events in the history of radioactive fallout, notably the commencement of nuclear weapons testing in 1954 and the period of peak fallout in 1963–1964. A general disadvantage of these techniques, noted by Oldfield *et al.* (1980b), is that the transport of fission products to the southern hemisphere is subject to a time-lag, and that their concentrations in these regions are often close to the detection limits.

Caesium-137 dating was developed for lake sediments by Krishnaswami, *et al.* (1971), Pennington *et al.* (1973) and Ritchie *et al.* (1973), and has also been used for dating peat (Aaby and Jacobsen, 1979; Oldfield *et al.*, 1979) and ice (Jaworowski *et al.*, 1975). The distribution of ^{137}Cs in ice, corresponding to specific fallout events, is shown in Fig. 10. In less than ideal conditions, however, inconsistencies can be observed between the observed and expected distributions of ^{137}Cs. First, a large terrigenous input to lake sediments can result in a time lag between the peak fallout and the peak input of ^{137}Cs to the sediments (Müller *et al.*, 1980; Davis *et al.*, 1984). The remobilization of ^{137}Cs in peat and aquatic sediments is a major drawback in the use of this method (Battarbee, 1978; Aaby and Jacobsen, 1979; Oldfield *et al.*, 1979; Tracy and Prantl, 1983), although Clymo (1983) has proposed that the particulate fraction of ^{137}Cs in peat may provide an accurate record of atmospheric fallout.

Other products of atomic weapons testing can be used to provide reference horizons, including ^{239}Pu, ^{240}Pu, ^{90}Sr, ^{3}H and ^{55}Fe. Work on the application of $^{239+240}$Pu dating for marine sediments and peat (Koide *et al.*, 1975; Oldfield *et al.*, 1979) has suggested that terrigenous transport to sediments and remobilization may detract from the usefulness of this method, and ^{90}Sr and tritium have been shown to be mobile under some circumstances in aquatic sediments and peat (Lerman and Lietzke, 1975; Gorham and Hofstetter, 1971; Aaby and Jacobsen, 1979).

(d) Geomagnetic techniques. Geomagnetic chronologies depend on the occurrence in natural deposits of magnetic minerals, originating either from

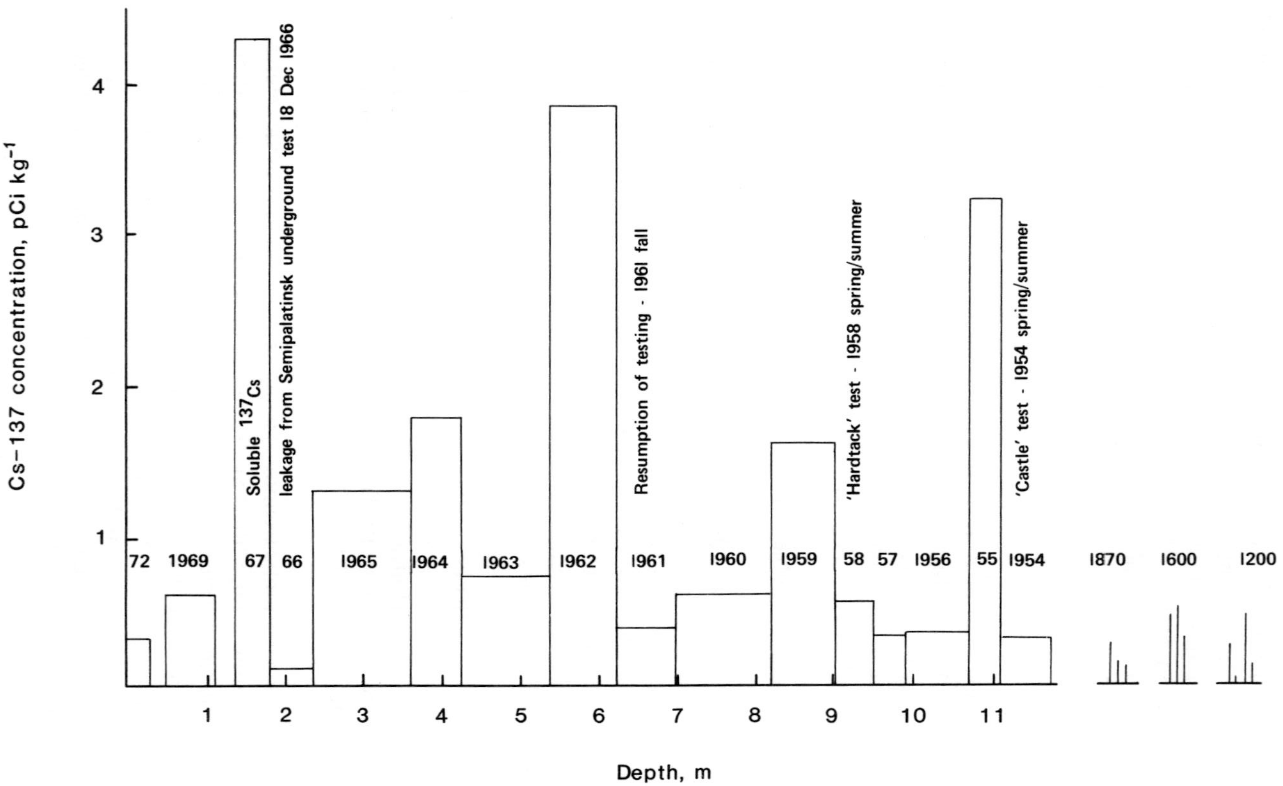

Fig. 10. Distribution of ^{137}Cs concentration with depth in a profile from Storbreen Glacier, southern Norway. Redrawn from Jaworowski *et al.* (1975).

ombrogenous or terrigenous input, or *in situ* formation. These techniques are only applicable to minerogenic deposits. The broad heading "geomagnetic techniques" encompasses a number of different chronologies which can be divided into two groups.

Palaeomagnetic dating is based on the preservation in mineral deposits of a fossil record of changes in the direction and intensity of the earth's magnetic field over at least the last 10 m.y. This record—the natural remanent magnetism (NRM) of a deposit—reflects changes which range in periodicity from 10^6 years (polarity reversals), to 10^4 years (partial reversals) and 10^2 years (secular changes in declination and inclination). The long periodicity of even this last type of record, however, produces a time-scale which is usually too coarse for use in pollution monitoring.

A second group of geomagnetic methods is based on small-scale temporal variations in the amounts and types of magnetic minerals in deposits, and on the specific properties which these confer—namely, magnetic susceptibility (the magnetizability of the sample), and the various functions of isothermal remanent magnetization, IRM (the degree of magnetic moment produced by an applied magnetic field). The variations in these properties are usually highly site-specific and are not necessarily attributable to either regular seasonal, or singular historical events. The susceptibility and IRM profiles of sediments, therefore, form the basis of a floating chronology which is used to correlate synchronous horizons in cores from any one site (e.g. Oldfield *et al.*, 1983). Problems can arise occasionally from the transformation of magnetic minerals in response to changing redox potential, which can deform peaks in the magnetic record (Oldfield *et al.*, 1978b; Hilton and Lishman, 1985).

This discussion shows that aquatic sediments, which often comprise significant minerogenic and organogenic fractions, are amenable to a wide range of dating techniques, but that problems frequently arise as a result of fluctuating conditions in the sedimentary environment, and large terrigenous inputs. There are fewer methods which are applicable to ice deposits, but these are less prone to error. As a non-layered, predominantly organogenic deposit, peat presents the greatest challenges in dating, and there are no satisfactory methods for dating the most recent deposits, those accumulated over the last *c.* 50 years.

C. Interpretation

The discussion in Section III.A has shown that the build-up of the stratigraphic sequence of heavy metals in ice, peat and aquatic deposits is accessible to heavy metal inputs from diverse sources in their geochemical cycles. The interpretation of the heavy metal record therefore involves, first, the quantitative assessment of the total heavy metal input, and second, a discriminatory assessment of the different components of this total metal load.

1. Quantitative Assessment

Snow and ice deposits in Antarctica present virtually no problems in quantitative assessment, as the concentrations of heavy metals in freshly fallen snow closely resemble those in the low-altitude aerosol (see Section III.A). Therefore, the heavy metal ice record accurately reflects historical changes in atmospheric heavy metal concentrations. It follows, also, that there is a linear relationship between the deposition rate of metals and the accumulation rates of snow across Antarctica, as demonstrated by Boutron *et al.* (1972).

In other regions, however, heavy metals are "concentrated" during atmospheric scavenging, and during their subsequent accumulation by deposits. Therefore, to be able to make meaningful comparisons between data from different deposits and different areas, it is necessary to derive heavy metal flux rates from the concentrations in successive sediment slices. In its simplest form, flux rate, F (dimensions $\mathsf{M}_m \cdot \mathsf{L}_d^{-2} \cdot \mathsf{T}^{-1}$), is defined as:

$$F = R \cdot D \cdot C$$

where: R is the accumulation rate of the deposit, d (dimensions $\mathsf{L}_d \cdot \mathsf{T}^{-1}$); D is the bulk density of the deposit (dimensions $\mathsf{M}_d \cdot \mathsf{L}^{-3}$); and C is the concentration of heavy metal m (dimensions $\mathsf{M}_m \cdot \mathsf{M}_d^{-1}$, where M denotes dry mass throughout). This basic formula was used by Livett *et al.* (1979) and Aaby *et al.* (1979) in peat studies. The latter authors made an additional numerical adjustment to the bulk density term to correct for swelling of the peat, which was frozen prior to slicing.

When dry density, ρ, is used rather than bulk density, a porosity function, φ, is incorporated in the formula:

$$F = R(1 - \varphi) \cdot \rho C$$

as given by Hamilton-Taylor (1979).

Few authors have attempted to estimate the error attached to calculated flux rates. Ochsenbein *et al.* (1983) suggested that the calculation of error should take into account uncertainties in the estimates of all the quantities used in the flux rate formula, which could yield an error of as much as 30%.

In geochemical monitoring using peat and ice deposits, heavy metal flux rates derived from a single sample or core are usually assumed to be broadly representative of flux rates over the adjoining area as a whole. In aquatic systems, however, sedimentation rate is rarely uniform over the whole basin, and the vertical sediment–heavy-metal burden tends to vary proportionately with the total depth of sediment (Section III.A); therefore, metal flux rates derived from a single core are usually invalid (Edgington and Robbins, 1976). Evans and Rigler (1980), Muhlbaier and Tisue (1981) and Dillon and Evans (1982) showed that multiple coring over the entire lake basin area, to sample zones of shallow, intermediate and deep sediment accumulation, can yield a whole-lake metal burden, from which the mean heavy metal burden can be derived, as illustrated in Fig. 11.

2. *Discriminative Assessment*

All deposits receive variable inputs of heavy metals from natural and anthropogenic sources, and it is essential in geochemical monitoring to distinguish between these different fractions. The simplest approach, where a complete profile is analysed, and when it can be assumed that the anthropogenic input is increasing while the natural input remains constant, is to define all heavy metal below the inflection point in the profile as "natural", and heavy metal above that point as "natural + anthropogenic". The inflection point is determined either subjectively or statistically (Evans *et al.*, 1983; Johnson *et al.*, 1986). In many instances, however, the assumptions

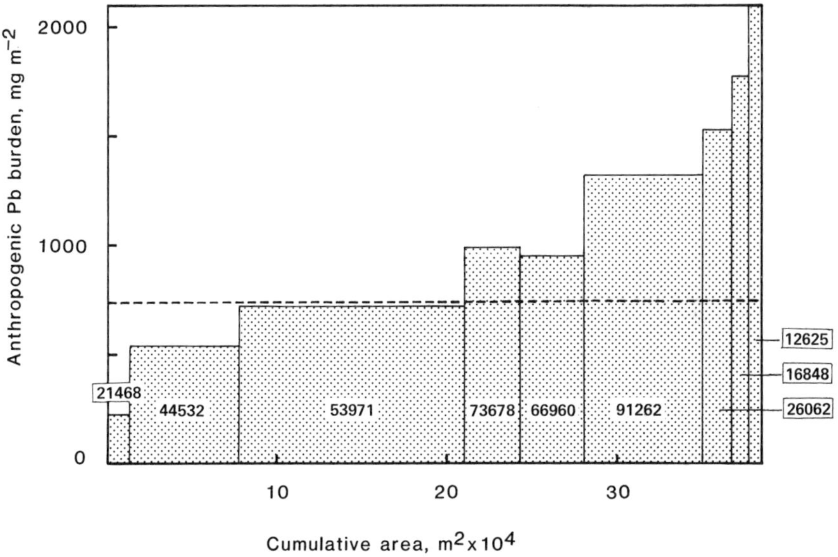

Fig. 11. Variation of total anthropogenic lead burden with coreable area of the basin, Jerry Lake, Ontario. The histogram bars represent the total sediment accumulation of, successively, 0–4 m, 4–8 m ... 32–36 m depths. Total lead burden, g, in the sediment column in different parts of the lake basin are shown on the diagram. Dashed line shows mean sediment lead burden (733 mgm^{-2}). Data of Dillon and Evans (1982).

underlying this simple approach are not valid, or a complete heavy metal profile is not analysed. In these circumstances other approaches must be adopted.

(a) Crustal enrichment. If it is assumed that the main source of non-pollutant heavy metal in a deposit is crustal, i.e. derived from weathered rock and soil dust, then the concentrations of so-called crustal reference elements can enable natural and anthropogenic heavy metals to be distinguished.

This approach has proved to be particularly valuable in polar investigations, where total heavy metal concentrations are often at the detection limit, and where it is often difficult to obtain long firn profiles spanning pre- and post-industrial periods (see Section IV). The crustal enrichment factor, EF_{crust}, of a metal, X, is defined as:

$$EF_{crust} = \frac{[X/x]_{snow}}{[X/x]_{crust}}$$

where x is the crustal reference element, and the ratio $[X/x]_{crust}$ is taken from a standard table of element abundance in crustal material (e.g. Taylor, 1964). The most commonly used crustal elements are silicon (e.g. Murozumi *et al.*, 1969) and aluminium (e.g. Boutron, 1979b; Dick and Peel, 1985). Duce *et al.* (1975) proposed a guideline of $EF_{crust} > 10$ to distinguish metals whose concentrations are significantly enriched above crustal levels in the polar atmosphere.

The significance of crustal heavy metal inputs to polar regions, where heavy metal inputs from other sources are very low, was emphasized by Boutron (1979a, 1980). He showed that the higher flux of heavy metals (13- to 91-fold) in Greenland compared with Antarctica is at least partly attributable to the much larger extent of ice-free areas in Greenland, whilst volcanic emissions can make significant contributions to heavy metal fluxes in the Antarctic on a short-term basis. Crustal enrichment has been used by several workers (e.g. Barrie and Vet, 1984; Laird *et al.*, 1986) to evaluate heavy metal concentrations in seasonal snow samples.

Calculations of crustal enrichment are rarely employed in peat studies, though Aaby and Jacobsen (1979) and Schell *et al.* (1986) showed that crustal sources can contribute significantly to loadings of copper, nickel and vanadium in some rural areas.

In investigations of aquatic sediments, "normalization", or crustal correction of heavy metal concentrations, is usually considered essential in order to account for the often large terrigenous inputs to the sediments. Taking aluminium as crustal reference element, the value of $[X/Al]_{crust}$ is usually calculated from the pre-cultural section of the profile, or from topsoil within the lake catchment (Hamilton-Taylor, 1979). The normalized, or excess, concentration of heavy metal, X, in a recent sediment horizon is then calculated as follows:

$$[X]_{excess} = [X]_{total} - [X]_{crust'}$$

where $[X]_{crust'}$, the crustal contribution to metal in a recent horizon, is derived from:

$$[X]_{crust'} = [Al]_{crust'} \cdot [X/Al]_{crust}$$

In lake sediments the values of $[X]_{crust}$ and $[X]_{crust'}$ are often widely disparate as a result of changes in the rate of erosion in the catchment from, for example, human disturbance. In a study of northern New England Lakes, Norton (1986) observed that vanadium was associated, not only with certain pollutant emissions, but also with enhanced erosion within the catchments. He therefore emphasized the importance of making a crustal correction to separate these fractions, which were both linked with human activity. Recently, Hilton *et al.* (1985) have developed a model to describe heavy metal profiles in terms of "background", $[X]_b$, "erosional", $[X]_a$, and "pollutant", $[X]_p$, fractions, as follows:

$$[X]_{total} - [X]_p = \beta[\text{clay}]_{total} + \alpha$$

where

$$\beta[\text{clay}]_{total} = [X]_a$$

and

$$\alpha = [X]_b$$

The unknown quantities α and β were determined by means of a successive approximation regression technique, allowing $[X]_p$ to be calculated. The authors showed that this model distinguished between those metals which were supplied to the lake in solution (e.g. cadmium and zinc) and those which covaried with erosional inputs as well as anthropogenic inputs (e.g. copper, nickel, lead), as shown in Fig. 12. The authors stressed, however, that the model does not provide a means for distinguishing direct atmospheric inputs of heavy metal from heavy metals deposited first to the catchment and then transported to the sediments.

(b) Stable lead isotope analysis. This approach is based on the occurrence in rocks and minerals of four stable isotopes of lead: ^{206}Pb, ^{207}Pb and ^{208}Pb, which are radiogenic; and ^{204}Pb, which is non-radiogenic and is thought to be the primaeval lead isotope (Russell and Farquar, 1960). The respective ratios of these four isotopes in lead-bearing rocks and minerals are specific to the age and type of rock. Generally, rocks of a younger geological age contain a greater proportion of radiogenic to non-radiogenic lead, and a higher concentration of ^{206}Pb, specifically, than older rocks. Thus, since industrial

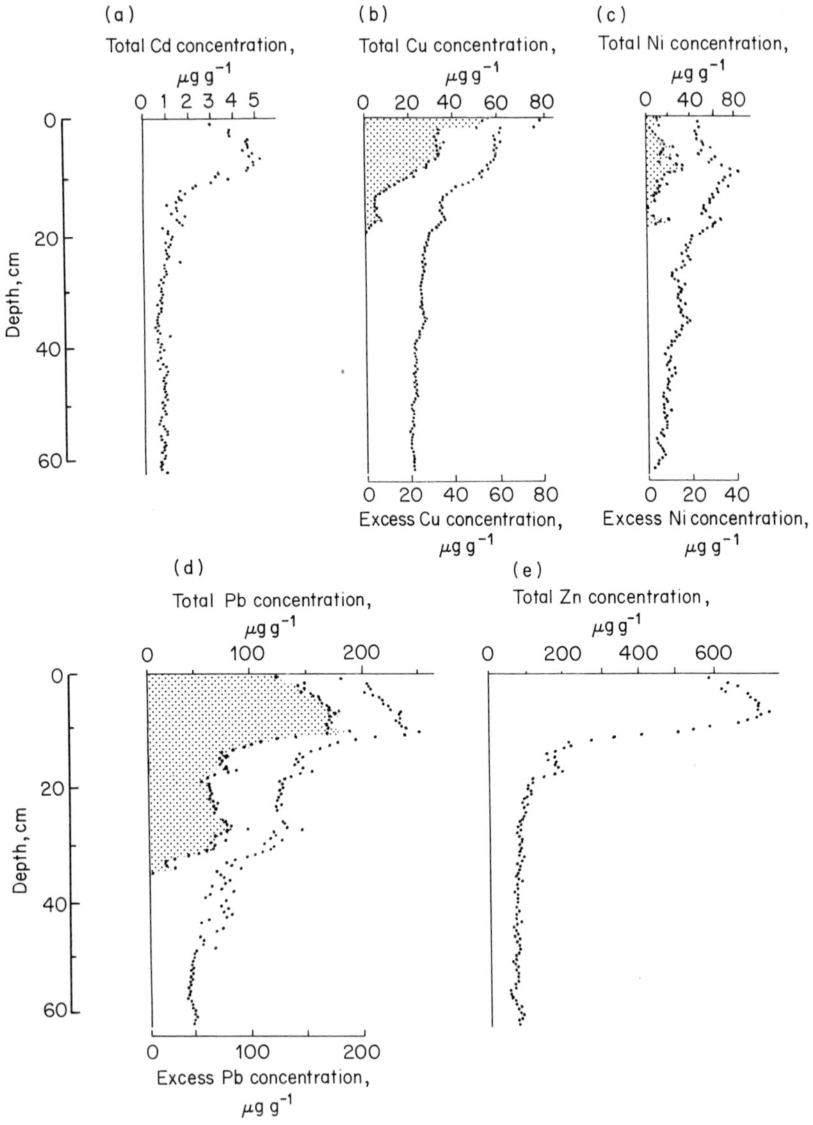

Fig. 12. Total (unshaded) and excess (shaded) concentrations of: (a) cadmium; (b) copper; (c) nickel; (d) lead; and (e) zinc in a sediment profile from Blelham Tarn, UK (see text for explanation). Redrawn from Hilton *et al.* (1985).

and natural lead in the environment often have different geological origins, it is theoretically possible to distinguish between lead from these two sources on the basis of isotope ratios (Chow *et al.*, 1975). The discrimination of different pollutant sources by this technique is discussed in Section IV.

In practice, stable isotope analysis has been proved useful in the detection of early pollutant inputs to sediments in remote areas (e.g. Shirahata *et al.*, 1980; Martin, 1985), or where terrigenous inputs are high (e.g. Ng and Patterson, 1982). Petit (1974) showed that this technique can also enable pollutant contribution to total lead in sediments to be quantified, if the isotope ratios of atmospheric and natural lead in the region are known. The proportion of atmospheric lead, X, was derived as follows:

$$X = \frac{\alpha_{i,s} - \alpha_{i,n}}{\alpha_{i,a} - \alpha_{i,n}}$$

where: $\alpha_{i,s}$, $\alpha_{i,n}$, and $\alpha_{i,a}$ are the isotopic ratios of lead in surface sediments, crustal material, and the atmosphere, respectively.

(c) Physico-chemical heavy metal partitioning. The theory behind this approach is that metals entering aquatic systems from different sources are incorporated into the sediments in different ways. As discussed in Section III.A, heavy metals deposited from the atmosphere to the water surface tend to be in fine-particulate form and are recruited to the sediments by means of scavenging, preferentially by fine-grained sediments. Accordingly, Filipek and Owen (1979) proposed that an inverse relationship between heavy metal concentrations and sediment particle size is indicative of a large autochthonous (pollutant) input.

It is also believed that metals from different sources show characteristic chemical associations in sediments. For example it was discussed in Section III.A that co-precipitation with Fe/Mn oxides is often the ultimate fate of heavy metals in permanent sediments, whereas inwashed metal, which is already complexed to organic matter or bound within crystalline lattices, is likely to retain these associations. These respective fractions of heavy metals in sediments can be separated by sequential chemical analysis, such as the scheme produced by Tessier *et al.* (1979), which is outlined in Table 4. This type of technique has been used by a number of workers (e.g. Patchineelam and Förstner, 1977; Farmer, 1978; Vuorinen *et al.*, 1986), and their results typically show that the excess fraction of metal in recent horizons is usually released in Stages 1–3, whilst the lower, constant fraction of metal is leached in Stages 4 and 5. Nevertheless, Tessier *et al.* (1979) emphasized that these categories of metal are only operationally defined, and that there may be some overlap between leachants. Also, some associations of heavy metals in sediments, e.g. car exhaust particles in road drainage, and ombrogenous

Table 4
Sequential chemical extraction of aquatic sediments (from Tessier *et al.*, 1979).

Stage	Reagent	pH	Probable chemical association of metal in leachate
1	$MgCl_2$	7·0	Exchangeable, sorbed to Fe/Mn oxides, hydroxides and organic matter
2	NaOAc/HOAc	5·0	Carbonate-bound
3	$NH_2OH.HCl/HOAc$	2·0	Coprecipitated with Fe/Mn oxides
4	H_2O_2/HNO_3	2·0	Complexed with organic matter
5	$HF + HClO_4$	< 1·0	Residual, lattice-bound

metals complexed to organic matter in sewage effluent, might confound this classification.

Attempts have been made to adapt the sequential analysis shown in Table 4 to peat, with some success (e.g. Jones, 1987). The chemical forms and cycling of heavy metals within peat are, however, so different from those in aquatic sediments that a rather different approach is probably needed.

This discussion suggests that aquatic sediments present greater interpretative problems in geochemical monitoring than do either peat or ice deposits. Heavy metal in sediments has diverse origins, and the ombrogenous fraction may not be separable. In many circumstances, aquatic sediments can be regarded as only semi-quantitative monitors of atmospheric heavy metal deposition. In comparison, the interpretation of the heavy metal record in peat and ice deposits is more straightforward. The main problems with ice analysis, which are practical rather than theoretical, are discussed by MARC (1985) and Boutron (1987).

IV. PRACTICAL APPLICATIONS

The discussion in the earlier sections of this paper has indicated that, within certain theoretical constraints, the analysis of aquatic sediments, peat and ice holds considerable potential for the monitoring of atmospheric heavy metal deposition. The extent to which this potential has been realized is explored in the following sections.

There is an extensive treatment of the broader aspects of geochemical monitoring in the literature, exemplifying the uses primarily of aquatic sediments in monitoring heavy metal inputs to the ecosystem from many sources; these are comprehensively reviewed by MARC (1985). This paper, however, will be concerned with a critical assessment of the contribution of geochemical monitoring to our understanding of airborne heavy metals in the environment. Because such investigations have focused largely on North-west Europe and North America, records from these two regions are discussed to illustrate the historical development of pollution and the dispersal of pollutants in industrialized countries, and the importance of different sources. World-wide aspects of atmospheric pollution are discussed lastly by reference to the oceanic and polar regions.

A. Historical Perspectives

1. Atmospheric Pollution in North-west Europe from Roman Times to AD 1900

Probably the longest continuous stratigraphic records of atmospheric heavy metal deposition are those from British peats, reflecting the environmental impact of sporadic industrial activity over almost 2000 years. Figure 13(a), (b) shows heavy metal profiles from two peat sites associated with early heavy metal smelting—the Gordano Valley, near Bristol (Martin *et al.*, 1979) and Featherbed Moss, Derbyshire (Lee and Tallis, 1973). At both sites the profiles exhibit short, but distinct episodes of elevated heavy metal concentrations which were dated by pollen analysis and radiocarbon, respectively, to between 1500 and 2000 years BP. This evidence suggests that the Roman smelting operations close to these sites produced a measurable pollutant influx to the local environment. Later small increases in heavy metal deposition, dated to the thirteenth–fifteenth centuries, also occur in these two profiles; these were thought to represent the revival of the lead industry in the Mediaeval Period. Heavy metal enrichments dating back to the Middle Ages were also detected in profiles from Glenshieldaig, North-west Scotland (Livett *et al.*, 1979) and from Lake Windermere, Cumbria (Aston *et al.*, 1973), which suggests that metal smelting in Britain in the Middle Ages was already beginning to pollute the regional aerosol.

The latter part of the eighteenth century marked the beginning of the major period of prosperity of the lead industry in Britain. At sites near to functioning lead smelters, heavy metal profiles closely mirror the fluctuations in emissions of the smelters, in terms of deposition to the immediate surroundings. This is particularly well illustrated by the profile from Moor House, Cumbria (Livett *et al.*, 1979), a site close to the lead smelting district of Alston Moor, but now remote from industrial influences. Figure 14 (a) shows that the changes in lead concentration in the profile parallel the rising prosperity and subsequent decline of the local lead industry in the eighteenth and nineteenth centuries (implied from the fluctuations in numbers of men

(a)

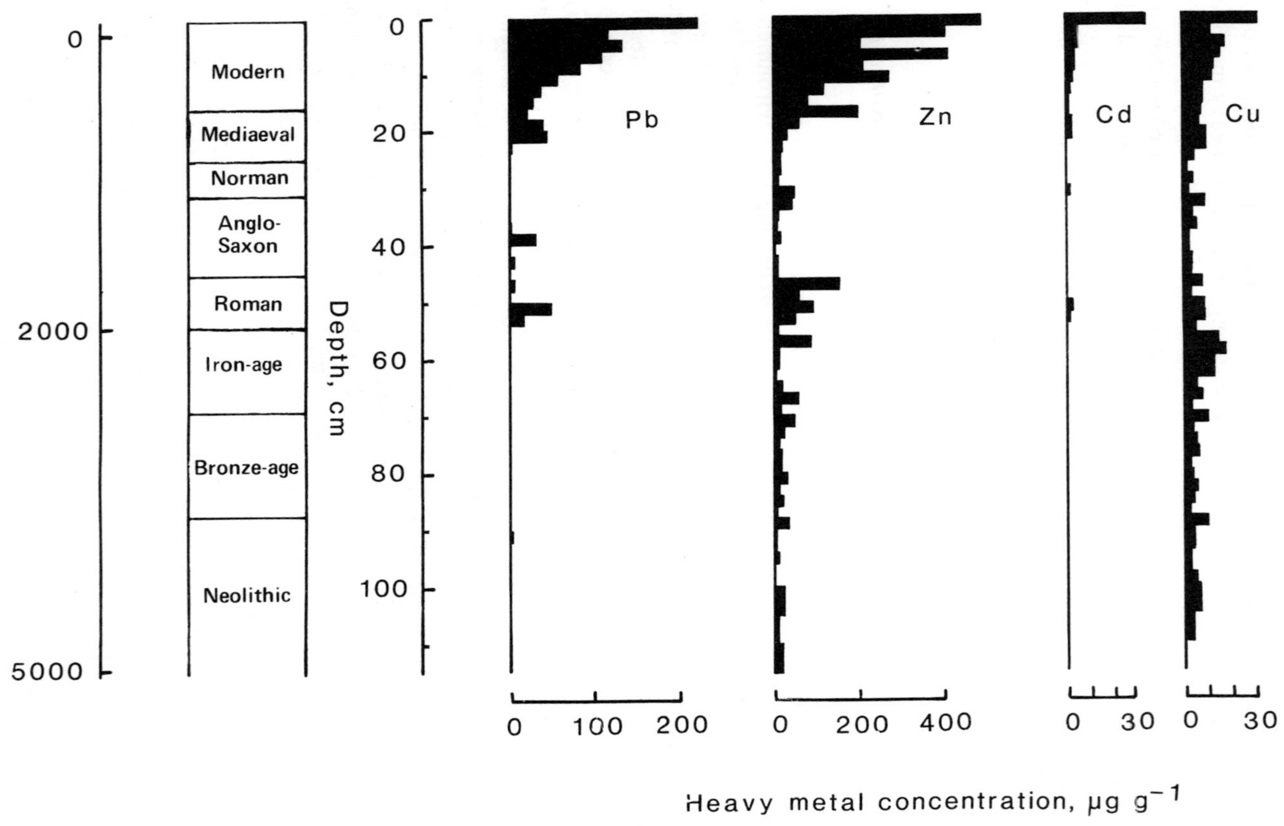
Years BP
0
2000
5000
Modern
Mediaeval
Norman
Anglo-Saxon
Roman
Iron-age
Bronze-age
Neolithic
Depth, cm
0
20
40
60
80
100
Pb
Zn
Cd
Cu
0 100 200
0 200 400
0 30
0 30
Heavy metal concentration, μg g⁻¹

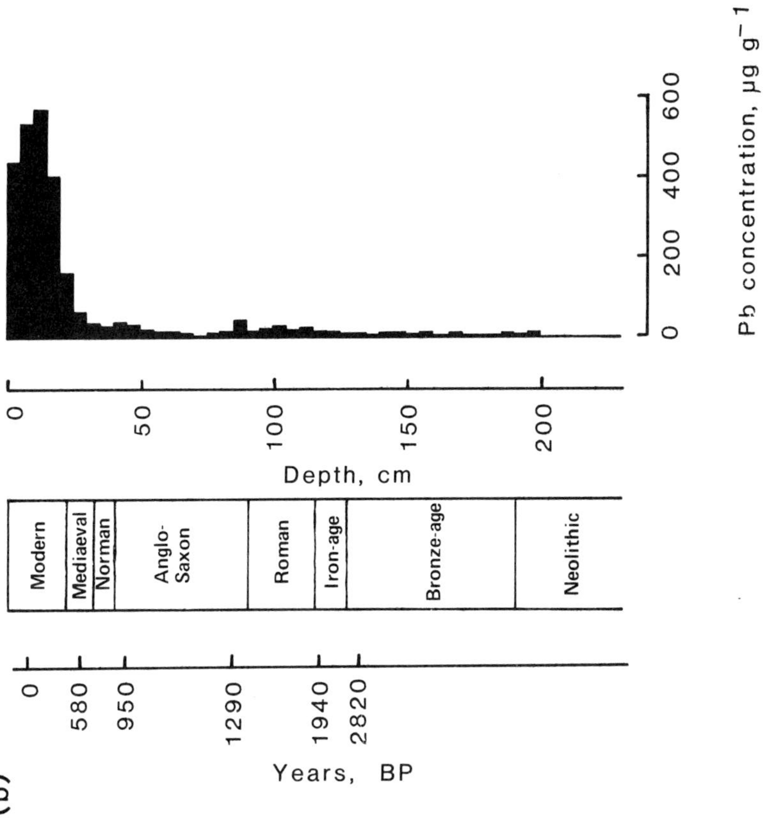

Fig. 13. Distribution of heavy metals with depth in peat profiles from (a) Gordano Valley (redrawn from Martin *et al.*, 1979); and (b) Featherbed Moss (redrawn from Lee and Tallis, 1973). Approximate duration of cultural periods is shown. Time-scale for the Gordano Valley profile is based on pollen-analytical evidence, and for the Featherbed Moss profile on radiocarbon dates calibrated according to Klein *et al.* (1982).

employed at the smelter over that period—Fig. 14(b)). A close correspondence between implied smelter emissions and heavy metal deposition in the vicinity of the smelter was also reported in profiles from Grassington Moor, West Yorkshire (Livett *et al.*, 1979) and Ringinglow Bog, South Yorkshire (Livett, 1982).

The severe impact of these smelters on their local environment is also demonstrated by the contemporary heavy metal deposition rates in their vicinity. Table 5 lists some examples of eighteenth century lead and zinc deposition rates calculated from peat profiles, and for comparison includes present-day values recorded in industrial areas by conventional deposition

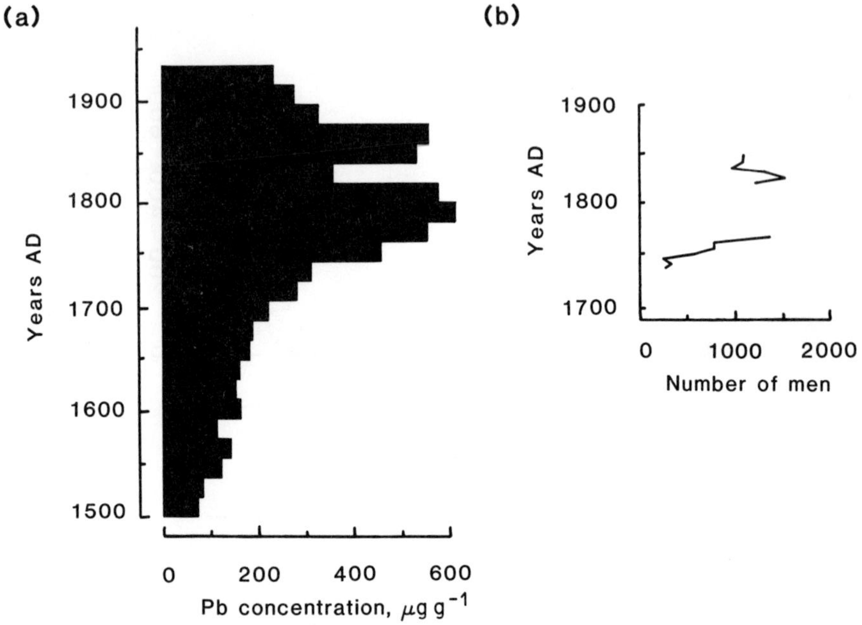

Fig. 14. Lead concentration in peat at Moor House, UK, in relation to local lead smelting; (a) concentration of lead in peat profile, dated using pollen-analytical and documentary evidence (c.f. Fig. 9); and (b) number of men employed at Alston Moor lead smelter from AD 1738 to AD 1844 (data of Hunt, 1970). Redrawn from Livett *et al.* (1979).

gauges. The similarity between the two sets of lead values suggests that by the eighteenth century sites such as Moor House and Ringinglow Bog could already have been classified as "polluted" by present-day standards. Peat-derived rates of zinc deposition, however, are substantially lower than corresponding present-day values, which is consistent with the occurrence of zinc only as an incidental pollutant in lead smelter emissions. This discrepancy could indicate also that there had been some loss or relocation of zinc in the peat profile (see Section III). This is supported by the poor correspondence between present-day peat-derived and deposition-gauge-derived zinc values for Shetland.

Profiles from Glenshieldaig (Livett *et al.*, 1979) and Shetland (Livett, 1982) suggest that lead smelter emissions resulted in a widespread enhancement of atmospheric lead deposition in Britain in the seventeenth and eighteenth centuries.

The environmental impact of the Industrial Revolution in Britain, which began in the eighteenth century, rapidly overtook that of the declining lead industry. The scale of the increase in heavy metal deposition which occurred at the height of urban/industrial expansion in the nineteenth century is most evident in regions such as the southern Pennines that were directly affected by the rapid growth of the nearby industrial towns of Lancashire and Yorkshire. Investigations by Livett (1982) and Lee and Tallis (1973) show a rapid two- to three-fold increase in rates of lead and zinc deposition on Featherbed Moss, Derbyshire, and Ringinglow Bog, South Yorkshire—already enhanced above background levels by local lead smelter emissions—that can be attributed to the almost exponential expansion of the nearby towns. By 1900, heavy metal concentrations in the peat were enriched by about 20-fold over background concentrations. Contemporary deposition rates of lead and zinc calculated from peat profiles are given in Table 5.

Table 5

Maximum eighteenth century, maximum nineteenth century and present-day bulk deposition rates of lead and zinc in Europe.

Site	Heavy metal deposition rate[a] ($\mu g\ cm^{-2}\ yr^{-1}$)		Source
	Pb	Zn	
(a) *Eighteenth century*			
Buxton, England[p]	2·1	—	Lee and Tallis (1973)
Featherbed Moss, England[p]	3·0	—	Livett (1982)
Grassington Moor, England[p]	3·6	1·1	Livett *et al.* (1979)
Moor House, England[p]	6·5	1·3	Livett *et al.* (1979)
Ringinglow Bog, England[p]	3·6	3·5	Livett (1982)
(b) *Nineteenth century*			
Glenshieldaig, Scotland[p]	0·12	—	Livett *et al.* (1979 and unpublished)
Moor House, England[p]	5·7	0·3	Livett *et al.* (1979 and unpublished)
Ringinglow Bog, England[p]	6·5	7·3	Livett (1982)
Draved Mose, Denmark[p]	0·5	0·4	Aaby and Jacobsen (1979)
Lake Mirwart, Belgium[a]	0·001	—	Oldfield *et al.* (1980a)
Lake Sorvalampi, Finland[a]	0·001	0·004	Tolonen and Jaakkola (1983)
(c) *Present day* (*c.* 1973–1976)			
Baglan, Wales[g]	5·5	13·7	Pattenden (1975)
Port Talbot, Wales[g]	3·5	18·3	Pattenden (1975)
Trebanos, Wales[g]	2·0	5·5	Pattenden (1975)
Collafirth, Shetland[g]	< 1·0	5·5	Cawse (1977)
Fladdabister, Shetland[p]	0·7	1·3	Livett (1982)

[a]Deposition rates determined from: [a]aquatic sediments, [g]bulk deposition gauges, [p]peat deposits.

Recognition of the long, and latterly extreme, exposure of the southern Pennines to atmospheric pollution prompted the suggestion by Conway (1949), and more recently by other workers (e.g. Tallis, 1965; Ferguson and Lee, 1980; Lee *et al.*, 1988), that this factor could have been at least partly responsible for the well-documented degeneration of the blanket mire vegetation of this area. Examination of the exact chronology of this change and of the enhanced deposition of heavy metals at Ringinglow Bog (Livett, 1982) tentatively supports this hypothesis. Chambers *et al.* (1979) have proposed that a similar link exists between the increase in atmospheric pollution and the degradation of the vegetation of the South Wales uplands.

The widespread impact of the Industrial Revolution all over rural Britain is revealed in the heavy metal record from, for example, Glenshieldaig, north-west Scotland (Livett *et al.*, 1979), Lake Windermere, Wastwater and Ennerdale Water in Cumbria (Aston *et al.*, 1973; Hamilton-Taylor, 1979, 1983), Loch Lomond, Scotland (Farmer *et al.*, 1980), and Lough Neagh, Northern Ireland (Rippey *et al.*, 1982). At many of these sites two- to three-fold increases in a range of metals (Pb, Zn, Cu, Cd, Ni, Hg) occurred, which dated back to the eighteenth and nineteenth centuries and represented the earliest detectable enrichments.

The growth of industry and spread of pollution in continental North-west Europe in the nineteenth century paralleled that in Britain, and was most severe in the central region. Sediment records from the Baltic Sea (Erlenkeuser *et al.*, 1974; Müller *et al.*, 1980) show an up to five-fold increase in the deposition of heavy metals from the major onset of pollution in 1830 to the end of the century, which paralleled industrial growth in Western Europe. Air pollutants were apparently transported efficiently southwards and eastwards from the industrialized central regions, as enhanced heavy metal deposition dating back to 1850 has been recorded at sites in south Belgium (Oldfield *et al.*, 1980a; Petit *et al.*, 1984), the Polish Tatra Mountains (Jaworowski, 1968), the French and Swiss Alps (Vernet and Favarger, 1982) and the French Massif Central (Martin, 1985). Heavy metal enrichments at these sites, however, were between one and two orders of magnitude lower than in the Baltic sediments.

Heavy metal records from Southern Scandinavia suggest that these regions may have received long-range airborne pollutants from Britain as well as from industrial central Europe. A lead profile from Draved Mose, Denmark (Aaby *et al.*, 1979) showed a late-1700s onset of anthropogenic deposition and a three-fold increase before 1900. The enhanced deposition of zinc, cadmium, copper, nickel, mercury and vanadium at the same site appears to have occurred during the nineteenth century (Aaby and Jacobsen, 1979; Madsen, 1981), as shown in Fig. 15; the large sampling intervals used preclude a more precise dating.

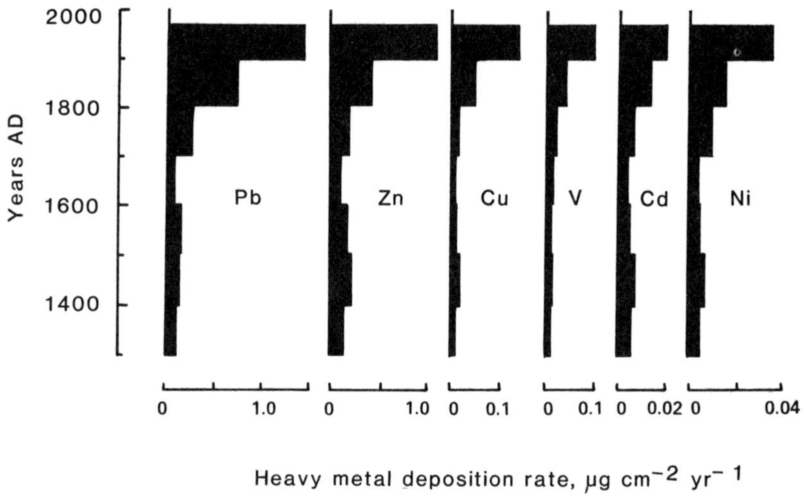

Fig. 15. Calculated deposition rates of heavy metals in the peat at Draved Mose, Denmark, over the last 600 years. The time-scale is based on Pb-210 dating. Data of Aaby and Jacobsen (1979).

There is evidence of the long-range transport of heavy metals from Britain during the eighteenth and nineteenth centuries to more northerly, remote regions of Scandinavia as well. An ice profile from the Jotunheimen Mountains, southern Norway (Jaworowski *et al.*, 1975), revealed a slight but detectable increase in lead and cadmium concentrations as early as the late eighteenth century, although the major increase in concentrations occurred in the nineteenth century horizons. Enhanced lead deposition dating to between 1770 and 1880 has also been reported in lake sediments from southern Norway and southern Finland (Norton and Hess, 1980; Davis *et al.*, 1983; Tolonen and Jaakkola, 1983). The wide range of these onset dates is attributed largely to variable terrigenous inputs to these sediments, rather than to any geographical variation. In most instances, enrichments of other heavy metals—zinc, copper, cadmium—in northern Scandinavia during the nineteenth century post-dated those of lead, or were absent, which is consistent with the relatively lower emission rates of these metals from coal combustion. Mercury enrichments, however, were easier to detect, because background, terrigenous inputs to sediments were lower in most instances; accordingly, trends of mercury deposition in Scandinavia (Björklund *et al.*, 1984) followed trends in lead deposition, which is consistent with the view that mercury is a general indicator of industrialization (Aston *et al.*, 1973).

Table 5 shows that heavy metal deposition rates in Scandinavia, rural

Table 6
Present-day and maximum rates of anthropogenic[a] atmospheric heavy metal deposition in Britain.

Site		Deposition rate [b] ($\mu g\ cm^{-2}\ yr^{-1}$)				Source
		Pb	Zn	Cu	Cd	
Fladdabister, Shetland[p,1]						Livett (1982)
	present-day	0·63	0·9	0·18		
	maximum (date)	—	1·8 (1910)	—		
Glenshieldaig, Scotland[p,1]						Lee and Tallis (1973 and unpublished)
	present-day	0·15	0·4	0·0		
	maximum	—	—	—		
Loch Lomond, Scotland[a,2]						Farmer *et al.* (1980)
	present-day	2·51	8·84	0·97	0·02	
	maximum	—	—	—	—	
Lough Neagh, N. Ireland[a,2]						Rippey *et al.* (1982)
	present-day	7·2	16·4	6·1	0·0	

present-day	2·0	0·5	0·17	0·0	Livett (1982)
maximum (date)	5·7 (1830)	—	—	—	
Lake Windermere, England[a,2]					Hamilton-Taylor (1979)
present-day	9·9	16·9	1·8		
maximum (date)	25·1 (1900)	47·9 (1950)	2·8 (1950)		
Grassington Moor, England[p,1]					Lee and Tallis (1973 and unpublished) and Livett *et al.* (1979)
present-day	1·3	0·42	0·25		
maximum (date)	3·6 (1870)	0·9 (1890)	—		
Featherbed Moss, England[p,1]					Lee and Tallis (1973 and unpublished)
present-day	3·5	1·7	0·8		
maximum (date)	4·8 (1850)	—	0·9 (1850)		
Ringinglow Bog, England[p,1]					Livett (1982)
present-day	3·9	4·4			
maximum (date)	6·5 (1930)	7·3 (1940)			
Buxton, England[p,1]					Lee and Tallis (1973) and Livett *et al.* (1979)
present-day	2·2	0·4	0·4		
maximum (date)	3·0 (1880)	0·9 (1880)	0·5 (1880)		

[a] Anthropogenic rates determined by: [1]subtraction of pre-cultural from post-cultural rate; and [2]sequential chemical analysis (see Section III).
[b] Deposition rates determined from: [a]aquatic deposits; and [p]peat deposits.
—, maximum deposition rate occurs at the present-day.

central Europe and Scotland were still very low throughout the nineteenth century by present-day standards, and between one and three orders of magnitude lower than at rural sites in northern England.

2. Twentieth Century Trends in Atmospheric Heavy Metal Deposition in Britain and Continental Europe

The theory that atmospheric heavy metal concentrations in Europe, having increased rapidly in the nineteenth century, might have started to decline in the present century, was first investigated by Rühling and Tyler (1968, 1969), using herbarium moss samples. Later results by Rühling and Tyler (1984) and by Lee and Tallis (1973) confirmed that in certain localities in Britain and Scandinavia present-day heavy metal deposition rates are now significantly lower than they were at various times in the past. To what extent these observations constitute a real trend over Europe as a whole can be assessed by continuous stratigraphical records spanning the last 80–100 years.

The stratigraphical evidence for Britain is summarized in Table 6, which lists present-day and maximum anthropogenic deposition rates of lead, zinc, copper, cadmium and mercury at ten rural sites in northern regions. Geographically, the sites fall into three categories: firstly, the two northern Scottish sites which are isolated from any urban/industrial influence; secondly, rural sites which are marginally influenced by industrial emissions (Grassington Moor, Lough Neagh, Moor House); and thirdly, sites not more than 30 km from industrial towns (Loch Lomond, Lake Windermere, and the three southern Pennine sites). Table 6 shows that this geographical grouping corresponds well with present-day deposition rates of lead, copper, and, to a lesser extent, zinc. On the basis of recent trends in pollution, as revealed by the dates of maximum deposition rates, however, two different groupings emerge. At the remote sites, and at Loch Lomond and Lough Neagh, heavy metals show a steady and maintained increase to the top of the profile where maximum deposition rates occurred. In contrast, at Moor House, Grassington Moor, Lake Windermere and the high-background sites, deposition rates of some or all of the metals measured were higher during the late nineteenth or early twentieth centuries, than at the present day. At Moor House and Grassington Moor this trend reflects the cessation of smelting locally, followed by a return to a more typically "rural" level of heavy metal deposition. The Lake District was also associated with lead smelting in the nineteenth and early twentieth centuries; thus the apparently recent decline in heavy metal deposition in Lake Windermere could be related to trends in local emissions. Alternatively, the subsurface accumulation of heavy metal could be a product of diagenesis, or be at least partly enhanced by terrigenous inputs (see Section III). The same explanation obviously does not apply to the recent decrease in heavy metal deposition at the three southern Pennine sites, which have had a continuous exposure to industrial emissions up to the present day. The most likely explanation—that there has been a real reduction in urban/industrial heavy metal emissions in these areas over the last 50–60 years—is discussed further in Section IV.B.

The consensus of evidence from recent stratigraphic records is therefore that concentrations of all heavy metals in the rural British aerosol are still increasing; at marginal rural sites, however, a decline in deposition is apparent—largely because these areas had experienced such severe air pollution during the nineteenth century.

Table 7 summarizes the stratigraphic evidence relating to recent trends in heavy metal deposition rates/concentrations in deposits in Continental North-west Europe. A slow, steady increase in the deposition of most metals throughout the twentieth century has been observed in Northern and Central Scandinavia; at some sites, increases in zinc and copper represent the earliest enrichments of these metals. Further south, in regions on the periphery of the industrialized areas (Denmark, Poland, Switzerland and France) post-1900 increases in lead deposition of up to six-fold have been observed (e.g. Draved Mose: Aaby *et al.*, 1979). Increases of other metals, however, were not usually so marked. In southern Belgium (Thomas *et al.*, 1984) there was no change in the concentrations of zinc, copper and cadmium, and at Mont Blanc (Briat, 1978; Batifol and Boutron, 1984) results were conflicting regarding the presence of anthropogenic enrichments of copper and zinc. Indeed a pronounced decline in heavy metal concentrations was observed in Lake Constance and the Baltic Sea. A possible link between this trend and the declining consumption of coal in Western Europe is discussed further in Subsection IV.B.

As mentioned above, mercury enrichments at most sites coincided with the general period of heavy metal increase. Major enhancements of this metal in atmospheric deposition are, however, a feature of the post-1900 sedimentary record and are associated with the expansion of the cellulose and chlor-alkali industries after the turn of the century. Because these industries are often sited in rural localities, considerable increases in mercury deposition have been recorded from many rural Scandinavian sites (e.g. Tolonen and Jaakkola, 1983; Björklund *et al.*, 1984). The observation of Fredriksson and Qvarfort (1973) of no historical trend in mercury concentrations in Lake Öjesjön, central Sweden, therefore seems to be anomalous. The failure to detect mercury enrichments at this site could possibly be a function of the large sampling intervals adopted and consequent loss of fine resolution, or of large terrigenous inputs to the sediments which would have masked a trend in atmospheric deposition.

The prevailing trend in heavy metal deposition in Continental Europe currently appear to be upward, therefore, as in many rural areas of Britain.

Table 7

Geochemical evidence for trends [a] in atmospheric heavy metal deposition in Continental Europe since AD 1900.

Site	Pb	Zn	Cu	Cd	Ni	Hg	V	Source
Storbreen Glacier, S. Norway[i]	+			+				Jaworowski *et al.* (1975)
Southern Norway lakes[a]	+	+	+					Norton and Hess (1980). Davis *et al.* (1983)
Southern Finland lakes[a]	+ +	+	+	+	+	+ +		Tolonen and Jaakkola (1983)
Lake Öjesjön, Sweden[a]						0		Fredriksson and Qvarfort (1973)
Central Sweden lakes[a]						+		Björklund *et al.* (1984)
Draved Mose, Denmark[p]	+ +	+	+	+	+	+	+	Aaby and Jacobsen (1979), Aaby *et al.* (1979), Madsen (1981)
Tatra Mountains, Poland[i,*]	+ +							Jaworowski (1968)
Kieler Bucht, W. Baltic[a]	+/−	+/−	+/−	+	−			Erlenkeuser *et al.* (1974), Müller *et al.* (1980)
Lake Constance, Germany[a]	+/−	+/−	+/−	+/−	+	+		Müller *et al.* (1977)
Lake Mirwart and Lake Willerzie, Belgium[a]	+ +	−	−	−				Oldfield *et al.* (1980a), Thomas *et al.* (1984)
Lake Annecy, Lake Bourget and Lake Léman, France/Switzerland[a]	+ +		+			+		Vernet and Favarger (1982)
Mont Blanc, France[i,*]	+ +	−	0	+			+	Briat (1978), Batifol and Boutron (1984)
Pavin Crater Lake, France[a]	+ +	+	+					Martin (1985)

[a] Geochemical evidence provided by the analysis of: [a]aquatic sediments; [i]ice deposits; and [p]peat deposits. Trends denoted by: +, increase; + +, increase of > two-fold; +/−, increase followed by levelling or decrease; *, change occurred after 1950; 0, no anthropogenic enrichment.

There is little evidence to support the view of Rühling and Tyler of a recent amelioration in air pollution in Scandinavia.

3. The Industrial Revolution in North America

North America has had a very short history of industrialization compared with Continental Europe, and particularly Britain. The developments which took place in the nineteenth century resulted in the introduction of anthropogenic heavy metals into a previously unpolluted air-mass. Geochemical investigations have focused on two aspects of this development: firstly, the history and magnitude of pollution on a local/regional scale; and secondly, the dispersal of atmospheric pollutants throughout the remote aerosols of the subcontinent.

(a) Local/regional aspects of pollution. The spread of industry in North America, and particularly the USA, can be related broadly to the westerly movement of settlers which started in the early 1900s. The chronology of this progression is summarized in Table 8.

The East Coast region of the USA was the first to undergo development, and this history is reflected in the sediment records of two reservoirs near New Haven, Connecticut, and Narragansett Bay, Rhode Island (Bertine and Mendeck, 1978; Goldberg *et al.*, 1977). In the first of these studies, the onset of heavy metal enrichments was dated at between 1830 and 1850, and the rapid scale of the increase was found to correlate closely with the documented population growth of the nearby town of New Haven. The assemblage of metals which were enriched—lead, zinc, copper and cadmium—are usually thought to be diagnostic of miscellaneous urban/industrial emissions (e.g. Erlenkeuser *et al.*, 1974). The pollution record from Narragansett Bay showed a similar development and scale of pollution, except that the sediment profile was not long enough to confirm the early 1800s onset of pollution. Atmospheric inputs of metals to the sediments were also thought to have been augmented by direct inputs of coal and coke particles from ships. This must partly explain the very high heavy metal deposition rates which were calculated for this site (Pb, 168 $\mu g\ cm^{-2}\ yr^{-1}$; Zn, 330 $\mu g\ cm^{-2}\ yr^{-1}$; Cu, 260 $\mu g\ cm^{-2}\ yr^{-1}$).

The industrial development of the Great Lakes region (encompassing parts of Ontario, Michigan, Illinois, Indiana, Wisconsin and Minnesota), has been the subject of a number of geochemical investigations. Sediment records from Lake Michigan indicate that the earliest enrichments of lead and other metals in this region date back to 1820–1830. A detailed study by Edgington and Robbins (1976), in which a model was constructed to account for the historical trend in lead concentrations in the sediments (see Fig. 21(b), Section IV.B), indicated that miscellaneous emissions, dominated by fossil

Table 8
Geochemical evidence[a] for the chronology of industrialization of North America.

Site	Date	Metal(s)	Source
Narragansett Bay, RI	pre–1860	Pb, Zn, Cu	Goldberg *et al.* (1977)
New Haven, CT	1830–1850	Pb, Zn, Cu	Bertine and Mendeck (1978)
South-eastern Ontario[1]	1830–1850	Pb, Zn, Cu, Cd	Edgington and Robbins (1976), Kemp and Thomas (1976), Farmer (1978), Nriagu *et al.* (1979), Goldberg *et al.* (1981), Evans and Dillon (1982)
South-eastern Ontario[1]	1900	Hg	Thomas (1972), Kemp *et al.* (1974)
South-central Ontario[2]	1870	Pb, Zn, Cd	Kemp *et al.* (1978), Johnson *et al.* (1986)
South-central Ontario[2]	1900	Hg	Kemp *et al.* (1978)
South-central Ontario[2]	1940	Cu, Ni	Kemp *et al.* (1978)
Sudbury, Ontario	1900	Pb, Cu, Ni	Nriagu *et al.* (1982)
South-western Ontario[3]	1890	Pb, Zn, Cd, Hg	Kemp *et al.* (1978), Johnson (1987)
Southern Wisconsin	1850	Pb, Zn, Cu, Cd	Iskandar and Keeney (1974)
Northern Wisconsin	1900	Hg	Syers *et al.* (1973)
Northern Minnesota	1880	Hg	Meger (1986)
Lake Washington	1880–1900	Pb, Zn, Cu, Hg	Barnes and Schell (1973), Crecelius and Piper (1973)
Southern California	1900–1930	Pb, Hg	Chow *et al.* (1973), Young *et al.* (1973), Bruland *et al.* (1974), Ng and Patterson (1982)
Southern California	1940	Zn, Cu, Cd, V	Bruland *et al.* (1974), Bertine and Goldberg (1977)

[a] Evidence from the analysis of aquatic sediments.
[1] Includes Lakes Erie, Michigan and Ontario and Found Lake; [2] includes Lake Huron and Turkey Lakes; [3] includes Lake Superior.

fuel combustion products, from the highly industrialized southern shore of the lake were responsible for the early inputs of lead. The link between combustion products and heavy metal enrichments in Lake Michigan sediments was also explored by Goldberg *et al.* (1981). They characterized carbonaceous material of different kinds in different sediment strata, and showed that the particles produced from the burning of coal were particularly closely correlated with the occurrence of heavy metal enrichments.

The history of heavy metal deposition in the other four lakes in the chain has been broadly outlined by Kemp and Thomas and their co-workers, who showed that the onset of local heavy metal pollution was synchronous with the successive movement westwards and northwards of the early settlers—marked by the *Ambrosia* pollen horizon (see Fig. 16). The settlement of the various lake environs took place in the following order: Lakes Erie and Ontario (south-eastern Ontario), 1850; Lake Huron (south-central Ontario), 1870; Lake Superior (south-western Ontario), 1890. In most instances, simultaneous increases in the concentrations of lead, zinc, copper and cadmium were observed; in Lake Superior direct deposits of copper mine tailings were thought to have masked early atmospheric inputs (Kemp *et al.*, 1978). The only estimate of metal deposition rate for this early industrial period is the value of 0·2 μg Pb cm^{-2} yr^{-1} calculated by Evans and Dillon (1982) for the Muskoka–Haliburton region of Southern Ontario, and this is of the same order as nineteenth century lead deposition rates recorded for rural localities in Europe (Table 5(b)). Generally, nineteenth century heavy metal enrichments over background concentrations were lower in Lakes Huron and Superior (< 2·0) than in Lakes Erie and Ontario (2–3), which is consistent with the lesser degree of urbanization of these more northerly districts. Throughout the Great Lakes region, mercury enrichments dated from *c.* 1900. As in Europe, this probably represents the first major emissions of the metal from the paper and chemical industries; earlier inputs associated with fossil fuel combustion were apparently not detected. There was no significant anthropogenic deposition of either nickel or vanadium in these regions during the nineteenth century.

The stratigraphic record from lakes in the Sudbury district of Ontario relates, not to general urban/industrial development as described above, but specifically to the growth of the local nickel/copper smelting industry. Nriagu *et al.* (1982), in an analysis of sediment profiles from a number of small lakes in this area, reported that anthropogenic deposition of copper and nickel dated back to *c.* 1890, when commercial exploitation of the ores commenced. In most of the lakes, enrichments of lead and zinc were dated to slightly later than those of copper and nickel, in line with the lower concentrations of these metals in the ores.

Geochemical evidence for the history of pollution in regions to the west of

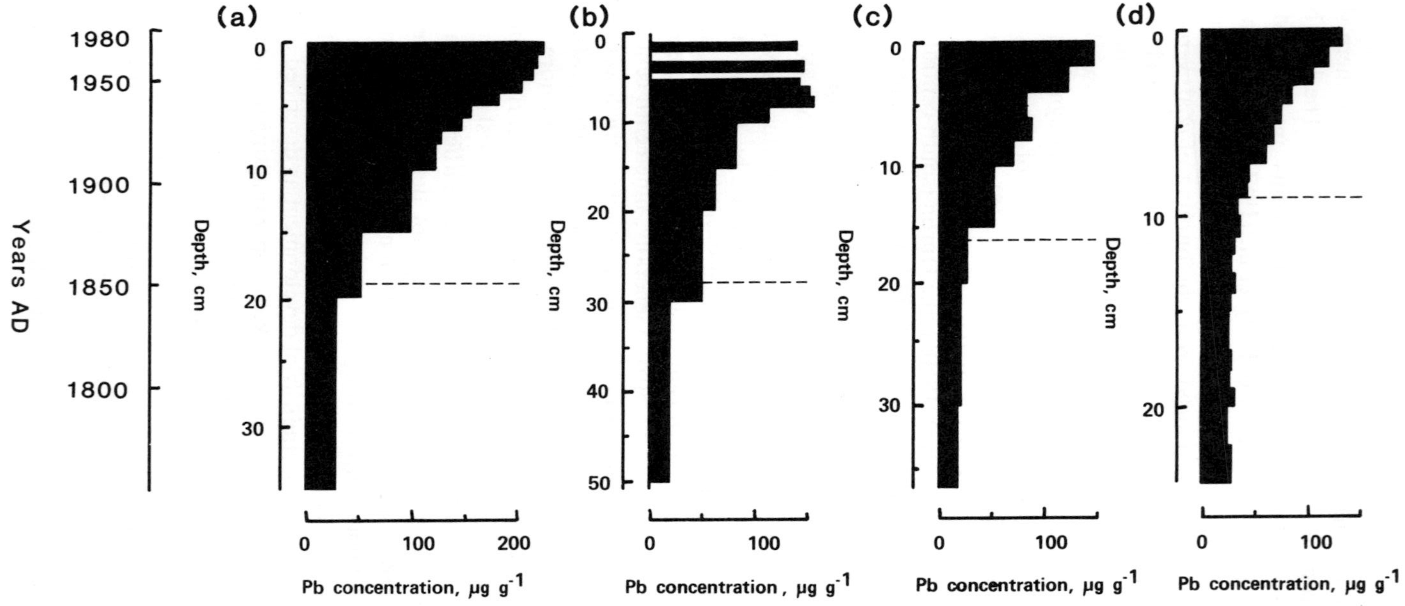

Fig. 16. Concentration of lead in sediment profiles from: (a) Lake Ontario; (b) Lake Erie; (c) Lake Huron; and (d) Lake Superior. The time-scale and the position of the *Ambrosia* horizon (broken line) were determined by pollen analysis. Redrawn from Kemp and Thomas (1976) and Kemp *et al.* (1978).

the Great Lakes is rather scanty. A number of small lakes in southern and northern Wisconsin have been investigated by Syers *et al.* (1973) and Iskandar and Keeney (1974), but the considerable degree of human influence in some of these watersheds (for example, logging, sewage inputs, chemical treatments) diminishes the value of the sedimentary records. Also, the apparently high rate of supply of most heavy metals in terrigenous materials, especially to the southern lakes, made interpretation difficult. The records from the southern lakes indicated that enhanced inputs of lead, zinc and copper date back to *c.* 1850, when the area was first settled, but in the northern lakes atmospheric inputs were not detectable until 1900, which was thought to be indicative of long-range transport rather than local settlement. The record of atmospheric deposition of mercury to the southern lakes was, in most instances, distorted by inputs of sewage; in the northern lakes, however, the enhanced deposition of mercury was detected at around the turn of the century (Syers *et al.*, 1973), as it was further to the north-west, in northern Minnesota (Meger, 1986). In the latter region, total concentrations of mercury in the nineteenth century lake sediments were very low (*c.* 0·4–0·7 $\mu g\ g^{-1}$), to which terrigenous inputs made the largest contribution (*c.* 75%). Meger (1986) concluded, therefore, that the excess mercury enrichments must have originated from remote sources, and that this atmospheric fraction was effectively "magnified" by a large input from the catchment.

The industrial development of the north-west coastal region of the USA began in the late nineteenth century and was associated particularly with the growth of the base metal smelting industry. Sediment cores from Lake Washington (Barnes and Schell, 1973; Crecelius and Piper, 1973) show that the first increase in lead deposition in the district reflected the early settlement of the area in 1880, and that the subsequent two-fold increase in lead concentrations by the turn of the century (resulting in lead concentrations as high as 100 $\mu g\ g^{-1}$) dated back to 1890, when the Tacoma lead smelter commenced work. The atmospheric deposition of mercury in the region was similarly detectable from the late-nineteenth to early-twentieth centuries (Barnes and Schell, 1973).

One of the last regions in the USA to undergo urban/industrial development was the Southern California coastal belt, and this development was based, not on coal as elsewhere in the USA, but largely on oil and light industry. Geochemical evidence of the impact of this industrial development on atmospheric deposition in this region is provided by the sediments of the inshore and offshore coastal basins, but interpretation of the pollution record is complicated by the presence of sewer outfalls and the effects of coastal transport processes (Ng and Patterson, 1982). These disadvantages are offset, however, by a high rate of sedimentation which helps to preserve the heavy metal record, and by the occurrence of "biogenic varves" which facilitates dating (see Section III). Sediment profiles from the three inner basins—Santa Barbara, Santa Monica and San Pedro—showed a major increase in the deposition of lead and mercury which dated back to 1910–1920 (Chow *et al.*, 1973; Young *et al.*, 1973; Bruland *et al.*, 1974), whilst enrichments of cadmium, copper, nickel, vanadium and zinc dated back only to *c.* 1950 (Bruland *et al.*, 1974). More recently, Ng and Patterson (1982) carried out a particularly detailed study of these sediments (Fig. 17), in which sequential analysis and stable lead isotope analysis were used to distinguish early inputs of industrial lead, which occurs in the "easily-leached" fraction and has a lower radiogenicity (and hence a lower $^{206}Pb/^{207}Pb$ ratio) than "natural" lead (see Section III). The authors were able to confirm that lead pollution in the region dated from *c.* 1910, but concluded that sewage inputs would have masked atmospheric inputs to the sediments in all but the Santa Barbara Basin.

The outer basins, which are isolated from all direct heavy metal inputs in run-off and sewage, had originally seemed to be more promising with respect to atmospheric monitoring. In a study of the sediments of one of these—the San Clemente Basin—Bertine and Goldberg (1977) found, however, that bioturbation occurred within the top 10 cm of the profile, and thus greatly reduced the degree of resolution of the recent part of the heavy metal record.

(b) Regional baselines of atmospheric pollution. Lead, more than any other heavy metal, can be regarded as a general indicator of human influence on atmospheric deposition in the industrialized countries of the world. Therefore, the early beginning of atmospheric pollution in North America can be defined by the onset of enhanced lead deposition at isolated sites. Geochemical evidence for this baseline is summarized in Table 9.

Perhaps the most notable feature of the data in Table 9 is the considerable degree of intra-site variation in baseline dates, particularly in northern New England and the Adirondacks. Norton and his co-workers give the earliest baseline for this region—1850—which suggests that it was only 20–30 years before long-range pollutants from the industrialized East Coast of the USA reached detectable levels in the atmosphere of this isolated region. It is, nevertheless, reasonable to assume that this early date is correct, since Norton and his co-workers used crustal corrections to distinguish between terrigenous and atmospheric inputs of heavy metals, and the cores were dated accurately with a lead-210 chronology. The profiles of Galloway and Likens (1979) yielded a much later baseline date of 1950, perhaps because their sediment cores were of insufficient length to enable the background concentration of lead to be estimated satisfactorily, and thus to allow the first enrichments of lead to be detected. Large fluctuations in background concentrations were also a problem in the investigation by Heit *et al.* (1981),

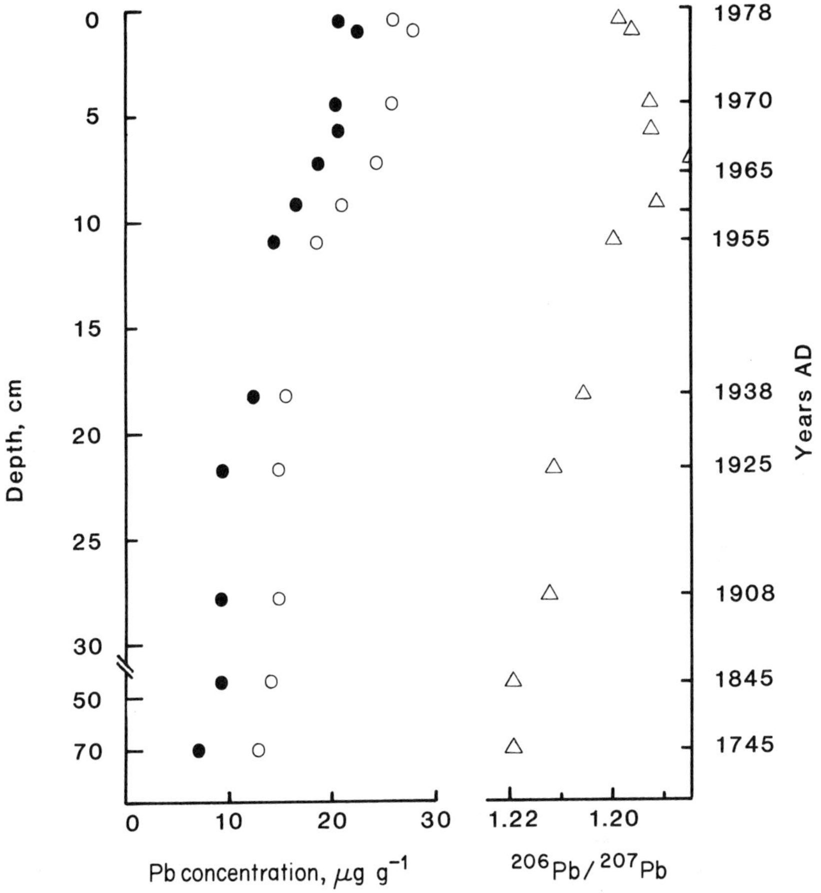

Fig. 17. Concentration of total lead (open circles) and leached lead (closed circles) and $^{206}Pb/^{207}Pb$ ratio in a sediment profile from the Santa Barbara Basin, southern California. Time-scale based on varve-chronology and Pb-210 dating. Data of Ng and Patterson (1982).

and these workers did not use crustal corrections. Furthermore, their use of the caesium-137 dating technique in isolation means that back-extrapolation would have been necessary in order to identify sediment horizons of 1900 and earlier. In the study of Big Heath Bog, Maine, by Norton (1987), the late baseline date appears to have been a function of a more fundamental problem—relocation of lead in the profile—as discussed by the author.

In studies of lake sediments from Ontario and southern and central Quebec a similarly wide variation in baseline dates is apparent—from 1850 to

Table 9
Baselines for anthropogenic lead deposition in isolated rural regions of North America.

Site	Date	Source
Adirondack Lakes, New York	1850	Norton (1986)
	1950	Galloway and Likens (1979), Heit *et al.* (1981)
Northern New England lakes	1820–1870	Hanson and Norton (1982), Hanson *et al.* (1982), Davis *et al.* (1983), Kahl *et al.* (1984), Norton (1986)
Big Heath Bog, Maine	1880	Norton (1987)
Algonquin Provincial Park, Ontario	1900	Wong *et al.* (1984)
Southern Quebec lakes	1940	Ouellet and Jones (1983)
Central Quebec lakes	1850–1855	Evans and Rigler (1985)
Rocky Mountain lakes	1855–1905	Baron *et al.* (1986)
Thompson Canyon Pond, Sierra Nevada, California	1850	Shirahata *et al.* (1980)

the early- to mid-1900s. It is tempting to accept the early date given by Evans and Rigler (1985), since these workers undertook a rigorous analysis of total lead burdens to account for the effects of sediment focusing (see Section III), and the onset dates were obtained by integration over the whole lake areas. The authors attributed the lead enrichments in the sediments to long-range transport from the industrial regions on the southern shore of Lake Michigan, which almost certainly invalidates the later dates reported from more southerly regions of Quebec, and the Algonquin Provincial Park, Ontario. The source of error in the studies by Ouellet and Jones (1983) and Wong *et al.* (1984) is likely to have been inadequate estimation of background concentration, and the poor dating precision for the pre-1900 horizons provided by the caesium-137 method.

Lastly, studies of lake sediments from the Rocky Mountains and the Sierra Nevada gave baselines of 1855–1900 and 1850, respectively, for remote western USA. The study by Shirahata *et al.* (1980) was particularly meticulous: lead in the sediments was apportioned to "natural" and "anthropogenic" fractions using sequential chemical analysis and stable lead isotope analysis, and a very precise dating scheme was constructed in which lead-210 dates were substantiated by a "biogenic varve" chronology (see Fig. 23(b)). The authors concluded that 1850 could be considered to be a pollution baseline for the whole North American subcontinent, on account of the extreme isolation of the site.

5. Twentieth Century Trends in Atmospheric Heavy Metal Deposition in North America

The discussion in the previous sections has shown that lead was dispersed rapidly throughout the subcontinent following the beginning of industrialization, but that the anthropogenic deposition of other heavy metals was much lower, and sometimes not detectable during the nineteenth century. The geochemical data in Table 10 show, broadly, that lead deposition has continued to increase—in some instances by two- or three-fold—during the twentieth century, and that the first detectable enrichments of other heavy metals have occurred within this period, although some regional heterogeneity exists.

In the industrialized parts of north-eastern USA (Rhode Island and Connecticut), and in the south-eastern Great Lakes region, the early 1900s were marked by increases in the rates of deposition—sometimes of up to five-fold—of lead, zinc, copper and cadmium. Muhlbaier and Tisue (1981) showed that if the deposition rate of cadmium to the southern basin of Lake Michigan continued to increase, even slightly, the concentration of dissolved cadmium in the lake water could reach a toxic level within 100 years. In many of the Great Lakes regions, however, the early 1900s increase in deposition has been followed by a levelling-off or decrease, which has been attributed to changes in fuel usage (see Section IV.B). This phenomenon is also apparent at some rural sites (e.g. northern New England and Quebec), but often lead concentrations alone have continued to increase. Norton (1986) and other workers have discussed the problems associated with the interpretation of very recent trends in heavy metal deposition, and have pointed out that a change in concentration which occupies only the top few layers of a deposit need not reflect a real change in deposition, but equally could be attributable a remobilization, particularly in lakes which have become acidified (see Section III). This illustrates that the interpretation of recent trends is one of the most difficult aspects of geochemical monitoring using aquatic sediments or peat.

At all the other rural sites listed in Table 10 (namely sites in Pennsylvania, the Adirondacks, Western Ontario, Wisconsin, Quebec and Southern California), the evidence points to a continued, steady increase in the deposition of most heavy metals up to the present day, in accordance with general urban/industrial expansion. The recent trends in the deposition of nickel and vanadium, however, are associated with more specific activities—metal smelting and oil combustion, respectively.

Table 10
Geochemical evidence for trends[a] in atmospheric heavy metal deposition in North America since AD 1900.

Site	Pb	Zn	Cu	Cd	Ni	Hg	V	Source
Narragansett Bay, RI[a,*]	+	+	+	+	0		0	Goldberg *et al.* (1977)
New Haven Reservoirs, CT[a]	++/−	++/−	++	+/−	+		+/−	Bertine and Mendeck (1978)
Spruce Flats Bog, PA[p]	++	++		++			0	Schell *et al.* (1986)
Northern New England lakes[a,*]	++	+/−		+/−			+	Hanson and Norton (1982), Kahl *et al.* (1984), Norton (1986)
Adirondack lakes, NY[a]	++	++	+	++	0	0	?	Galloway and Likens (1979), Heit *et al.* (1981), Norton (1986)
South-eastern Ontario[a,1]	++/−	++/−	+/−	+/−	+/−	+	?	Edgington and Robbins (1976), Kemp and Thomas (1976), Farmer (1978), Goldberg *et al.* (1981), Evans and Dillon (1982), Christensen and

								Johnson *et al.* (1986)
South-western Ontario[a,3]	+ +	+	+	+	0	+	0	Kemp *et al.* (1978)
Sudbury, Ontario[a]	+ +	+	+ +	+	+ +			Nriagu *et al.* (1982)
Algonquin Provincial Park, Ont.[a]	+ +	+ +	+	+	+			Wong *et al.* (1984)
Northern Wisconsin[a] and Northern Minnesota[a]	+ +	+	+	+	+	+		Syers *et al.* (1973), Meger (1986)
Southern and Central Quebec[a]	+ +	+/−	+		+	+		Ouellet and Jones (1983), Evans and Rigler (1985)
Rocky Mountain lakes[a]	+/−	0	0	+	0	0	0	Heit *et al.* (1984), Baron *et al.* (1986)
Thompson Canyon Pond, CA[a]	+ +							Shirahata *et al.* (1980)
Inner coastal basins, CA[a]	+ +	+	+	+	0	+	+	Young *et al.* (1973), Bruland *et al.* (1974)
Outer coastal basins, CA[a]	+	+	0	0			+	Bertine and Goldberg (1977)

[a]Geochemical evidence provided by the analysis of: [a]aquatic sediments; and [p]peat deposits. Trends denoted by: +, increase; + +, increase of > two-fold; +/−, increase followed by levelling or decrease; *, change occurred after 1950; 0, no anthropogenic enrichment; ?, inconclusive.

[1]includes Lakes Erie, Michigan and Ontario and Found Lake; [2]includes Lake Huron and Turkey Lakes; [3]includes Lake Superior.

B. Present-day Sources of Atmospheric Heavy Metal Pollutants

The respective contributions of different pollution sources to the atmospheric heavy metal burden have been changing continually throughout history and, as discussed in the previous section, have diversified greatly during the twentieth century. In the geochemical monitoring of pollution sources one of two approaches is usually adopted: either the investigation of the dispersal of heavy metals from a point source to the surrounding area, or the assessment of the respective contributions of a number of sources to heavy metal deposition in an urban or rural area.

1. The Dispersal of Pollutants from Point Sources

The effect of a point source on atmospheric deposition is a function of the dispersal of pollutants to the surrounding area; this can be demonstrated most clearly in isolated rural areas. Classic examples of this situation are the base-metal smelting complexes of Sudbury, Ontario (nickel, copper) and Flin Flon, Manitoba (zinc). In the Sudbury region, monitoring is facilitated by the presence of numerous small lakes in the vicinity of the smelters. The atmospheric dispersal of metals away from the point source was investigated by the measurement of the flux to the surface sediments of lakes situated at different distances from the main smelter (Nriagu *et al.*, 1982). Figure 18 shows that nickel deposition decreases sharply up to a distance of *c.* 10 km from the smelter, but declines much more gradually thereafter. The extent of the influence of the smelter emissions on the atmospheric heavy metal burden in the surrounding area has not been quantified, but it is significant that nickel enrichments, attributed to the Sudbury smelters, have been detected in the sediments of Lake Huron and a number of small lakes in central Ontario over 100 km to the east (Kemp *et al.*, 1978; Johnson *et al.*, 1986), but not yet in the sediments of Lake Superior (Kemp *et al.*, 1978). The recent historical aspect of the dispersal of pollutants from the smelters to the surrounding district is interesting. Nickel and copper deposition to the lakes in the Sudbury area which are farthest from the smelters (Lakes Nelson and Windy) increased sharply after 1970, corresponding to the construction of the 381 m-high "superstack", which almost certainly would have facilitated the long-distance transport of heavy metals (Nriagu *et al.*, 1982). Whether the "superstack" actually fulfils its purpose, which is to alleviate pollution in the immediate vicinity, cannot be assessed with certainty, as the heavy metal records in the lakes closest to the smelters are now affected by acidification (Nriagu and Rao, 1987; see also Section III).

The Flin Flon smelting complex in remote western Manitoba is a similar "outdoor laboratory", where the atmospheric impact of zinc smelting has been assessed in isolation from a background of miscellaneous emissions.

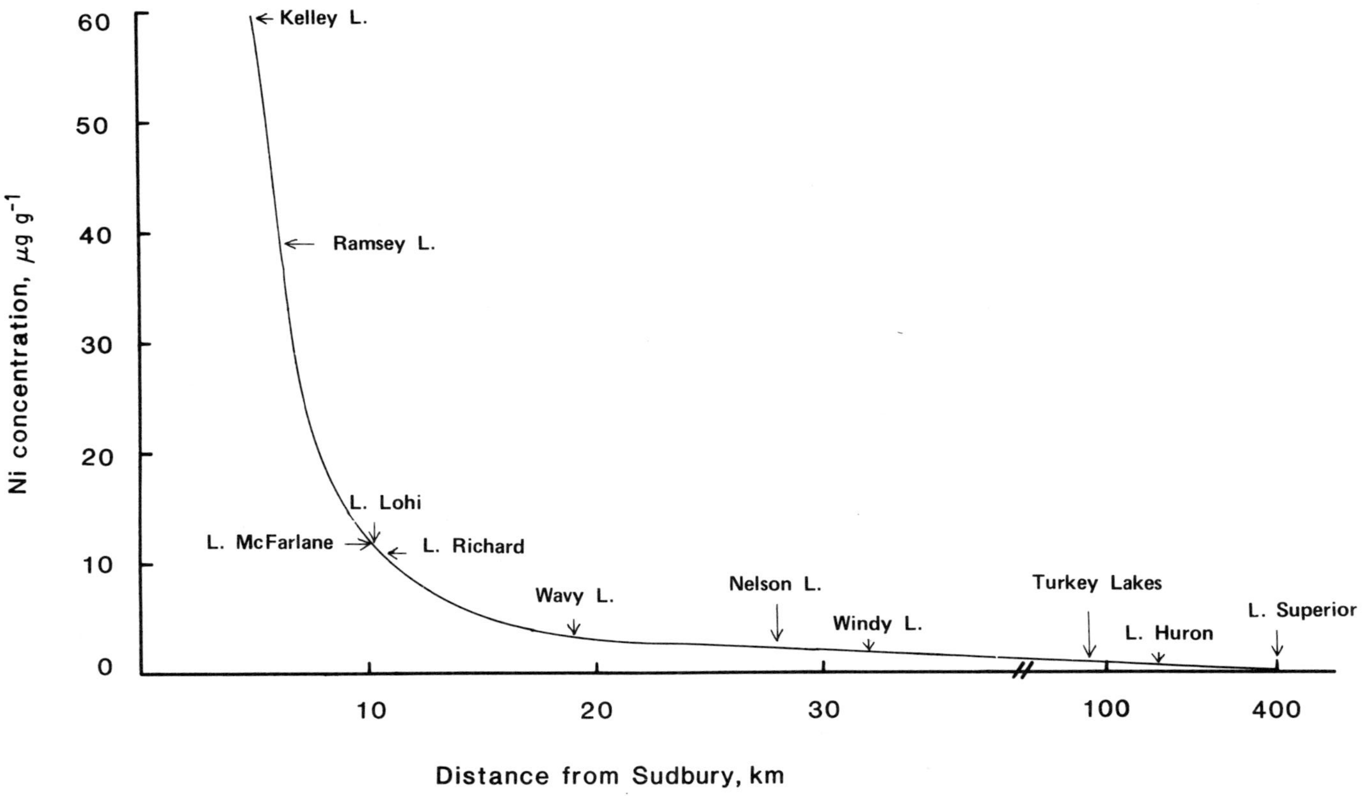

Fig. 18. Concentration of nickel in the surface sediments of lakes in relation to distance from Sudbury, Ontario. Data of Kemp *et al.* (1978), Nriagu *et al.* (1982) and Johnson (1987). Note change of scale between 30 and 100 km.

Following a series of analyses of the winter snowpack in the area, Franzin *et al.* (1979) were able to construct a mathematical model to describe the decrease in deposition with distance from the smelter, taking account of meteorological variables. They calculated that the smelter was the predominant source of zinc, copper and cadmium in the local environment, and that its measureable influence extended over an area of 250 000 km^2. A comparison of the respective ratios of copper, cadmium, lead and zinc in emissions from the smelter and deposition to the surrounding area suggested that, whilst most of the zinc was deposited in the immediate vicinity of the smelter, a significant fraction of the other metals was transported further afield. In a subsequent snowpack survey, however, Shewchuk (1985) reported that enrichments of not only lead, but also of zinc from the Flin Flon smelters could be detected in atmospheric deposition 100 km away in northern Saskatchewan. A long-term snowpack survey recently completed by Phillips *et al.* (1986), in which sampling was carried out over four consecutive years, showed that zinc deposition decreased 20-fold after 1983, when electrostatic precipitators were installed at the smelter.

2. The Contributions of Multiple Sources to Heavy Metal Deposition

Ever since Patterson (1965) drew attention to the rapidly increasing concentrations of lead in the environment, the relative importance of motor vehicles and fossil fuel combustion as sources of atmospheric lead (and other metals) has been a subject of debate. The question has gained significance recently, as attention has been focused on the scale of the emissions of other pollutants (e.g. NO_x) from these two respective sources. The emissions inventory for lead (Nriagu, 1979) suggests that cars may be the more significant source of lead pollution globally, but this masks significant regional heterogeneity.

Elemental ratios are now widely used in direct air-sampling as source tracers—for example, Br/Pb for motor-vehicle exhaust (Sturges and Harrison, 1986), Fe/Mg for coal-fired power plants (Parekh and Husain, 1987), and V/As and V/Se for oil-fired power plants (Husain, 1986). The chemical characteristics of some of these elements, however, preclude their use as tracers in geochemical monitoring. The approaches most often used, therefore, involve the comparison of temporal trends in deposition with documentary evidence relating to trends in emissions.

More specific, supplementary evidence is provided by the analysis of related pollutants such as coal fly-ash and polycyclic aromatic hydrocarbons (PAHs), a group of compounds which includes some analogues that can be specifically related to the high-temperature processes involved in coal- rather than petroleum-combustion. Stable lead-isotope analysis (described in Section III) can also provide valuable supporting evidence relating to sources of lead in the environment, both natural and anthropogenic.

In Britain and Continental Europe many sites have been heavily polluted by industrial emissions since the eighteenth and nineteenth centuries, and current trends in heavy metal deposition have accordingly been attributed to changing patterns in coal usage during the latter part of the twentieth century.

It was shown in Section IV.A that a recent decline in heavy metal deposition rate has taken place at many moderately polluted rural sites in northern Britain. This is exemplified by the heavy metal record from Ringinglow Bog (Yorkshire), where the still actively growing upper peat horizons (*c.* $1{\cdot}3\,\mathrm{mm\,yr^{-1}}$) provide a detailed record of recent trends in deposition (Livett, 1982). The profile from this site (Fig. 19(a)) shows a marked decrease in lead deposition which was dated to post-1930, and which corresponds closely with the decrease in consumption of coal for primary energy over the same period (Fig. 19(b)). The effect of the continued increase in the use of coal for secondary energy would almost certainly have been offset by controls imposed on particulate emissions since the 1960s.

In industrialized regions of Europe deposition rates of a suite of metals characteristic of coal combustion—lead, zinc, copper and cadmium—have continued to increase during the twentieth century. Examples include the Baltic Coastal region and Lake Constance, where Erlenkeuser *et al.* (1974) and Müller *et al.* (1977, 1980) have attributed surface enrichments of between two- and seven-fold in the sediments to the continued increase in consumption of coal in the countries bordering this region. Erlenkeuser *et al.* (1974) showed that the concentration of heavy metal-rich fly-ash in the sediments would be alone sufficient to account for the observed enrichments of heavy metals. Additionally, Müller *et al.* (1977) showed that trends in heavy metal deposition in the post-1900 sediments of Lake Constance were correlated with enrichments of the particular PAHs which are associated with coal combustion. It is interesting to note, however, that the heavy metal profiles in the sediments of both the Baltic Sea and Lake Constance actually show a decline in deposition since the late 1970s. This trend, if not a product of diagenesis, could be a consequence of the decrease in coal usage in Western Europe over the last few decades (Fig. 20(a)). The authors unfortunately did not comment on this feature.

The evidence for trends and sources of lead deposition in more isolated areas of Europe is less conclusive, and thus open to various interpretations. In isolated regions of northern Britain, e.g. northern Scotland and Shetland, the increase in lead deposition since 1900 appears to have been very gradual (Section IV.A); Lee and Tallis (1973) therefore concluded that the impact of

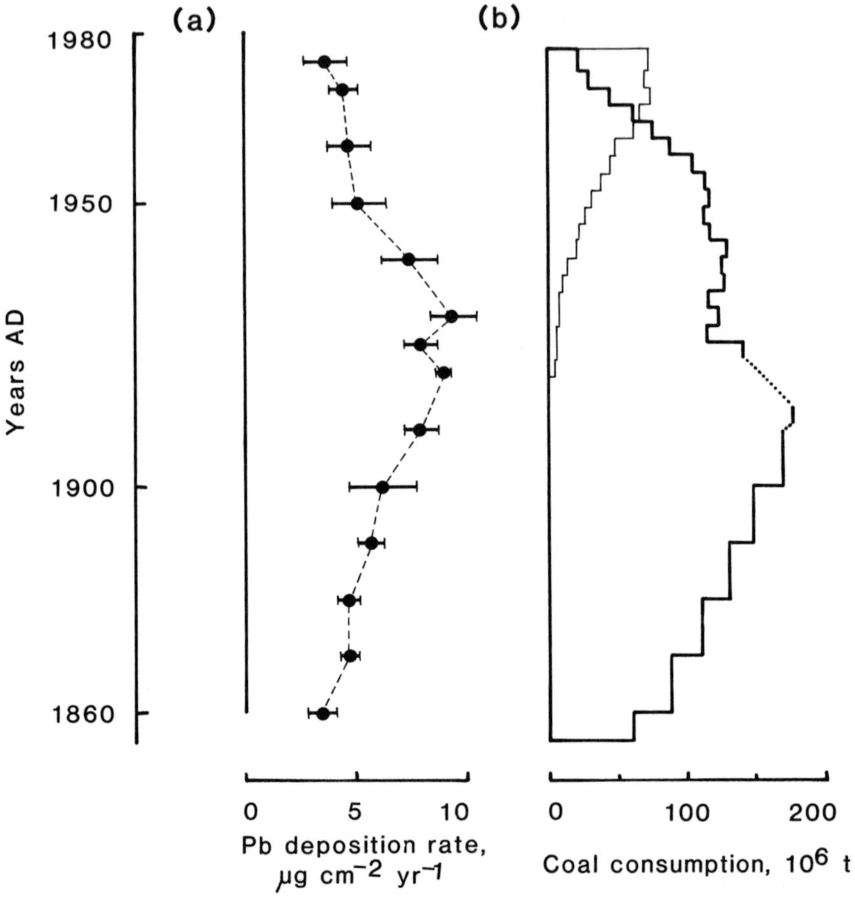

Fig. 19. Lead deposition in Northern England in relation to trends in fossil fuel consumption: (a) calculated mean lead deposition rates and standard errors for five peat profiles from Ringinglow Bog; the time-scale is based on pollen-analytical and documentary evidence (data of Livett, 1982 and unpublished); (b) combustion of coal for primary energy (bold line) and secondary energy (feint line) in the UK from 1860 to 1975 (data of Jevons, 1915; Ministry of Power, 1968; Department of Energy, 1978).

Fig. 20. Lead deposition in Central Europe in relation to trends in fossil fuel consumption: (a) coal consumption from 1895 to 1980 in Eastern Europe (E), Western Europe (W), and the two combined (T); (data of Putnam, 1953; United Nations, 1952, 1976, 1981); (b) concentration of lead in glacier ice dated from 1890 to 1968 at Great Mieguszowiecki Cirque (open circles) and Groto Lodowa (closed circles); mean lead determinations from short sections of the profile at the two sites denoted by feint and bold lines, respectively. Redrawn from Jaworowski (1968).

(a)

(b)

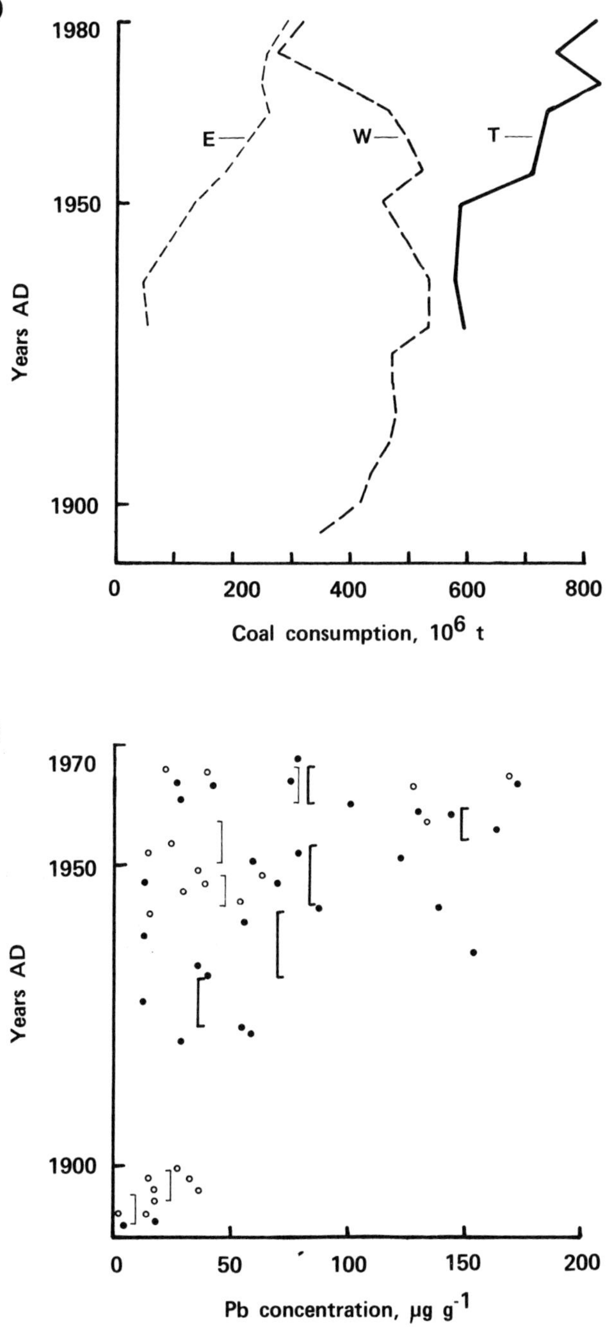

1980
1950
1900
Years AD
E
W
T
0
200
400
600
800
Coal consumption, 10^6 t
1970
1950
1900
Years AD
0
50
100
150
200
Pb concentration, μg g^{-1}

motor vehicle emissions in these regions is still negligible. In Continental Europe, in contrast, there has generally been a more marked increase in heavy metal deposition in the rural areas. For example, Jaworowski (1968) reported an increase of three-fold in lead concentrations in the post-1900 horizons of two glaciers in the Polish Tatra Mountains (Fig. 20(b)), and related this trend to the documented rise in coal combustion in Eastern Europe over the same period (Fig. 20(a)). Aaby *et al.* (1979) observed a similar trend in lead deposition at Draved Mose in Denmark, but attributed it to the increase in production of tetra-ethyl lead in Western Europe since 1950. They also argued that fine-particulate motor vehicle emissions would be more efficiently transported to high altitudes than would coarse coal fly-ash particles, and hence that the lead enrichments reported by Jaworowski (1968) in the Tatras were likely to be of automotive rather than industrial origin. Indeed, at another high-altitude site—Mont Blanc—post-1950 increases in the concentrations, not only of lead, but also of cadmium and vanadium, in a snow profile were interpreted as evidence of the long-distance transport of emissions from petrol and heating oil combustion (Briat, 1978). These examples illustrate that the matching of historical trends is a very subjective approach to source identification, and one which is unlikely to produce unequivocal conclusions.

Firmer positive evidence in favour of the importance of motor vehicle emissions in rural regions of Europe has been provided by stable lead isotope analysis. Petit (1974) and Petit *et al.* (1984) conducted lead isotope analyses of cores from three small south Belgian lakes (Lakes Mirwart, Vielsalm and Willerzie), and observed a progressive shift in the ratios in the post-1900 sediment horizons, which implied that natural, terrigenous lead was increasingly being diluted with older, less radiogenic lead of pollutant origin (see Section III). Further up the profile, the lead was even less radiogenic, and at the surface isotopically closely resembled anti-knock lead (which is generally of greater geological age than is fossil fuel lead). This evidence tied in well with the heavy metal profiles from the same lakes (Oldfield *et al.*, 1980a; Thomas *et al.*, 1984), which showed a three-fold increase in lead deposition after 1960, but a decrease in deposition of copper, cadmium and zinc, supporting the theory that industrial emissions have declined, and automotive emissions increased in importance in the region. Evidence of a similar nature has been produced from a study of Pavin Crater Lake in the French Massif Central by Martin (1985). Two lead profiles showed a gradual increase in concentrations dating from the early to the mid-1800s, followed by a steeper increase from 1900 to the present day. The accompanying stable isotope ratios indicated that anthropogenic lead, first of an industrial origin, and then of a vehicular origin, had been deposited to the lake.

In North America as in Britain, many regions which have been heavily polluted with lead, zinc, copper and cadmium since the nineteenth century have recently experienced a decline in atmospheric deposition which has been linked with trends in coal usage. In sediment cores from Lake Michigan, Goldberg *et al.* (1981) observed a steep increase in heavy metal deposition after 1920, followed by a decrease since *c.* 1960. The distribution of carbonaceous material in the sediments, identified by Griffin and Goldberg (1979, 1983) as coal combustion residues, implied that the historical trends in heavy metal deposition could be related at the beginning to the increasing use of coal for electricity generation in this region, and latterly to reduced emissions from power plants after the 1960s when electrostatic precipitators were installed.

Similar, recent decreases in heavy metal deposition as a response to emissions controls have been reported at other localities in the Great Lakes regions (e.g. Kemp and Thomas, 1976), but there is also evidence that in some instances the potential improvement in air quality has been more than offset by increased pollution from motor vehicles. This situation was demonstrated by Edgington and Robbins (1976) in a detailed analysis of sediment cores from the southern basin of Lake Michigan. They used information relating to emissions from local sources to construct a model to account for the historical trends in lead flux to the sediments (Figs 21(a),(b)), and showed that, whereas the installation of flue "scrubbers" in local power plants had effectively reduced emissions to one-third of their pre-1960 level, the simultaneous increase in the use of cars locally had maintained the upward trend in lead deposition.

The balance which can exist between a miscellany of emissions in an urban area was illustrated by a study of two reservoirs in the New Haven area, Connecticut, by Bertine and Mendeck (1978). It was established by heavy metal analyses of the sediments that coal combustion had made the most important contribution to pollution in the area from the mid-nineteenth to the early twentieth century (see Section IV.A), and it was assumed, therefore, that the post-1960 decline in lead deposition in the sediments of one of the reservoirs—Lake Whitney—was attributable to the changeover from coal to oil burning at the local electricity generating station. Stable lead isotope analysis showed, however, that after a steady decrease in $^{206}Pb/^{207}Pb$ ratio since the mid-nineteenth century, when "natural" lead was progressively diluted by "industrial" lead, by the 1950s the isotope ratio in the sediments more closely resembled "gasoline" lead than "coal" lead, as shown in Fig. 22. The authors concluded, therefore, that just prior to 1960, motor vehicle emissions had been the predominant lead source, and that the post-1960 decline in deposition was a result of the diversion of motor traffic away from the lake-shore in 1966 rather than a direct effect of industrial emissions controls implemented at about the same time.

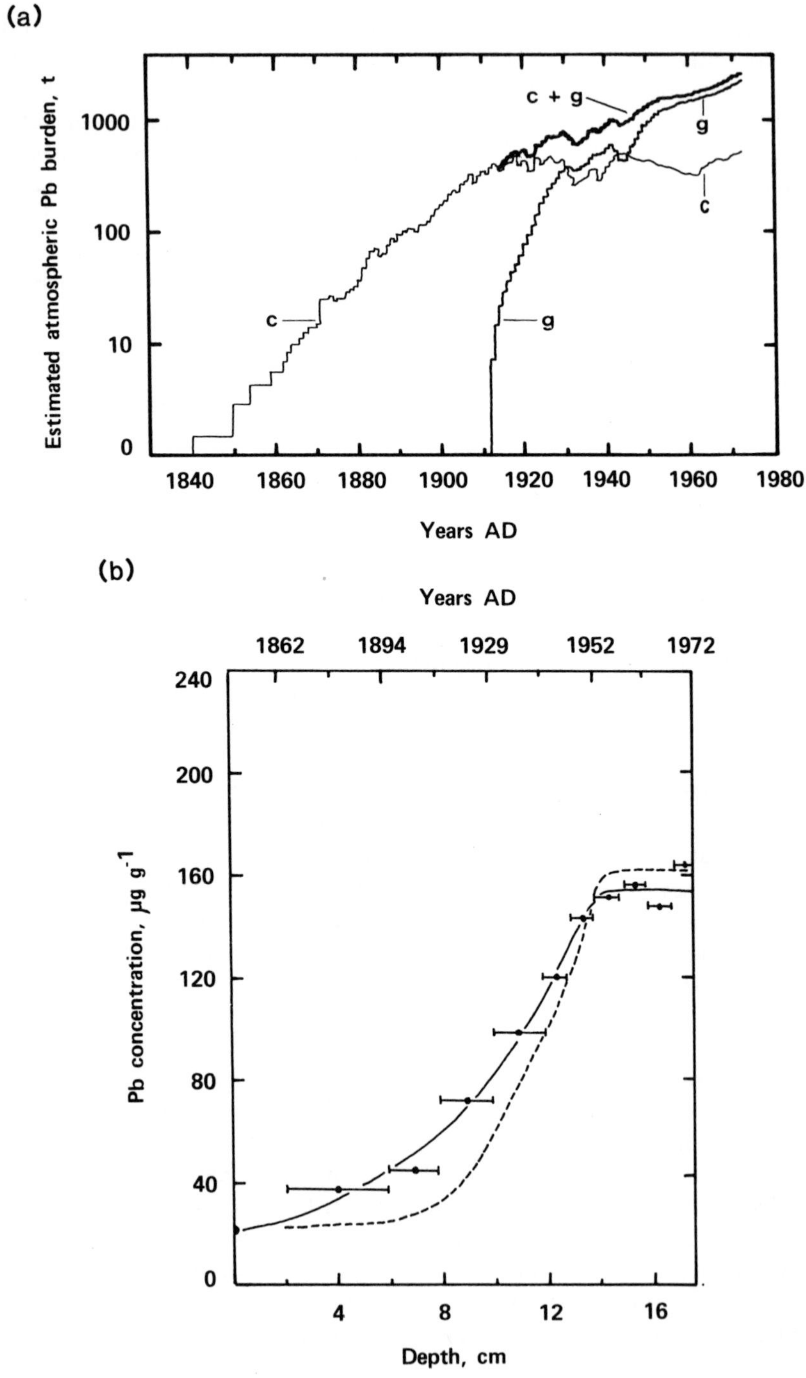

(a)
Estimated atmospheric Pb burden, t
1000
100
10
0
c + g
g
c
c
g
1840
1860
1880
1900
1920
1940
1960
1980
Years AD
(b)
Years AD
1862
1894
1929
1952
1972
240
200
160
120
80
40
0
Pb concentration, μg g⁻¹
4
8
12
16
Depth, cm

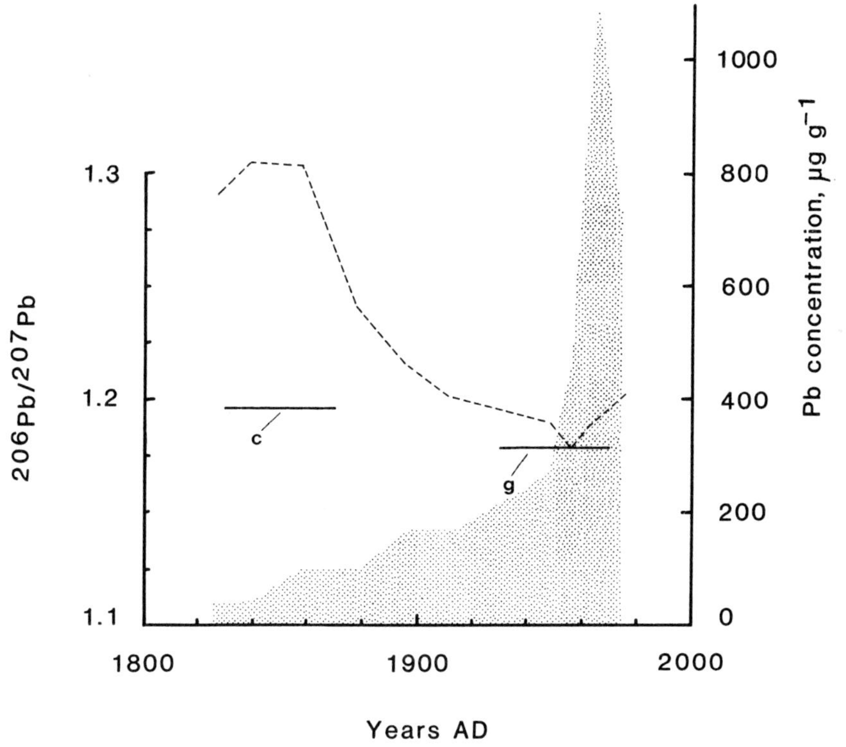

Fig. 22. Total lead concentration (shaded area) and stable lead isotope ratios (broken line) in a Pb-210 dated sediment profile from Lake Whitney, New Haven, Connecticut. The solid horizontal lines show the stable lead isotope ratios in present-day samples of coal (c) and gasoline (g). Data of Bertine and Mendeck (1978).

Fig. 21. Sources of lead in the southern basin of Lake Michigan. (a) Estimated atmospheric loading from coal burning (c), gasoline burning (g), and the two combined (c + g) from 1840 to 1970. (b) Concentration of lead with depth (closed circles) and standard errors (horizontal bars), in a Pb-210 dated sediment profile, showing the least-squares fit of source-functions for coal and gasoline (solid curve) and gasoline only (broken curve). Redrawn from Edgington and Robbins (1976).

Geochemical evidence from isolated rural areas in North America points unequivocally to a widespread increase in the deposition of lead (and other metals) since the early to mid-1900s (Table 10), but as in Europe, this trend is open to more than one explanation. Analyses of Adirondack lake sediments (Galloway and Likens, 1979; Heit *et al.*, 1981) have revealed a major increase in an assemblage of metals associated with coal combustion (lead, zinc, copper and cadmium) since 1950, and more specific evidence from the analysis of PAHs led Heit *et al.* (1981) to conclude that the metal enrichments originated from the long-distance transport of industrial emissions, facilitated by the construction of taller smoke stacks in the mid-1900s. There is an equally strong body of evidence, however, in favour of motor vehicles as the major source of lead enrichments in rural regions (e.g. Davis *et al.*, 1982; Ouellet and Jones, 1983; Baron *et al.*, 1986; Schell, 1986), and stable lead isotope analysis, chiefly from the western USA (summarized in Table 11), substantiates this point of view.

In the Southern California coastal belt, where heavy industry has never played a dominant part in the regional economy (see Section IV.A), it has been suggested that the present-day atmospheric lead burden chiefly originates from vehicular emissions. Chow *et al.* (1973) reported a progressive decline in the isotope ratios of lead from the pre- to the post-industrial sediment horizons of the coastal basins, and showed that ratios in the present-day horizons were similar to those reported for contemporary gasoline samples. In a later study of sediment profiles from the same region, Ng and Patterson (1982) observed a post-1970 upturn in the radiogenicity of deposited lead (see Fig. 17). This was thought to reflect the beginning of the use of the anomalously radiogenic lead from the Missouri orefield for the manufacture of anti-knock compound in the mid-1960s. The same characteristic feature was observed by Shirahata *et al.* (1980) in a profile from Thompson Canyon Pond, in the Sierra Nevada Mountains, California (Fig. 23(a)), and the steep increase in lead deposition, dated to after 1920, corresponded to the documented increase in gasoline consumption in the USA over the same period (Fig. 23(b)). These results pointed firstly, to the predominantly vehicular source of the lead enrichments, and secondly, to the rapid response of the airmass and the sediments in this remote region to country-wide trends in the chemistry of pollutant emissions. In another very isolated region, the Lassen Volcanic Park, California, Chow and Johnstone (1965) showed that the isotope ratios of freshly fallen snow closely matched those of gasoline sold in the region, and therefore concluded that motor vehicle exhausts were making a virtually 100% contribution to the atmospheric lead burden in this region.

The various lines of evidence which can help to identify the sources of heavy metals in atmospheric deposition all point to the importance of motor

Table 11
Isotopic composition of lead in atmospheric deposition[a], fossil fuel and lead ore in the USA.

Site and Material	Date	Lead isotope ratio 206/204	206/207	206/208	Source
(a) *Polluted*					
Lake Whitney, New Haven, CT[a]	1975	18·77	1·190	0·485	Bertine and Mendeck (1978)
Lake Whitney, New Haven, CT[a]	Pre-industrial	20·50	1·291	0·506	Bertine and Mendeck (1978)
Gasoline, Boston, MA	1964	18·45	1·179	0·482	Bertine and Mendeck (1978)
Coal, Boston, MA	—	18·77	1·196	0·484	Bertine and Mendeck (1978)
(b) *Intermediate*					
Santa Barbara Basin, CA[a]	1970	18·73	1·189	0·481	Chow *et al.* (1973)
Santa Barbara Basin, CA[a]	Pre-industrial	19·17	1·219	0·487	Chow *et al.* (1973)
Santa Barbara Basin, CA[a]	1977	18·66	1·199	0·486	Ng and Patterson (1982)
Santa Barbara Basin, CA[a]	Pre-industrial	18·98	1·219	0·486	Ng and Patterson (1982)
Gasoline, Los Angeles, CA	1968	18·08	1·155	0·476	Chow *et al.* (1973)
Lead ore, MO	—	21·75	1·385	0·534	Chow *et al.* (1975)
(c) *Remote*					
Lassen Park, CA[s]	*c.* 1963	18·01	1·144	0·469	Chow and Johnstone (1965)
Gasoline, Los Angeles, CA	1964	17·92	1·145	0·473	Chow and Johnstone (1965)
Thompson Canyon, CA[a]	1972	18·42	1·188	0·485	Shirahata *et al.* (1980)
Thompson Canyon, CA[a]	Pre-industrial	19·39	1·239	0·496	Shirahata *et al.* (1980)

[a] Atmospheric lead analysed in: [a]aquatic sediments; and [s]snow.

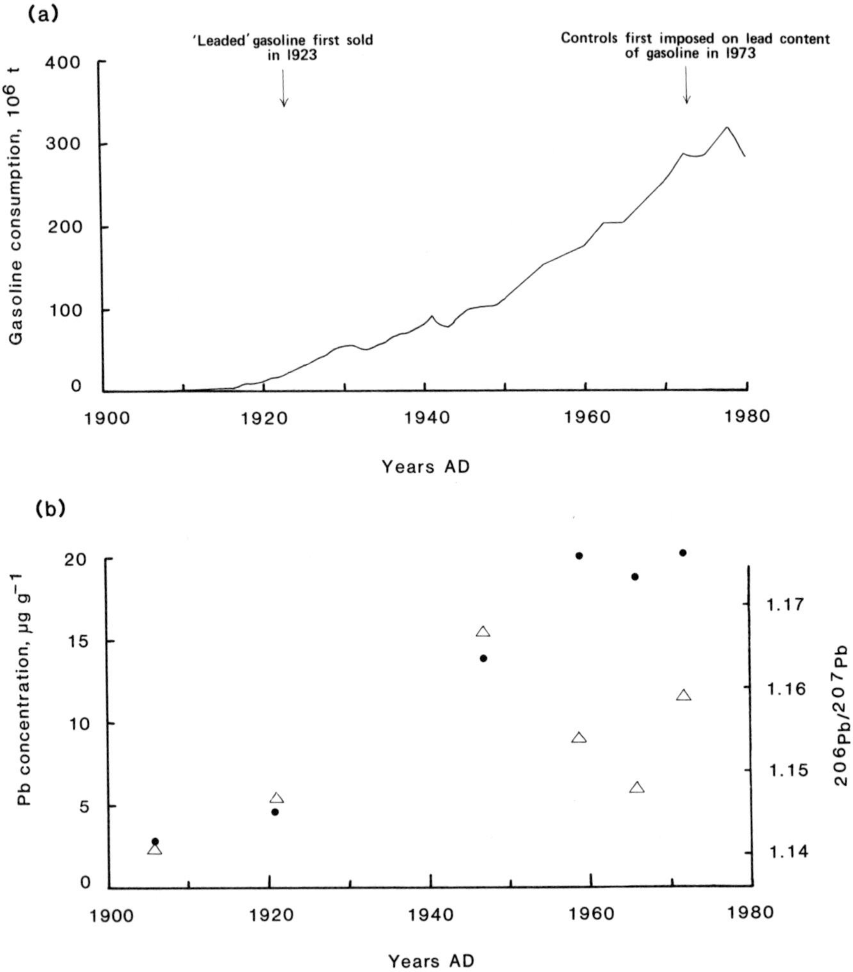

Fig. 23. Lead deposition in rural USA in relation to trends in vehicle emissions: (a) gasoline consumption in the USA from 1900 to 1980 (data of United Nations, 1952, 1976, 1981; Williamson *et al.*, 1963); (b) concentration of atmospheric, anthropogenic lead (closed circles) and $^{206}Pb/^{207}Pb$ ratio (triangles) in a varve- and Pb-210 dated sediment profile from Thompson Canyon Pond, Sierra Nevada Mountains, California. Data of Shirahata *et al.* (1980).

vehicle exhausts in both Europe and the USA. In urban areas in both continents this influence is balanced to some extent against industrial and domestic emissions; the direction of trends in these regions in the future will depend on patterns of fuel usage and the efficacy of emissions controls. In rural and isolated regions, however, particularly in the USA, automobile emissions have predominated over industrial emissions as a source of lead since the 1950s at least, and are therefore chiefly responsible for the continuing increases in lead deposition observed in these regions. In this context, it is interesting to note that there is as yet no firm geochemical evidence from the USA of the observed decrease in atmospheric lead concentration following the enforced reduction in the anti-knock content of gasoline in the 1970s (e.g. Eisenreich *et al.*, 1986). This is probably a function of the in-built time-lag in natural ecosystems (Schell, 1986), and the difficulties in the interpretation of the chemistry of the surface layers of deposits, as discussed in Section III.

C. Present World-wide Occurrence of Atmospheric Heavy Metal Pollutants

1. Heavy Metal Deposition Rates in North-west Europe, North America and the Remote Regions

Table 12 summarizes the geochemical evidence for the variation in deposition rates of the metals lead, zinc, copper, cadmium, nickel, mercury and vanadium in, on the one hand, two major industrialized regions, and on the other hand, three remote regions of the world. The sites listed as "polluted" are less than 5 km from a large urban or industrial complex, while the regions classified as "rural" are at least 20 km distant from such pollution sources. For some of these sites the anthropogenic fraction of total heavy metal deposition has been calculated, but for most of the sites it has been established that a significant fraction of the total metal influx is pollutant in origin. The criteria applied to the designation of "remote" sites, and the sources of the metals in atmospheric deposition at these sites, is discussed below. The deposition rates in Table 12 span approximately seven orders of magnitude; rates for lead and zinc correspond roughly with the three-tier categorization of sites, as follows:

$$\text{polluted} > 100\,\mu\text{g cm}^{-2}\,\text{yr}^{-1} > \text{rural} > 0{\cdot}1\,\mu\text{g cm}^{-2}\,\text{yr}^{-1} >> \text{remote}$$

These boundaries agree reasonably well with the tripartite classification devised by Galloway *et al.* (1982), which was based on a large number of

Table 12

Variation in present-day atmospheric heavy metal deposition rates in North-west Europe, North America and the remote regions.

Locality	Deposition rate[a]							Source[b]
	Pb	Zn	Cu	Cd	Ni	Hg	V	
(a) *Remote*			($ng\ cm^{-2}\ yr^{-1}$)					
Antarctica (South Pole)[t]	0·01							6
Antarctica (Dome C)[t]	0·12	0·11	0·12	0·02				5
Greenland[t]	5·97–7·85	9·16–10·0	1·65–1·89	0·07–0·37				2, 5
North-east Atlantic Ocean[t]	1·1							40
(b) *Rural*			($\mu g\ cm^{-2}\ yr^{-1}$)					
Northern Scandinavia[t]	0·08–0·55	0·77–1·0	0·11–0·25	0·47	0·06	0·04–0·05	0·31	21, 35, 36, 39
Northern Britain/Southern Scandinavia[t]	0·15–9·90	0·42–16·9	0·04–1·8	0·02–0·09				1, 11, 14, 17, 27, 28, 29
Southern Belgium[t]	2·19–12·0	11·97	0·47	0·15				33, 38
Northern/Western Ontario, North America[a]	0·10–1·50	0·10–2·50	0·10–0·6	⩽0·10	⩽0·50	< 0·01	0·01–0·2	8, 9, 10, 22, 23, 25, 26, 34
South-eastern Ontario, North America[a]	1·19–23·2	6·90–59·3	1·80–12·0	0·08–0·36		0·05–0·14		12, 13, 25, 31
Northern New England, USA[a]	0·70–5·56	2·73–6·50	0·07	0·05–0·12	⩽0·04	0·01	0·4	15, 19, 20,

(c) *Polluted*			($\mu g\ cm^{-2}\ yr^{-1}$)						
Ramsey Lake, Sudbury, Ontario, Canada[t]	2·7	4·8	30·1		38·6				32
Whitney Reservoir, New Haven, Conn., USA[a]	238·0	39·0		0·4	17·0			22·0	4
Narragansett Bay, RI USA[a]	168·0	330·0	260·0						16
Flin Flon, Manitoba, Canada[t]	5·7	382·0	12·3	0·24					37

[a]Deposition rates are: [t]total, or [a]anthropogenic.
[b]Published sources:

1 Aaby and Jacobsen (1979)
2 Appelquist *et al.* (1978)
3 Bertine and Goldberg (1977)
4 Bertine and Mendeck (1978)
5 Boutron (1979a)
6 Boutron (1982)
7 Bruland *et al.* (1974)
8 Christensen and Chien (1981)
9 Christensen and Goetz (1987)
10 Edgington and Robbins (1976)
11 Erlenkeuser *et al.* (1974)
12 Evans and Dillon (1982)
13 Farmer (1978)
14 Farmer *et al.* (1980)
15 Galloway and Likens (1979)
16 Goldberg *et al.* (1977)
17 Hamilton-Taylor (1979)
18 Hamilton-Taylor (1983)
19 Hanson and Norton (1982)
20 Heit *et al.* (1981)
21 Jaworowski *et al.* (1975)
22 Johnson (1987)
23 Johnson *et al.* (1986)
24 Kahl *et al.* (1984)
25 Kemp and Thomas (1976)
26 Kemp *et al.* (1978)
27 Livett (1982)
28 Livett *et al.* (1979)
29 Madsen (1981)
30 Ng and Patterson (1982)
31 Nriagu *et al.* (1979)
32 Nriagu *et al.* (1982)
33 Oldfield *et al.* (1980a)
34 Pakarinen and Gorham (1983)
35 Pakarinen *et al.* (1980)
36 Pakarinen *et al.* (1983)
37 Phillips *et al.* (1986)
38 Thomas *et al.* (1984)
39 Tolonen and Jaakkola (1983)
40 Veron *et al.* (1987)

determinations from standard deposition gauges as well as from natural deposits.

Table 12 shows that there is a dearth of geochemical data from polluted sites. Of the four listed, New Haven and Narragansett Bay are typical, in that they experience high deposition rates of three or more heavy metals, reflecting emissions from a miscellany of pollution sources, for example motor vehicle exhausts, electricity generation, and domestic incineration. In contrast, the deposition spectra at Sudbury and Flin Flon are the product of emissions from single point sources (base metal smelters), and are characterized by disproportionately high deposition rates for one or two particular metals (as discussed in Section IV.B).

The "rural" category is highly heterogeneous, and includes both sites which are directly in the path of pollution-bearing airstreams (e.g. the southern basin of Lake Michigan, Lake Erie, the southern Pennines, the Baltic Sea), and sites which are remote from such influences, by virtue of altitude (e.g. the Tatra Mountains, Thompson Canyon) or of horizontal distance (e.g. northern Norway, Shetland). Barrie *et al.* (1987) have discussed the problems inherent in the classification of sites when there are so many environmental factors, as well as location, which influence heavy metal deposition in any particular area.

The heterogeneous nature of the "rural" category highlights broad geographical trends in heavy metal deposition. In North-west Europe a north/south gradient is apparent, from the highly industrialized regions of central Europe, north-west England and the Baltic seaboard, to the predominantly rural areas of northern Britain and Scandinavia, where lead and zinc deposition rates are lower by a factor of about 100. In North America the highest regional levels of atmospheric heavy metals occur in the south and east of the Great Lakes region, which is influenced by emissions from the Upper Ohio Valley, and in the East Coast region and New England. A gradient in deposition rates of lead and zinc of about 20-fold is apparent in a westerly direction, and of about 200-fold in a northerly direction.

Heavy metal deposition rates in the remote regions cover nearly three orders of magnitude. Rates measured in Greenland are most similar to those recorded for Northern Scandinavia; indeed, Ross and Granat (1986) observed that heavy metal concentrations in snow in the isolated west of Sweden were similar to those measured in Greenland snow. In all the remote regions, nevertheless, total deposition rates are substantially lower than "natural" or "pre-cultural" fluxes calculated for rural localities in Europe and North America (e.g. 0·05–1·1 $\mu g\ cm^{-2}\ yr^{-1}$: Kemp *et al.*, 1978; Aaby *et al.*, 1979). This prompts the question of whether any fraction of such a small total heavy metal influx to the remote regions can be attributed to pollution; it is this question which is examined in the next section.

2. Heavy Metal Deposition in Remote Regions

One of the most important aspects of geochemical monitoring concerns the investigation of the extent to which heavy metal pollutants have pervaded the entire global atmosphere. Regions chosen for this aspect of geochemical monitoring must be remote from major sources such as Europe and North America. Three such regions are considered in this section: the North Atlantic Ocean; and the North and South Polar regions. The mid-ocean waters generally are protected from direct continental heavy metal drainage by the continental shelf sediments (which act as a filter), and so receive only airborne inputs. Amongst these, the North Atlantic is probably the most polluted ocean as it lies in the path of the westerlies which bring polluted air from North America (Patterson, 1987). The North and South Polar regions are shielded for a large part of the year against the influx of polluted air from the mid-latitudes by the cold air masses of the Arctic and Antarctic fronts, respectively. Some pollution reaches the Arctic in the winter when the Arctic air mass extends further south and covers some of the source areas (Barrie, 1986). Antarctica, by comparison, is much more isolated from prospective pollution sources in the southern hemisphere and is relatively cut off from the polluted air masses of the northern hemisphere by the tropospheric equatorial barrier. A significant route of pollutant input to the Antarctic is therefore probably via the stratospheric global circulation system (Newell, 1971). Thus, of these three regions, the atmosphere of the Antarctic is probably the most representative of a "global mean".

(a) North Atlantic Ocean sediments. The investigation of mid-ocean sediments is one of the newest areas in geochemical research, and has developed from the observation that the upper layers of the water column in several mid-ocean areas are now contaminated with lead (Patterson, 1987).

In a recent investigation Veron *et al.* (1987) analysed lead in two short sediment cores from sites situated 250 km and 150 km from the nearest continental areas (Stations 7 and 8, respectively). As shown in Fig. 24, the lead concentration over the greater part of the two profiles constituted stable background levels of *c.* 2–4 $\mu g\ g^{-1}$ and 5–8 $\mu g\ g^{-1}$, respectively, which were in accordance with the relative composition of silicate and carbonate of the two cores. In the surficial layers (< 1 cm) of the sediments, however, lead concentration rose sharply to *c.* 15 $\mu g\ g^{-1}$ and 20 $\mu g\ g^{-1}$, respectively. The authors showed that these surface peaks were significantly in excess of the concentrations which would be expected on the basis of the mineralogical composition, and therefore concluded that the excess metal was anthropogenic. Calculations of the budgets of lead in Atlantic waters and sediments suggested that most of the airborne lead emitted to the North Atlantic region since 1750 could be accounted for by these enrichments.

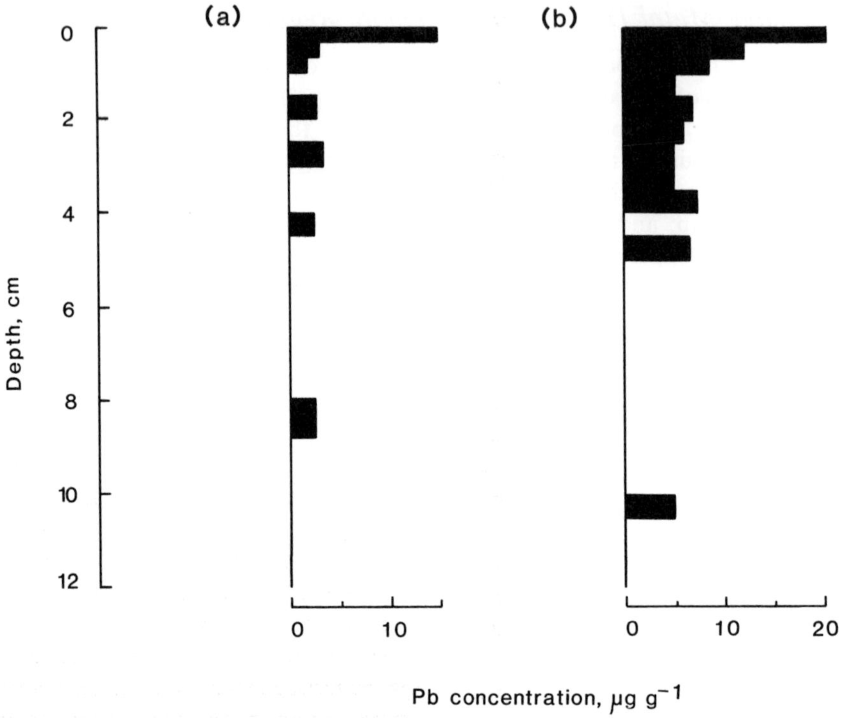

Fig. 24. Concentration of lead with depth in sediment profiles from (a) Station 7; and (b) Station 8 in the North-east Atlantic Ocean. Redrawn from Veron *et al.* (1987).

(b) Polar ice deposits. Advances in the understanding of sources and trends of heavy metals in the polar regions have been rapid over the last two decades. In particular, work has shown that analytical precision and contamination control are of prime importance in view of the very low concentrations of heavy metals (of the order of 1 pg g^{-1}) in these deposits: for this reason, much of the earlier work is now recognized to be flawed, and thus will be summarized only briefly in the following discussion. A fuller account of this early work is given by MARC (1985) and Boutron (1987).

The pioneers in Arctic and Antarctic pollution research were Murozumi *et al.* (1969), and their early conclusions (still held to be valid) were that the atmosphere of the North, and possibly also the South Polar regions are now contaminated by lead. These conclusions were based on chemical analyses of firn and ice cores from Camp Century and Camp Tuto, Greenland and New Byrd Station, Antarctica. The series of lead profiles from Greenland, spanning the years 800 BC to AD 1965 revealed that lead concentrations were already elevated above the prehistoric level by AD 1753. A further gradual increase occurred during the eighteenth and nineteenth centuries, followed by a three-fold increase from 1933 to 1965 (Fig. 25). The authors related this trend to the history of lead production and consumption in the northern hemisphere since 1750, and estimated that the present-day concentration of lead in the Greenland atmosphere is enriched by a factor of *c*. 500 over the crustal concentration. The results from Antarctica were less clear-cut. There was no systematic increase in concentrations of lead in samples spanning the years 1700 to 1966, but surface concentrations (only one-tenth of those in Greenland) were enriched over the mean crustal value by a factor of *c*. 100.

During the decade following the work of Murozumi *et al.*, a number of investigations were undertaken in Greenland and Antarctica to try to establish the nature and magnitude of anthropogenic enrichments, if these existed. The results were, however, very variable and inconclusive with respect to both surface and prehistoric concentrations of lead and other metals. In Antarctica no convincing historical trend of lead concentrations could be obtained, despite the apparently high enrichments of heavy metals which were consistently reported.

The turning point in polar research came in the early 1980s, initially with the work of Boutron and Patterson and their colleagues. These workers were the first to tackle the problem of contamination in deep-drilled firn and ice cores, and did so by using a meticulous analytical procedure which enabled the extent of the contamination to be assessed, and the true level of heavy metal to be determined. Ng and Patterson (1981) concluded that the original value of lead reported by Murozumi *et al.* (1969) for prehistoric Greenland ice had been valid, but that most of the subsequent results of other workers were affected by gross contamination by a factor of up to 10^6 for highly permeable firn cores. The findings from Antarctica, however, were inconclusive. Following a study by Boutron (1982) which showed that local contamination from scientific stations could affect surface snow for considerable distances, Boutron and Patterson (1983) undertook the analysis of surface snow cores from a contamination-free site in Adelie Land, East Antarctica, to try to establish the presence of a historical trend in concentrations using the techniques pioneered by Ng and Patterson (1981). The results were ambiguous; there appeared to be an underlying trend in concentrations over the last two centuries, but this was clouded by considerable variability.

Further light was shed on the nature of the variability of lead inputs to Antarctic snows by the approach of Peel and Wolff and their co-workers. By means of a series of simultaneous analyses of contemporary snow and air samples and of short snow cores at a remote, uncontaminated site in the Antarctic Peninsula, Landy and Peel (1981) and Peel and Wolff (1982) demonstrated that large-amplitude fluctuations in heavy metal deposition, of between two- and ten-fold, can occur over a day-to-day time-scale, and over

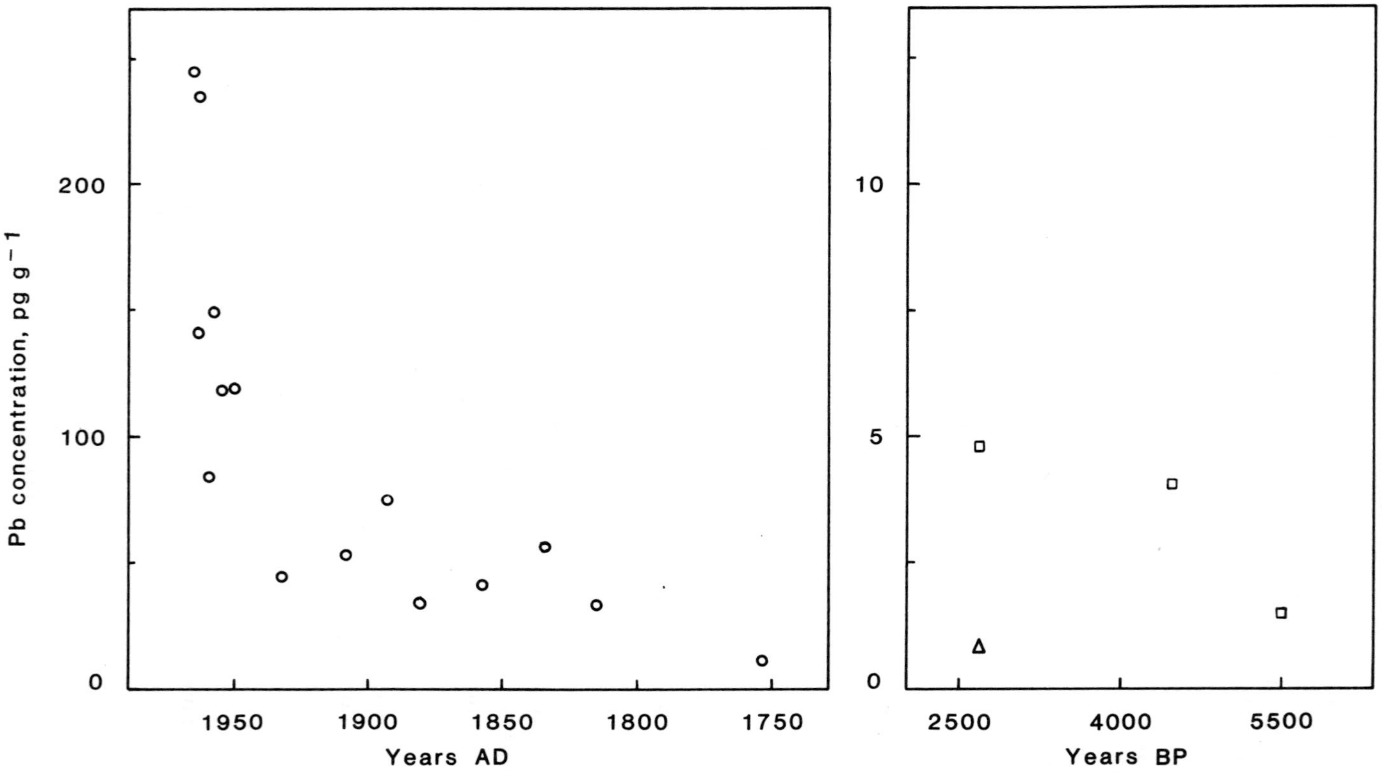

Fig. 25. Lead concentration in a stratigraphically-dated snow and firn profile from Camp Century and Virgin Site, Greenland (open circles: data of Murozumi *et al.*, 1969), and in stratigraphically-dated ice horizons from Camp Century (squares: data of Ng and Patterson, 1981) and Camp Tuto (triangle: data of Murozumi *et al.* 1969). Redrawn from Boutron (1987).

periods of months or years. These trends were considered to be a true reflection of variations in aerosol composition that are, in turn, a product of the changes in the air circulation systems which transport impurities to the Antarctic atmosphere. Peel and Wolff (1982) proposed that these meteorological processes might be the major determinants of trends in heavy metal deposition that had previously been ascribed to an anthropogenic source.

As a result of the findings of these studies, polar pollution research in the last five years (1982–1987) has been directed largely towards a definition of the magnitude and sources of the fluctuations in natural heavy metal inputs to the Antarctic, and towards obtaining reliable determinations of heavy metal concentrations in surface snows from contamination-free sites.

Prehistoric fluctuations in lead deposition in Antarctica have recently been investigated by Boutron and Patterson (1986) and Boutron *et al.* (1987) in two deep ice cores from Vostok and Dome C, respectively. The cores date back to 155 000 yr BP, covering the time-span from the penultimate glacial, through the last interglacial and glacial periods, into the early Holocene Period. The results, summarized in Fig. 26, show that wide fluctuations in lead deposition in the Antarctic have occurred over the last 155 000 years, and that these are linked with the contrasting climatic regimes represented, namely, the glacial periods, when large amounts of crustal material were transported to the ice-sheet, and interglacials, when lead deposition was much lower and originated mainly from volcanic eruptions and other undetermined sources. Lead determinations of recent snow and firn from Adelie Land (a remote region of Eastern Antarctica) by Boutron and Patterson (1983, 1987) show that the more recent horizons contain lead concentrations which are certainly higher than at other times in the Holocene Period, thus implying that there has been a recent anthropogenic lead input (Fig. 26). Partitioning of present-day lead concentrations into their respective "crustal", "marine" and "other" fractions (Boutron and Patterson, 1987) substantiates this conclusion. Boutron (1987) emphasizes, nevertheless, that such a conclusion would have to be supported by reliable analyses of continuous profiles covering at least the last few centuries.

The most reliable extant data for lead concentrations in Greenland are still those of Murozumi *et al.* (1969) and Ng and Patterson (1981), as summarized in Fig. 25. The composite historical trend shows that fluctuations in the flux of natural lead to the snow existed during prehistoric times, but that these fluctuations are exceeded by the very high concentrations determined in recent snow by Murozumi *et al.* (1969). These data, therefore, constitute fairly firm evidence of the presence of anthropogenic lead in the north polar atmosphere.

The data for other heavy metals in the polar regions (summarized in Table 13) are still scanty, and no continuous time series exist. Comparison between

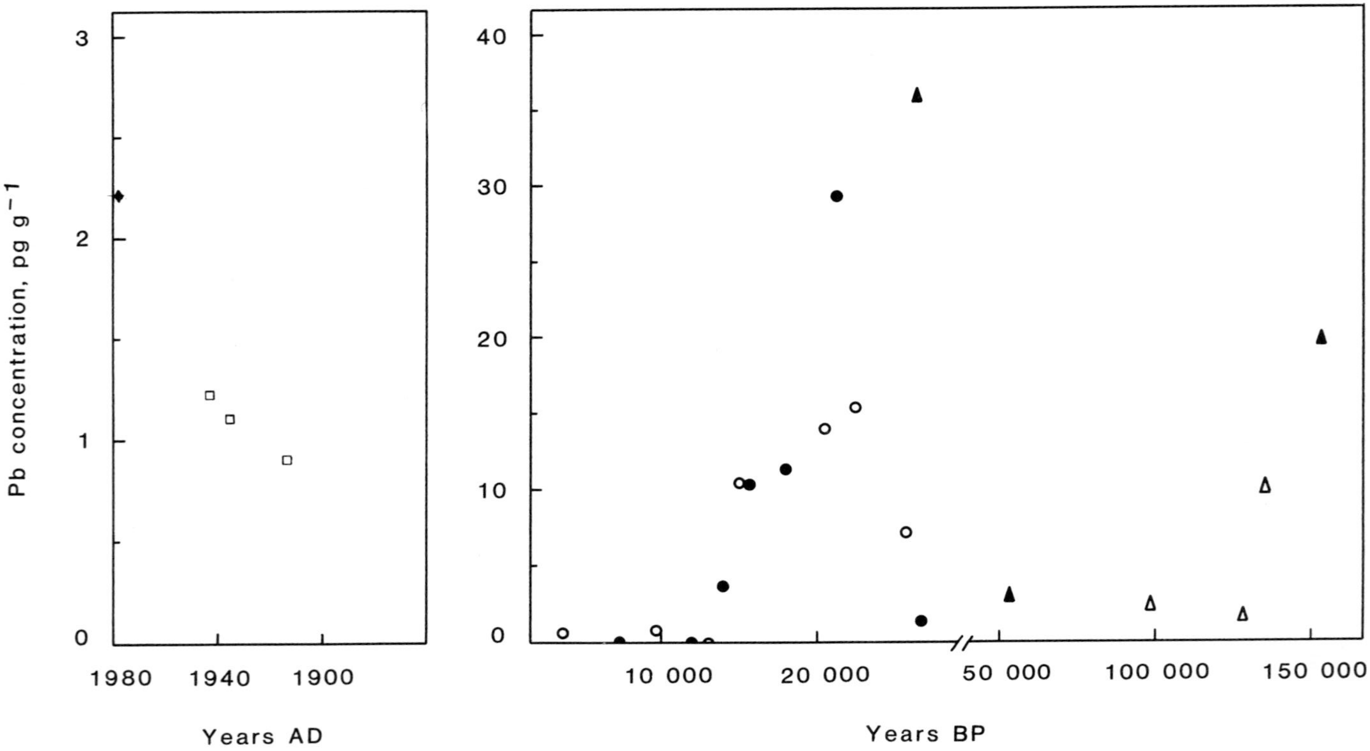

Fig. 26. Lead concentration in stratigraphically-dated Antarctic snow, firn and ice horizons from: Stake D80 (diamond: data of Boutron and Patterson, 1987), Stake D55 (squares: data of Boutron and Patterson, 1983), Vostok (triangles: data of Boutron *et al.* 1987) and Dome C (circles: data of Boutron and Patterson, 1986). Closed symbols denote determinations with known limits of accuracy, while open symbols denote determinations with uncertain limits of accuracy. Redrawn from Boutron (1987).

Table 13
Summary of reliable data for the concentrations of cadmium, copper, mercury and zinc in the polar regions.

Site	Date	Heavy metal concentration ($pg\ g^{-1}$)				Source
		Cd	Cu	Hg	Zn	
(a) *Greenland*						
Crete	1957–1971			7–12		Appelquist *et al.* (1978)
Crete	1727–1786			12–19		Appelquist *et al.* (1978)
Dye-3 Camp	1978–1979		30–100		140–580	Davidson *et al.* (1981)
Station Milcent	1971–1973	3–10			174–270	Herron *et al.* (1977)
T1–T46	1973–1974	0·7–33	30–100		97–1045	Boutron (1979b)
(b) *Antarctic*						
Dome C and South Pole	1925–1975	5	15–60		50	Boutron (1979a), Boutron and Lorius (1979), Boutron (1982)
Antarctic Peninsula (Spaatz Island)	1971–1979	1–10			50	Landy and Peel (1981), Peel and Wolff (1982)
Antarctic Peninsula (Spaatz Island and Plateau)	Recent	0·26	1·9		3·3	Wolff and Peel (1985b)
Adelie Land (Cap Prudhomme)	12 000 BP	2·6	17		60	Boutron *et al.* (1984)

prehistoric and recent concentrations of most metals shows that there are no significant differences which might constitute a historical trend. Wolff and Peel (1985b) have stressed the need to improve contamination control in the analysis of prehistoric ice samples, and to rationalize the large variability in surface concentrations which would obscure any existing historical trends.

(c) *Antarctic lake sediments.* The only published data for heavy metal concentrations in lake sediments in the Antarctic are those of Lyons *et al.* (1985). The study lakes—Don Juan Pond and Shea Sisters Lake—are closed basins in Southern and Northern Victoria Land, respectively, which are the main ice-free regions of Antarctica. Both lakes are fed almost exclusively by glacier and permafrost meltwater, which effectively magnifies atmospheric heavy metal inputs to the lakes.

In Don Juan Pond, normalized heavy metal concentrations fluctuated about the values: lead, 2 $\mu g\ g^{-1}$; zinc, 47 $\mu g\ g^{-1}$; and copper, 59 $\mu g\ g^{-1}$. The particularly low concentration of lead was attributed to remobilization from the surface sediments by leaching with brine. Higher normalized concentrations of heavy metals occurred in the sediments of Shea Sisters Lake, and the maximum values—lead, 70 $\mu g\ g^{-1}$; zinc, 95 $\mu g\ g^{-1}$; and copper, 70 $\mu g\ g^{-1}$—occurred just below the surface. This distribution was thought not to be connected with iron/manganese cycling, so was tentatively ascribed to an anthropogenic source. The authors pointed out, however, that heavy metal enrichments could have originated from local rather than global emissions, bearing in mind that the lakes are situated near some of the permanently settled coastal regions.

The conclusions from these investigations in remote regions are, broadly, that heavy metals, particularly lead, have attained a wide global distribution. Both the mid-latitude and high-latitude air-masses of the northern hemisphere are now significantly enriched with lead, by as much as several orders of magnitude over the natural level. Smaller enrichments may exist for other heavy metals, but these still have to be proved.

In Antarctica the picture is less clear. In certain areas there appears to be a recent elevation of lead concentrations, but it has not yet proved possible to determine whether this originates from local contamination, or is the result of long-distance transport from the mid-latitudes of the southern, or even the northern hemisphere.

V. CONCLUSIONS

In this review, the value of geochemical monitoring has been explored by reference to the theoretical constraints of the method, and in the context of its proven applications. A final evaluation can best be made in terms of the principal criteria of monitoring which have been set down by various authorities (e.g. Martin and Coughtrey, 1982; Henderson-Sellers, 1984; Stern *et al.*, 1984), and which include: (1) quantitative assessment; (2) spatial assessment; (3) temporal assessment; (4) source assessment; and (5) environmental impact assessment.

(1) With the exception of Antarctic snow deposits, which exactly mirror atmospheric concentrations of impurities, natural deposits accumulate heavy metals from the atmosphere with varying degrees of efficiency. Thus, concentrations of heavy metals in North Polar and temperate snow and ice deposits are invariably low, entailing practical analytical difficulties, whilst aquatic sediments, and particularly peat, concentrate atmospheric metals usually to well above the detection limits. The degree to which metals are retained in deposits also varies. Under well-defined conditions, ice is a fully conservative monitoring medium. In aquatic sediments and peat, however, heavy metal retention is always less than 100%, and under certain environmental conditions these deposits must be considered to be only semi-conservative sinks for heavy metals. The use of these deposits in monitoring therefore requires a detailed knowledge of aquatic and mire ecosystems. Additional problems are entailed in the interpretation of the information provided by aquatic sediments, as these are not exclusively ombrogenic.

Despite these complicating factors, it has been shown that quantitative assessment is possible in geochemical monitoring, and can be a useful tool in making inter-site comparisons, or predicting future trends. The relative shortage of this sort of information in the literature, however, suggests that the quantitative potential of geochemical monitoring is not usually fully exploited.

(2) Taken together, ice, peat and aquatic deposits are distributed widely on a global scale in all the main climatic and geographical regions, with the exception of some arid regions. Geochemical monitoring thus has a considerable potential for yielding information on the global distribution of heavy metals. This potential is only just beginning to be explored, and the majority of published works are still biased heavily towards North America and Europe.

On a small spatial scale, the scope of geochemical monitoring is limited by the irregular distribution of natural deposits. It is notable that built-up areas are very poorly represented in the literature.

(3) Most ice, peat and aquatic deposits date back far into the prehistoric era, so geochemical monitoring can provide a complete historical perspective of anthropogenic influence. Here again, the potential is

underexploited and very few studies extend back beyond the last 200 years.

The degree of detail of the historical record provided by geochemical monitoring varies considerably with the characteristic accumulation rates of the different deposits. In peat and aquatic deposits, resolution is rarely finer, and usually coarser than one year. Consequently, these deposits provide an integrated record of atmospheric deposition, in which short-term fluctuations are obscured. Recent depositional events are particularly poorly recorded in these deposits, as a result of physical and chemical processes specific to the surface horizons. The depositional record thus has an inbuilt time-lag. Ice deposits have considerably higher accumulation rates than peat and aquatic deposits, and consequently furnish detailed deposition records in which seasonal, or even daily events are represented and can be dated accurately. This advantage is often outweighed, however, by the practical difficulties entailed in obtaining long, continuous time sequences from ice deposits.

(4) Geochemical monitoring provides a number of indirect approaches to source assessment, all of which rely on the fundamental properties outlined in (1)–(3). A qualitative evaluation of pollution sources is implicit in most geochemical investigations in the discussion of spatial or temporal trends in heavy metal deposition. On the other hand, the quantitative, or semi-quantitative apportionment of heavy metals at a particular site is possible using additional information, for example, the concentration of related pollutants or indicator substances, or, in the case of lead, stable isotope ratios. The latter approach has particular potential in geochemical monitoring.

The quantitative assessment of the dispersal of pollutants from a point source relies on small-scale spatial assessment, which limits its application to geochemical monitoring (see (2) above). Most published studies of this type have used seasonal snow deposits.

(5) Geochemical monitoring invariably entails assessment of environmental impact, in its broad interpretation, because the monitoring media are part of natural ecosystems. The more specific effects of pollutants on organisms can only be assessed by peat and aquatic deposits which comprise a significant organogenic fraction. The best known example of such an assessment is the analysis of diatom remains in lake sediments to trace the onset of acidification or eutrophication. In peat deposits, complete individual organisms can rarely be distinguished in the macrofossil record so that impact assessment is on the scale of the community rather than the individual. This aspect of monitoring has received very little attention.

The conclusion from this evaluation is that, within its theoretical constraints, geochemical monitoring fulfils the basic criteria of monitoring outlined above. Its chief weaknesses are in quantitative assessment, where the interpretation of the depositional record is not straightforward, and in the lack of fine resolution provided generally in spatial and temporal assessment. In these spheres, conventional monitoring devices or biological monitors are often more satisfactory. The particular strengths of geochemical monitoring lie firstly, in the immediate relevance of the information to natural ecosystems—an advantage shared with biological monitors—and secondly, in the broad spatial and temporal scope of the information provided—unique to this method. Much of the potential of geochemical monitoring to shed light on the broad historical perspective of heavy metal pollution on a global scale has yet to be explored.

ACKNOWLEDGEMENTS

I am particularly grateful to Professor R. S. Clymo (Queen Mary College, London) for his meticulous criticism of part of this review, and to Dr J. H. Tallis (Manchester University) for his invaluable help and advice at all stages of the work. I also thank Dr A. H. Fitter and two anonymous referees for their thoughtful comments at a later stage of preparation.

REFERENCES

Aaby, B. and Jacobsen, J. (1979). Changes in biotic conditions and metal deposition in the last millenium as reflected in ombrotrophic peat in Draved Mose, Denmark. *Danm. geol. Unders., Årbog 1978*, 5–43.

Aaby, B., Jacobsen, J. and Jacobsen, O. S. (1979). Pb-210 dating and lead deposition in the ombrotrophic peat bog, Draved Mose, Denmark. *Danm. geol. Unders., Årbog 1978*, 45–68.

Andrews, J. (1979). The present Ice Age: Cenozoic. In: *The Winters of the World* (Ed. by B. John), pp. 173–218. Newton Abbot: David and Charles.

Appelquist, H., Jensen, K. O., Sevel, T. and Hammer, C. (1978). Mercury in the Greenland ice sheet. *Nature, Lond.* **273,** 657–659.

Arafat, N. and Nriagu, J. O. (1986). Simulated mobilization of metals from sediments in response to lake acidification. *Water Air Soil Pollut.* **31,** 991–998.

Aston, S. R., Bruty, D., Chester, R. and Padgham, R. C. (1973). Mercury in lake sediments: a possible indicator of technological growth. *Nature, Lond.* **241,** 450–451.

Ault, W. V., Senechal, R. G. and Erlebach, W. E. (1970). Isotopic composition as a natural tracer of lead in the environment. *Environ. Sci. Technol.* **4,** 305–313.

Baier, R. W. and Healy, M. L (1977). Partitioning and transport of lead in Lake Washington. *J. Environ. Qual.* **6,** 291–296.

Balistrieri, L. S., Brewer, P. G. and Murray, J. W. (1981). Scavenging residence times of trace metals and surface chemistry of sinking particles in the deep ocean. *Deep Sea Res.* **28A,** 101–121.

Balistrieri, L. S. and Murray, J. W. (1986). The surface chemistry of sediments from the Panama Basin: the influence of Mn oxides on metal adsorption. *Geochim. cosmochim. Acta* **50,** 2235–2243.

Barnes, R. S. and Schell, W. R. (1973). Physical transport of trace metals in the Lake Washington watershed. In: *Cycling and Control of Metals: Proceedings of an Environmental Resources Conference in Columbus, Ohio, sponsored by U.S. E.P.A./National Science Foundation*, pp. 45–53. Cincinnati, Ohio: National Environmental Research Centre.

Baron, J., Norton, S. A., Beeson, D. R. and Herrmann, R. (1986). Sediment diatom and metal stratigraphy from Rocky Mountain Lakes with special reference to atmospheric deposition. *Can. J. Fish. Aquat. Sci.* **43,** 1350–1362.

Barrie, L. A. (1986). Arctic air pollution: an overview of current knowledge. *Atmos. Environ.* **20,** 643–663.

Barrie, L. A. and Schemenauer, R. S. (1986). Pollutant wet deposition mechanisms in precipitation and fog water. *Water Air Soil Pollut.* **30,** 91–104.

Barrie, L. A. and Vet, R. J. (1984). The concentration and deposition of acidity, major ions and trace metals in the snowpack of the Eastern Canadian Shield during the winter of 1980–1981. *Atmos. Environ.* **18,** 1459–1469.

Barrie, L. A., Lindberg, S. E., Chan, W. H., Ross, H. B., Acrimoto, R. and Church, T. M. (1987). On the concentration of trace metals in precipitation. *Atmos. Environ.* **21,** 1133–1135.

Batifol, F. M and Boutron, C. (1984). Atmospheric heavy metals in high altitude surface snows from Mont Blanc, French Alps. *Atmos. Environ.* **18,** 2507–2515.

Battarbee, R. W. (1978). Observations on the recent history of Lough Neagh and its drainage basin. *Phil. Trans. R. Soc. B* **281,** 303–345.

Benninger, J. K., Aller, R. C., Cochran, J. K. and Turekian, K. K. (1979). Effects of biological sediment mixing on the ^{210}Pb chronology and trace metal distribution in a Long Island Sound sediment core. *Earth Planet. Sci. Lett.* **43,** 241–259.

Berg, van den, C., Merks, A. G. A. and Duursma, E. K. (1987). Organic complexation and its control of the dissolved concentrations of copper and zinc in the Scheldt Estuary. *Est. Coastal Shelf. Sci.* **24,** 785–797.

Bertine, K. K. and Goldberg, E. D. (1977). History of heavy metal pollution in the southern California coastal zone—a reprise. *Environ. Sci. Technol.* **11,** 297–299.

Bertine, K. K. and Mendeck, M. F. (1978). Industrialisation of New Haven, Conn., as recorded in reservoir sediments. *Environ. Sci. Technol.* **12,** 201–207.

Birks, H. J. B. and Birks, H. H. (1980). *Quaternary Palaeoecology*. London: Edward Arnold. 289 pp.

Björklund, I., Borg, H. and Johansson, K. (1984). Mercury in Swedish Lakes—its regional distribution and causes. *Ambio* **13,** 118–121.

Boggie, R., Hunter, R. F. and Knight, A. H. (1958). Studies of the root development of plants in the field using radioactive tracers. II. Communities growing in deep peat. *J. Ecol.* **46,** 629–639.

Boutron, C. (1979a). Past and present tropospheric fallout fluxes of Pb, Cd, Cu, Zn and Ag in Antarctica and Greenland. *Geophys. Res. Lett.* **6,** 159–162.

Boutron, C. (1979b). Trace element content of Greenland snows along an east–west transect. *Geochim. cosmochim. Acta* **43,** 1253–1258.

Boutron, C. (1980). Respective influence of global pollution and volcanic eruptions on the past variations of the trace metals content of Antarctic snows since 1880's. *J. geophys. Res.* **85C,** 7426–7432.
Boutron, C. (1982). Atmospheric trace metals in the snow layers deposited at the South Pole from 1928 to 1977. *Atmos. Environ.* **16,** 2451–2459.
Boutron, C. (1987). Atmospheric lead, cadmium, mercury and arsenic in Antarctic and Greenland recent snow and ancient ice. *Proceedings of Metals Cycling Workshop,* sponsored by S.C.O.P.E./U.N.E.P. (in press).
Boutron, C. and Lorius, C. (1979). Trace metals in Antarctic snows since 1914. *Nature, Lond.* **277,** 551–554.
Boutron, C. and Patterson, C. C. (1983). The occurrence of lead in Antarctic recent snow, firn deposited over the last two centuries and prehistoric ice. *Geochim. cosmochim. Acta* **47,** 1355–1368.
Boutron, C. and Patterson, C. C. (1986). Lead concentration changes in Antarctic ice during the Wisconsin/Holocene transition. *Nature, Lond.* **323,** 222–225.
Boutron, C and Patterson, C. C. (1987). Relative levels of natural and anthropogenic lead in recent Antarctic snow. *J. geophys Res.* **92,** 8454–8464.
Boutron, C., Echevin, M. and Lorius, C. (1972). Chemistry of polar snows: estimation of rates of deposition in Antarctica. *Geochim. cosmochim. Acta* **36,** 1029–1041.
Boutron, C., Leclerc, M. and Risler, N. (1984). Atmospheric trace metals in Antarctic prehistoric ice collected at a coastal ablation area. *Atmos. Environ.* **18,** 1947–1953.
Boutron, C., Patterson, C. C., Petrov, V. N. and Barkov, N. I. (1987). Preliminary data on changes of lead concentrations in Antarctic ice from 155 000 to 26 000 yr BP. *Atmos. Environ.* **21,** 1197–1202.
Bradley, R. S. (1985). *Quaternary Paleoclimatology; Methods of Paleoclimatic Reconstruction.* London: Allen & Unwin. 472 pp.
Briat, M. (1978). Evaluation of levels of Pb, V, Cd, Zn and Cu in the snow of Mont Blanc during the last 25 years. In: *Proceedings of the 13th International Colloquium on Atmospheric Pollution* (Ed. by M. M. Benarie), pp. 225–228. Amsterdam: Elsevier.
Bruland, K. W., Bertine, K. K., Koide, M. and Goldberg, E. D. (1974). History of heavy metal pollution in the Southern Californian coastal zone. *Environ. Sci. Technol.* **8,** 425–432.
Bunzl, K., Schmidt, W. and Sansoni, B. (1976). Kinetics of ion exchange in soil organic matter. IV. Adsorption and desorption of Pb^{2+}, Cu^{2+}, Cd^{2+}, Zn^{2+} and Ca^{2+} by peat. *J. Soil Sci.* **27,** 32–41.
Carignan, R. and Tessier, A. (1985). Zinc deposition in acid lakes: the role of diffusion. *Science, N.Y.* **228,** 1524–1526.
Cawse, P. A. (1977). A survey of atmospheric trace elements in the U.K.: results for 1976. Harwell: United Kingdom Atomic Energy Authority.
Chamberlain, A. C. (1960). Aspects of the deposition of radioactive and other gases and particles. *Int. J. Air. Pollut.* **3,** 63–88.
Chamberlain, A. C. (1986). Deposition of gases and particles on vegetation and soils. In: *Advances in Science and Technology.* (Ed. by A. H. Legge and S. V. Krupa), Vol. 18, pp. 189–209. New York: J. Wiley & Sons.
Chambers, F. M., Dresser, P. Q. and Smith, A. G. (1979). Radiocarbon dating

evidence on the impact of atmospheric pollution on upland peats. *Nature, Lond.* **282,** 829–831.

Chambers, L. A. (1976). The classification and extent of air pollution problems. In: *Air Pollution*, 3rd edn (Ed. by A. C. Stern), pp. 3–22. New York: Academic Press.

Chow, T. J. (1970). Lead accumulation in roadside soil and grass. *Nature, Lond.* **225,** 295–296.

Chow, T. J. and Johnstone, M. S. (1965). Lead isotopes in gasoline and aerosols of Los Angeles Basin, California. *Science, N.Y.* **147,** 502–503.

Chow, T. J., Bruland, K. W., Bertine, K., Soutar, A., Koide, M. and Goldberg, E. D. (1973). Lead pollution: records in Southern Californian coastal sediments. *Science, N.Y.* **181,** 551–552.

Chow, T. J., Snyder, C. B. and Earl, J. L. (1975). Isotope ratios of lead as pollutant source indicators. In: *Isotope Ratios as Pollutant Source and Behaviour Indicators*, pp. 95–108. Vienna: International Atomic Energy Authority.

Christensen, E. R. and Chien, N. K. (1981). Fluxes of arsenic, lead, zinc and cadmium to Green Bay and Lake Michigan sediments. *Environ. Sci. Technol.* **15,** 553–558.

Christensen, E. R. and Goetz, R. H. (1987). Historical fluxes of particle bound pollutants from deconvolved sedimentary records. *Environ. Sci. Technol.* **21,** 1088–1096.

Cline, J. T. and Chambers, R. L. (1977). Spatial and temporal distribution of heavy metals in lake sediments near Sleeping Bear Point, Michigan. *J. sedim. Petrol.* **47,** 716–727.

Clymo, R. S. (1963). Ion exchange in *Sphagnum* and its relation to bog ecology. *Ann. Bot.* **27,** 309–324.

Clymo, R. S. (1978). A model of peat bog growth. *Ecological Studies* **27,** 187–223.

Clymo, R. S. (1983). Peat. In: *Ecosystems of the World 4A. Mires: Swamp, Bog, Fen and Moor. General Studies* (Ed. by A. J. P. Gore), pp. 159–224. Amsterdam: Elsevier.

Clymo, R. S. (1984). The limits to peat bog growth. *Phil. Trans. R. Soc. B.* **303,** 605–654.

Clymo, R. S. and Mackay, D. (1987). Upwash and downwash of pollen and spores in the unsaturated surface layer of *Sphagnum* dominated peat. *New Phytol.* **105,** 175–183.

Conway, V. M. (1949). Ringinglow Bog, near Sheffield II. The present surface. *J. Ecol.* **37,** 148–170.

Corn, M. (1976). Aerosols and the primary air pollutants—nonviable particles, their occurrence, properties and effects. In: *Air Pollution*, 3rd Edn (Ed. by A. C. Stern), pp. 78–168. New York: Academic Press.

Crecelius, E. A. and Piper, D. Z. (1973). Particulate lead contamination recorded in sedimentary cores from Lake Washington, Seattle. *Environ. Sci. Technol.* **7,** 1053–1055.

Crisp, D. T. (1966). Input and output of minerals from an area of Pennine moorland: the importance of precipitation, drainage, erosion and animals. *J. appl. Ecol.* **3,** 327–348.

Damman, A. W. H. (1978). Distribution and movement of elements in ombrotrophic peat bogs. *Oikos* **30,** 480–495.

Davidson, C. I., Chu, L., Grimm, T. C., Nasta, M. A. and Qamoos, M. P. (1981). Wet and dry deposition of trace elements onto the Greenland ice sheet. *Atmos. Environ.* **15,** 1429–1437.

Davis, A. O., Galloway, J. N. and Nordstrom, D. K. (1982). Lake acidification: its effect on lead mobility in the sediments of two Adirondack lakes. *Limnol. Oceanogr.* **27,** 163–167.

Davis, R. B., Norton, S. A., Hess, C. T. and Brakke, D. F. (1983). Palaeolimnological reconstruction of the effects of atmospheric deposition of acids and heavy metals on the chemistry and biology of lakes in New England and Norway. *Hydrobiologia* **103,** 113–123.

Davis, R. B., Hess, C. T., Norton, S. A., Hanson, D. W., Hoagland, K. D. and Anderson, D. S. (1984). ^{137}Cs and ^{210}Pb dating of sediments from soft-water lakes in New England (USA) and Scandinavia, a failure of ^{137}Cs dating. *Chem. Geol.* **44,** 151–185.

Davison, W. (1985). Conceptual models for transport at a redox boundary. In: *Chemical Processes in Lakes* (Ed. by W. Stumm), pp. 31–53. New York: J. Wiley & Sons.

Deniseger, J., Austin, A., Roch, M. and Clark, M. J. R. (1986). A persistent bloom of the diatom *Rhizoselenia eriensis* (Smith) and other changes associated with decreases in heavy metal contamination in an oligotrophic lake, Vancouver Island. *Envir. exp. Bot.* **26,** 217–226.

Denny, P. and Welsh, R. P. H. (1979). Lead accumulation in plankton blooms from Ullswater, the English Lake District. *Environ. Pollut.* **18,** 1–10.

Department of Energy (1978). *Digest of United Kingdom Energy Statistics.* London: Government Statistical Services.

Dick, A. L. and Peel, D. A. (1985). Trace elements in Antarctic air and snowfall. *Ann. Glaciol.* **7,** 12–19.

Dickson, W. (1980). Properties of acidified waters. In: *Proceedings of an International Conference on the Ecological Impact of Acid Precipitation, S.N.S.F., Sandefjord, Norway* (Ed. by D. Drabløs and A. Tollan), pp. 75–83. Oslo: SNSF.

Dijk, van, H. (1971). Cation binding of humic acids. *Geoderma* **5,** 53–56.

Dillon, P. J. and Evans, R. D. (1982). Whole-lake lead burdens in sediments of lakes in southern Ontario, Canada. *Hydrobiologia* **91,** 121–130.

Dollard, C. J., Unsworth, M. H. and Harvey, M. J. (1983). Pollutant transfer in upland regions by occult precipitation. *Nature, Lond.* **302,** 241–243.

Duce, R. A. and Hoffman, E. J. (1976). Chemical fractionation at the air/sea interface. *Ann. Rev. Earth Planet. Sci.* **4,** 187–228.

Duce, R. A., Hoffman, G. L. and Zoller, W. H. (1975). Atmospheric trace metals at remote northern and southern hemisphere sites: pollution or natural? *Science, N.Y.* **187,** 59–61.

Durham, R. W., Joshi, S. R. and Allen, R. J. (1980). Radioactive dating of sediment cores from four contiguous lakes in Saskatchewan, Canada. *Sci. Total Environ.* **15,** 65–71.

Edgington, D. N. and Robbins, J. A. (1976). Records of lead deposition in Lake Michigan sediments since 1800. *Environ. Sci. Technol.* **10,** 266–274.

Eisenreich, S. J., Metzer, N. A., Urban, N. R. and Robbins, J. A. (1986). Responses of atmospheric lead to decreased use of lead in gasoline. *Environ. Sci. Technol.* **20,** 171–174.

El-Daoushy, F., Tolonen, K. and Rosenberg, R. (1982). Lead-210 and moss-increment dating of two Finnish *Sphagnum* hummocks. *Nature, Lond.* **296,** 429–431.

Elderfield, H. and Hepworth, A. (1975). Diagenesis, metals and pollution in estuaries. *Mar. Pollut. Bull.* **6,** 85–87.

Elgmork, K., Hagen, A. and Langeland, A. (1973). Polluted snow in Southern Norway during the winters 1968–1971. *Environ. Pollut.* **4,** 41–52.

Elzerman, A. W. and Armstrong, D. E. (1979). Enrichment of Zn, Cd, Pb and Cu in the surface microlayer of Lakes Michigan, Ontario and Mendota. *Limnol. Oceanogr.* **24,** 133–144.

Erlenkeuser, H., Suess, E. and Willkomm, H. (1974). Industrialization affects heavy metal and carbon isotope concentrations in recent Baltic Sea sediments. *Geochim. cosmochim. Acta* **38,** 823–842.

Evans, H. E., Smith, P. J. and Dillon. P. J. (1983). Anthropogenic zinc and cadmium burdens in sediments of selected Southern Ontario lakes. *Can. J. Fish. Aquat. Sci.* **40,** 570–579.

Evans, R. D. and Dillon, P. J. (1982). Historical changes in anthropogenic lead fallout in Southern Ontario, Canada. *Hydrobiologia* **91,** 131–137.

Evans, R. D. and Rigler, F. H. (1980). Calculation of the total anthropogenic lead in sediments of a rural Ontario lake. *Environ. Sci. Technol.* **14,** 216–218.

Evans, R. D. and Rigler, F. H. (1985). Long distance transport of anthropogenic lead as measured by lake sediments. *Water Air Soil Pollut.* **24,** 141–151.

Everard, M. and Denny, P. (1985a). Flux of lead in submerged plants and its relevance to a freshwater system. *Aquat. Bot.* **21,** 181–193.

Everard, M. and Denny, P. (1985b). Particulates and the cycling of lead in Ullswater, Cumbria. *Freshwater Biol.* **15,** 215–226.

Farmer, J. G. (1978). Lead concentration profiles in lead-210 dated Lake Ontario sediment cores. *Sci. Total Environ.* **10,** 117–127.

Farmer, J. G., Swan, D. S. and Baxter, M. S. (1980). Records and sources of metal pollutants in a dated Loch Lomond sediment core. *Sci. Total Environ.* **16,** 131–147.

Ferguson, N. P. and Lee, J. A. (1980). Some effects of bisulphite and sulphate on the growth of *Sphagnum* species in the field. *Environ. Pollut. A* **21,** 59–71.

Filipek, L. H. and Owen, R. M. (1979). Geochemical associations and grain-size partitioning of heavy metals in lacustrine sediments. *Chem. Geol.* **26,** 105–117.

Fischer, K., Dymon, J., Lyle, M., Soutar, A. and Rau, S. (1986). The benthic cycle of copper: evidence from sediment trap experiments in the eastern tropical North Pacific Ocean. *Geochim. cosmochim. Acta* **50,** 1535–1543.

Flegal, A. R. and Patterson, C. C. (1983). Vertical concentration profiles of lead in the Central Pacific at 15°N and 20°S. *Earth Planet. Sci. Lett.* **64,** 19–32.

Forrest, G. I. (1971). Structure and production of north Pennine blanket bog vegetation. *J. Ecol.* **59,** 453–479.

Förstner, U. (1982). Accumulative phases for heavy metals in limnic sediments. *Hydrobiologia* **91,** 269–284.

Förstner, U. and Wittmann, T. W. W. (1981). *Metal Pollution in the Aquatic Environment*, 2nd edn. Berlin: Springer-Verlag. 486 pp.

Franzin, W. G., McFarlane, G. A. and Lutz, A. (1979). Atmospheric fallout in the vicinity of a base metal smelter at Flin Flon, Manitoba, Canada. *Environ. Sci. Technol.* **13,** 1513–1521.

Fredriksson, I. and Qvarfort, U. (1973). The mercury content of sediments from two lakes in Dalarna, Sweden. *Geol. För. Stockh. Förh.* **95,** 237–242.

Frevert, T. (1985). Chemical limnology. In: *The Handbook of Environmental Chemistry* (Ed. by O. Hutzinger), Vol. 1, Part D, pp. 83–124. Heidelberg: Springer-Verlag.

Frevert, T. (1987). Heavy metals in Lake Kinneret (Israel) II. Hydrogen sulfide-dependent precipitation of copper, cadmium, lead and zinc. *Arch. Hydrobiol.* **109,** 1–24.

Frevert, T. and Sollmann, C. (1987). Heavy metals in Lake Kinneret (Israel) III.

Concentrations of iron, manganese, nickel, cobalt, molybdenum, zinc, cadmium, lead and copper in interstitial water and sediment dry weight. *Arch, Hydrobiol.* **109,** 181–205.

Galloway, J. N. and Likens, G. E. (1979). Atmospheric enhancement of metal deposition in Adirondack lake sediments. *Limnol. Oceanogr.* **24,** 427–433.

Galloway, J. N., Thornton, J. D., Norton, S. A., Volchok, H. L. and McLean, R. A. N. (1982). Trace metals in atmospheric deposition: a review and assessment. *Atmos. Environ.* **16,** 1677–1700.

Garrett, R. G. and Hornbrook, E. H. W. (1976). The relationship between zinc and organic content in centre-lake bottom sediments. *J. geochem. Explor.* **5,** 31–38.

Gendron, A., Silverberg, N., Sundby, B. and Lebel, J. (1986). Early diagenesis of cadmium and cobalt in sediments of the Laurentian Trough. *Geochim. cosmochim. Acta* **50,** 741–747.

Given, P. H. and Dickinson, C. H. (1975). Biochemistry and microbiology of peats. *Soil Biochem.* **3,** 123–212.

Gobeil, C., Silverberg, N., Sundby, B. and Cossa, D. (1987). Cadmium diagenesis in Laurentian Trough sediments. *Geochim. cosmochim. Acta* **51,** 589–596.

Goldberg, E. D. and Arrhenius, G. O. S. (1958). Chemistry of Pacific pelagic sediments. *Geochim. cosmochim. Acta* **13,** 153–212.

Goldberg, E. D., Gamble, E., Griffin, J. J. and Koide, M. (1977). Pollution history of Narragansett Bay as recorded in its sediments. *Estuar. Coast. Mar. Sci.* **5,** 549–561.

Goldberg, E. D., Hodge, V. F., Griffin, J. J., Koide, M. and Edgington, D. N. (1981). Impact of fossil fuel combustion on the sediments of Lake Michigan. *Environ. Sci. Technol.* **15,** 466–471.

Goodman, B. A. and Cheshire, M. V. (1976). The occurrence of copper–porphyrin complexes in soil humic acids. *J. Soil Sci.* **27,** 337–347.

Gore, A. J. P. and Olson, J. S. (1967). Preliminary models for accumulation of organic matter in an *Eriophorum*/*Calluna* ecosystem. *Aquilo, Ser. Botanica* **6,** 297–313.

Gorham, E. (1953a). A note on the acidity and base status of raised and blanket bogs. *J. Ecol.* **41,** 153–156.

Gorham, E. (1953b). Chemical studies on the soils and vegetation of waterlogged habitats in the English Lake District. *J. Ecol.* **41,** 345–360.

Gorham, E. and Hofstetter, R. H. (1971). Penetration of bog peats and lake sediments of tritium from atmospheric fallout. *Ecology* **52,** 898–902.

Gorham, E. and Swaine, D. J. (1965). The influence of oxidising and reducing conditions upon the distribution of some elements in lake sediments. *Limnol. Oceanogr.* **10,** 268–279.

Griffin, J. J. and Goldberg, E. D. (1979). Morphologies and origin of elemental carbon in the environment. *Science, N.Y.* **206,** 563–565.

Griffin, J. J. and Goldberg, E. D. (1983). Impact of fossil fuel combustion on sediments of Lake Michigan: a reprise. *Environ. Sci. Technol.* **17,** 244–245.

Hamilton-Taylor, J. (1979). Enrichments of zinc, lead and copper in recent sediments of Windermere, England. *Environ. Sci. Technol.* **13,** 693–697.

Hamilton-Taylor, J. (1983). Heavy metal enrichments in the recent sediments of six lakes in northwest England. *Environ. Technol. Lett.* **4,** 115–122.

Hamilton-Taylor, J., Willis, M. and Reynolds, C. S. (1984). Depositional fluxes of metals and phytoplankton in Windermere as measured by sediment traps. *Limnol. Oceanogr.* **29,** 695–710.

Hanson, D. W. and Norton, S. A. (1982). Spatial and temporal trends in the chemistry of atmospheric deposition in New England. In: *International Symposium on Hydrometeorology*, sponsored by American Water Resources Association, pp. 25–33.

Hanson, D. W., Norton, S. A. and Williams, J. S. (1982). Modern and palaeolimnological evidence for accelerated leaching and metal accumulation in soils in New England, caused by atmospheric deposition. *Water Air Soil Pollut.* **18,** 227–239.

Harding, J. P. C. and Whitton, B. A. (1978). Zinc, cadmium and lead in water, sediments and submerged plants of the Derwent Reservoir, Northern England. *Water Res.* **12,** 307–316.

Hart, B. T. (1982). Uptake of trace metals by sediments and suspended particulates: a review. *Hydrobiologia* **91,** 299–313.

Heirtzler, J. R. (1968). Sea floor spreading. *Scient. Am.* **21,** 60–70.

Heit, M., Tan, Y., Klusek, C. and Burke, J. C. (1981). Anthropogenic trace elements and polycyclic hydrocarbon levels in sediment cores from two lakes in the Adirondack acid lakes region. *Water Air Soil Pollut.* **15,** 441–464.

Heit, M., Klusek, C. and Baron, J. (1984). Evidence of deposition of anthropogenic pollutants in remote Rocky Mountain lakes. *Water Air Soil Pollut.* **22,** 403–416.

Henderson-Sellers, B. (1984). *Pollution of Our Atmosphere.* Bristol: Adam Hilger. 210 pp.

Henriksen, A. and Wright, R. F. (1978). Concentrations of heavy metals in small Norwegian lakes. *Water Res.* **12,** 101–102.

Herron, M. M., Langway, C. C. Jr, Weiss, H. V. and Cragin, J. H. (1977). Atmospheric trace metals and sulfate in the Greenland ice sheet. *Geochim. cosmochim. Acta* **41,** 915–920.

Hildebrand, E. E. and Blum, W. E. (1975). Fixation of emitted lead by soils. *Z. Pflanzenern. Bodenk.* **3,** 279–294.

Hilton, J. (1985). A conceptual framework for predicting the occurrence of sediment focusing and sediment redistribution in small lakes. *Limnol. Oceanogr.* **30,** 1131–1143.

Hilton, J. and Lishman, J. P. (1985). The effect of redox changes on the magnetic susceptibility of sediments from a seasonally anoxic lake. *Limnol. Oceanogr.* **30,** 907–909.

Hilton, J., Davison, W. and Ochsenbein, U. (1985). A mathematical model for analysis of sediment core data: implications for enrichment factor calculations and trace-metal transport mechanisms. *Chem. Geol.* **48,** 281–291.

Hilton, J., Lishman, J. P. and Allen, P. V. (1986). The dominant processes of sediment distribution and focusing in a small, eutrophic, monomictic lake. *Limnol. Oceanogr.* **31,** 125–133.

Hodges, L. (1977). *Environmental Pollution*, 2nd edn. New York: Holt, Rinehart & Winston. 496 pp.

Holdgate, M. W. (1979). *A Perspective of Environmental Pollution.* Cambridge: Cambridge University Press. 278 pp.

Hosker, R. P. Jr, (1986). Practical application of air pollutant deposition models—current status, data requirements, and research needs. In: *Advances in Science and Technology* (Ed. by A. H. Legge and S. V. Krupa), Vol. 18, pp. 505–567. New York: J. Wiley & Sons.

Hosker, R. P. Jr and Lindberg, S. E. (1982). Review: Atmospheric deposition and plant assimilation of gases and particles. *Atmos. Environ.* **16,** 889–910.

Hughes, M. K., Lepp, N. W. and Phipps, D. A. (1980). Aerial heavy metal pollution and terrestrial ecosystems. *Adv. Ecol. Res.* **11,** 218–327.

Hunt, C. J. (1970). *Lead Mines of the Northern Pennines in the Eighteenth and Nineteenth Centuries*. Manchester: Manchester University Press. 282 pp.
Husain, L. (1986). Chemical elements as tracers of pollutant transport to a rural area. In: *Toxic Metals in the Atmosphere* (Ed. by J. O. Nriagu and C. I. Davidson), pp. 295–317. New York: J. Wiley & Sons.
Hutchinson, G. E. (1957). *A Treatise on Limnology*, Vol. 1: *Geography, Physics and Chemistry*. New York: J. Wiley & Sons. 1015 pp.
Hvatum, O. Ø. (1971). Sterk blyopphopning i overflatesjiktet i myrjord. *Tek. Ukebl.* **27,** 40.
Imboden, D. M. and Schwarzenbach, R. P. (1985). Spatial and temporal distribution of chemical substances in lakes: modelling concepts. In: *Chemical Processes in Lakes* (Ed. by W. Stumm), pp. 1–30. New York: J. Wiley & Sons.
Imboden, D. M. and Stiller, M. (1982). The influence of radon diffusion on the ^{210}Pb distribution in sediments. *J. geophys. Res.* **87C,** 557–565.
Imeson, A. C. (1974). The origin of sediment in a moorland catchment with particular reference to the role of vegetation. In: *Fluvial Processes in Instrumented Watersheds* (Ed. by K. J. Gregory and D. E. Walling), pp. 59–72. London: Institute of British Geographers.
Ingri, J. and Pontér, C. (1986). Iron and manganese layering in recent sediments in the Gulf of Bothnia. *Chem. Geol.* **56,** 105–116.
Iskandar, I. K. and Keeney, D. R. (1974). Concentrations of heavy metals in sediment cores from selected Wisconsin lakes. *Environ. Sci. Technol.* **8,** 165–170.
Jackson, T. A. (1978). The biogeochemistry of heavy metals in polluted lakes and streams at Flin Flon, Canada, and a method for limiting heavy metal pollution of natural waters. *Environ. Geol.* **2,** 173–189.
Jaworowski, Z. (1968). Stable lead in fossil ice and bones. *Nature, Lond.* **217,** 152–153.
Jaworowski, Z., Bilkiewicz, J., Dobosz, E. and Wódkiewicz, L. (1975). Stable and radioactive pollutants in a Scandinavian glacier (radium-226, caesium-137, lead-210, uranium and cadmium). *Environ. Pollut.* **9,** 305–315.
Jevons, H. S. (1915). *The British Coal Trade*. London: Kegan, Paul, Trench, Trübner & Co. Ltd. 876 pp.
Johnson, M. G. (1987). Trace element loadings to sediments of fourteen Ontario lakes and correlations with concentrations in fish. *Can. J. Fish. Aquat. Sci.* **44,** 3–13.
Johnson, M. G., Culp, L. R. and George, S. E. (1986). Temporal and spatial trends in metal loadings to sediments of the Turkey Lakes, Ontario. *Can. J. Fish. Aquat. Sci.* **43,** 754–762.
Jones, J. M. (1987). Chemical fractionation of copper, lead and zinc in ombrotrophic peat. *Environ. Pollut.* **48,** 131–144.
Junge, C. E. (1963). *Air Chemistry and Reactivity*. New York: Academic Press. 382 pp.
Junge, C. E. (1977). Processes responsible for the trace content in precipitation. In: *Isotopes and Impurities in Snow and Ice*. Proceedings of the I.U.G.G. Symposium, Grenoble, 1975. *Int. Ass. Soc. Hydrol. Sci.* **118,** 63–77.
Kahl, J. S., Norton, S. A. and Williams, J. S. (1984). Chronology, magnitude and palaeolimnological record of changing metal fluxes related to atmospheric deposition of acids and metals in New England. In: *Acid Precipitation* (Ed. by O. P. Bricker), Vol. 7, pp. 23–37. Boston: Butterworth.
Kemp, A. L. W. and Thomas, R. L. (1976). Impact of man's activities on the chemical composition of the sediments of Lakes Ontario, Erie and Huron. *Water Air Soil Pollut.* **5,** 469–490.

Kemp, A. L. W., Anderson, T. W., Thomas, R. L. and Mudrochova, A. (1974). Sedimentation rates and recent sediment history of Lakes Ontario, Erie and Huron. *J. sedim. Petrol.* **44,** 207–214.

Kemp, A. L. W., Williams, J. D. H., Thomas, R. L. and Gregory, M. L. (1978). Impact of man's activities on the chemical composition of the sediments of Lakes Superior and Huron. *Water Air Soil Pollut.* **10,** 381–402.

Kennett, J. P. (1977). Cenozoic evolution of Antarctic glaciation, the Circumpolar-Antarctic Ocean, and their impact on global paleoceanography. *J. geophys. Res.* **82,** 3843–3860.

Klein, J., Lerman, J. C., Damon, P. E. and Ralph, E. K. (1982). Calibration of radiocarbon dates: Tables based on the consensus data of the workshop on calibrating the radiocarbon timescale. *Radiocarbon* **24,** 103–150.

Knight, A. H., Crooke, W. M. and Inkson, R. M. E. (1961). Cation-exchange capacities of tissues of higher and lower plants and their related uronic acid contents. *Nature, Lond.* **192,** 142–143.

Koide, M., Bruland, K. W. and Goldberg, E. D. (1973). Th-228/Th-232 and Pb-210 geochronology in marine and lake sediments. *Geochim. cosmochim. Acta* **37,** 1171–1187.

Koide, M., Griffin, J. J. and Goldberg, E. D. (1975). Records of plutonium fallout in marine and terrestrial samples. *J. geophys. Res.* **80,** 4153–4162.

Kononova, M. M. (1966). *Soil Organic Matter. Its Nature, its Role in Soil Formation and in Soil Fertility*, 2nd English edn, translated by T. Z. Nowakowski and A. C. D. Newman. Oxford: Pergamon Press. 450 pp.

Krishnaswami, Lal. D., Martin, J. M. and Meybeck, M. (1971). Geochronology of lake sediments. *Earth Planet. Sci. Lett.* **11,** 407–414.

Laird, L. B., Taylor, H. E. and Kennedy, V. C. (1986). Snow chemistry of the Cascade-Sierra Nevada Mountains. *Environ. Sci. Technol.* **20,** 275–290.

Landy, M. P. and Peel, D. A. (1981). Short-term fluctuations in heavy metal concentrations in Antarctic snow. *Nature, Lond.* **291,** 144–146.

Lee, J. A. and Tallis, J. H. (1973). Regional and historical aspects of lead pollution in Britain. *Nature, Lond.* **245,** 216–218.

Lee, J. A., Tallis, J. H. and Woodin, S. M. (1988). Acidic deposition and British upland vegetation. In: *Ecological Change in the Uplands* (Ed. by M. B. Usher and B. A. Thompson) (in press). Oxford: Blackwell Scientific.

Lee, R. E. Jr, Patterson, R. K., Crider, W. L. and Wagman, J. (1971). Concentration and particle size-distribution of particulate emissions in automobile exhaust. *Atmos. Environ.* **5,** 225–237.

Lerman, A. and Lietzke, T. A. (1975). Uptake and migration of tracers in lake sediments. *Limnol. Oceanogr.* **20,** 497–510.

Likens, G. E. and Davis, M. B. (1975). Post-glacial history of Mirror Lake and its watershed in New Hampshire, U.S.A.: and initial report. *Int. Ver. Theor. Angew. Limnol. Verh.* **19,** 982–993.

Liss, P. S. (1974). Chemistry of the sea surface microlayer. In: *Chemical Oceanography* (Ed. by J. P. Riley and G. Skirrow), pp. 193–243. New York: Academic Press.

Livett, E. A. (1982). *The Interaction of Heavy Metals with the Peat and Vegetation of Blanket Bogs in Britain.* Ph.D. thesis, University of Manchester. 378 pp.

Livett, E. A., Lee, J. A. and Tallis, J. H. (1979). Lead, zinc and copper analyses of British blanket peats. *J. Ecol.* **67,** 865–891.

Loring, D. H. (1976). Distribution and partition of cobalt, nickel, chromium and

vanadium in the sediments of the Saguenay Fjord. *Can. J. Earth Sci.* **13,** 1706–1718.
Lowe, J. J. and Walker, M. J. C. (1984). *Reconstructing Quaternary Environments.* London: Longman. 389 pp.
Lu, J. C. S. and Chen, K. Y. (1977). Migration of trace metals in interfaces of seawater and polluted surficial sediments. *Environ. Sci. Technol.* **11,** 174–182.
Lu, X., Johnson, W. K. and Wong, C. S. (1986). Seasonal replenishment of mercury in a coastal fjord by its intermittent anoxicity. *Mar. Pollut. Bull.* **17,** 263–267.
Lyons, W. B., Mayewski, P. A. and Chormann, F. H. Jr (1985). Trace-metal concentrations in sediments from two closed-basin lakes, Antarctica. *Chem. Geol.* **48,** 265–270.
McIntosh, A. W., Shephard, B. K., Mayes, R. A., Atchison, G. J. and Nelson, D. W. (1978). Some aspects of sediment distribution and macrophyte cycling of heavy metals in a contaminated lake. *J. Environ. Qual.* **7,** 301–305.
McIntyre, F. (1974). Chemical fractionation and sea surface microlayer processes. In: *The Sea* (Ed. by E. D. Goldberg), Vol. 5, pp. 245–299. New York: J. Wiley & Sons.
Mackereth, F. J. H. (1966). Some chemical observations on Post-glacial lake sediments. *Phil. Trans. R. Soc. B.* **250,** 165–213.
Madsen, P. P. (1981). Peat bog records of atmospheric mercury deposition. *Nature, Lond.* **293,** 127–130.
Maenhaut, W., Zoller, W. H., Duce, R. A. and Hoffman, G. L. (1979). Concentration and size-distribution of particulate trace elements in the South Polar atmosphere. *J. geophys. Res.* **84,** 2421–2431.
Mamane, Y. (1987). Air pollution control in Israel during the First and Second Century. *Atmos. Environ.* **21,** 1861–1863.
MARC (1985). *Historical Monitoring.* Monitoring and Assessment Research Centre, London University. 320 pp.
MARC (1986). *Biological Monitoring (Plants).* Monitoring and Assessment Research Centre, London University. 247 pp.
Martin, J.–M. (1985). The Pavin Crater Lake. In: *Chemical Processes in Lakes* (Ed. by W. Stumm), pp. 169–188. New York: J. Wiley & Sons.
Martin, M. H. and Coughtrey, P. J. (1982). *Biological Monitoring of Heavy Metal Pollution.* Barking: Applied Science Publishers. 475 pp.
Martin, M. H., Coughtrey, P. J. and Ward, P. (1979). Historical aspects of heavy metal pollution in the Gordano Valley. *Proc. Bristol Naturalists Soc.* **37,** 91–97.
Matsuda, K. and Ito, S. (1970). Adsorption strength of zinc for soil humus III. Relationship between stability constants of zinc–humic and fulvic acid complexes and the degree of humification. *Soil. Sci. Plant Nutr.* **16,** 1–10.
Matsumoto, E. and Wong, C. S. (1977). Heavy metal sedimentation in Saanich inlet measured with ^{210}Pb technique. *J. geophys. Res.* **82,** 5477–5482.
Mayes, R. A., McIntosh, A. W. and Anderson, V. L. (1977). Uptake of Cd and Pb by rooted aquatic macrophytes (*Elodea canadensis*). *Ecology* **58,** 1176–1180.
Meetham, A. R. (1981). *Atmospheric pollution. Its History, Origins and Prevention,* 4th Edn. Oxford: Pergamon Press. 232 pp.
Meger, S. A. (1986). Polluted precipitation and the geochronology of mercury deposition in lake sediments of northern Minnesota. *Water Air Soil Pollut.* **30,** 411–419.
Mellanby, K. (1972). *The Biology of Pollution.* London: Edward Arnold. 60 pp.
Mercer, J. H. (1984). Late Cainozoic glacial variations in South America south of the

equator. In: *Late Cainozoic Palaeoclimates of the Southern Hemisphere* (Ed. by J. C. Vogel), pp. 45–58. Rotterdam: A. A. Balkema.
Ministry of Power (1968). *Statistical Digest (1967)*. London: HMSO.
Moore, P. D. and Webb, J. A. (1978). *An Illustrated Guide to Pollen Analysis*. London: Hodder & Stoughton. 133 pp.
Muhlbaier, J. and Tisue, G. T. (1981). Cadmium in the southern basin of Lake Michigan. *Water Air Soil Pollut.* **15,** 45–59.
Müller, G., Grimmer, G. and Böhnke, H. (1977). Sedimentary record of heavy metals and polycyclic aromatic hydrocarbons in Lake Constance. *Naturwissenschaften* **64,** 427–431.
Müller, G., Dominik, J., Reuther, R., Malisch, R., Schulte, E., Acker, L. and Irion, G. (1980). Sedimentary record of environmental pollution in the western Baltic Sea. *Naturwissenschaften* **67,** 595–600.
Murozumi, M., Chow, T. J. and Patterson, C. C. (1969). Chemical concentrations of pollutant lead aerosols, terrestrial dusts and sea salts in Greenland and Antarctic snow strata. *Geochim. cosmochim. Acta* **33,** 1247–1294.
Newell, R. E. (1971). The global circulation of atmospheric pollutants. *Scient. Am.* **224,** 32–33.
Ng, A. and Patterson, C. C. (1981). Natural concentrations of lead in ancient Arctic and Antarctic ice. *Geochim. cosmochim, Acta* **45,** 2109–2121.
Ng, A. and Patterson, C. C. (1982). Changes of lead and barium with time in Californian off-shore basin sediments. *Geochim. cosmochim. Acta* **46,** 2307–2321.
Nieboer, E. and Richardson, D. H. S. (1980). The replacement of the nondescript term "heavy metals" by a biologically and chemically significant classification of metal ions. *Environ. Pollut. B* **1,** 3–26.
Nilsson, I. (1972). Accumulation of metals in spruce needles and needle litter. *Oikos* **23,** 132–136.
Norton, S. A. (1986). A review of the chemical record in lake sediments of energy-related air pollution and its effects on lakes. *Water Air Soil Pollut.* **30,** 331–345.
Norton, S. A. (1987). The stratigraphic record of atmospheric loadings of metals at ombrotrophic Big Heath Bog, Mt Desert Island, Maine, USA. In: *Effects of Atmospheric Pollutants on Forests, Wetlands and Agricultural Ecosystems* (Ed. by T. C. Hutchinson and K. M. Meema). Berlin: Springer-Verlag (in press).
Norton, S. A. and Hess, C. T. (1980). Atmospheric deposition in Norway during the last 300 years as recorded in S.N.S.F. lake sediments I. Sediment dating and chemical stratigraphy. In: *Proceedings of an International Conference on the Ecological Impact of Acid Precipitation*, S.N.S.F., Sandefjord, Norway. (Ed. by D. Drabløs and A. Tollan), pp. 268–269. Oslo: S.N.S.F.
Norton, S. A., Dubiel, R. F., Sasseville, D. R. and Davis, R. B. (1978). Palaeolimnologic evidence for increased zinc loading in lakes of New England, U.S.A. *Int. Ver. Theor. Angew. Limnol. Verh.* **20,** 538–545.
Nriagu, J. O. (1979). Global inventory of natural and anthropogenic emissions of trace metals to the atmosphere. *Nature, Lond.* **279,** 409–411.
Nriagu, J. O. and Gaillard, J. F. (1984). The speciation of pollutant metals in lakes near the smelters at Sudbury, Ontario. In: *Advances in Science and Technology* (Ed. by J. O. Nriagu), Vol. 15, pp. 349–374. New York: J. Wiley & Sons.
Nriagu, J. O. and Rao, S. S. (1987). Response of lake sediments to changes in trace metal emissions from the smelters at Sudbury, Ontario. *Environ. Pollut.* **44,** 211–218.
Nriagu, J. O. and Wong, H. K. T. (1986). What fraction of the total metal flux into lakes is retained in the sediments? *Water Air Soil Pollut.* **31,** 999–1006.

Nriagu, J. O., Kemp, A. L. W., Wong, H. K. T. and Harper, N. (1979). Sedimentary record of heavy metal pollution in Lake Erie. *Geochim. cosmochim. Acta* **43,** 247–258.
Nriagu, J. O., Wong, H. K. T. and Coker, R. D. (1982). Deposition and chemistry of pollutant metals in lakes around the smelters at Sudbury, Ontario. *Environ. Sci. Technol.* **16,** 551–560.
Ochsenbein, U., Davison, W., Hilton, J. and Haworth, E. Y. (1983). The geochemical record of major cations and trace metals in a productive lake. *Arch. Hydrobiol.* **98,** 463–488.
Oldfield, F., Appleby, P. G. and Battarbee, R. W. (1978a). Alternative ^{210}Pb dating: Results from the New Guinea Highlands and Lough Erne. *Nature, Lond.* **271,** 339–342.
Oldfield, F., Dearing, J., Thompson, R. and Garrett-Jones, S. E. (1978b). Some magnetic properties of lake sediments and their possible link with erosion rates. *Polskie Archwm. Hydrobiol.* **25,** 321–331.
Oldfield, F., Appleby, P. G., Cambray, R. S., Eakins, J. D., Barber, K. E., Battarbee, R. W., Pearson, G. R. and Williams, J. M. (1979). ^{210}Pb, ^{137}Cs and ^{239}Pu profiles in ombrotrophic peat. *Oikos* **33,** 40–45.
Oldfield, F., Appleby, P. G. and Petit, D. (1980a). A re-evaluation of lead-210 chronology and the history of total lead influx in a small South Belgian pond. *Ambio* **9,** 97–99.
Oldfield, F., Appleby, P. G. and Thompson, R. (1980b). Palaeoecological studies of lakes in the highlands of Papua New Guinea I. The chronology of sedimentation. *J. Ecol.* **68,** 457–477.
Oldfield, F., Barnowsky, C., Leopold, E. B. and Smith, J. P. (1983). Mineral magnetic studies of lake sediments. *Hydrobiologia* **103,** 37–44.
O'Sullivan, P. E. (1983). Annually-laminated lake sediments and the study of Quaternary environmental changes—a review. *Quat. Sci. Rev.* **1,** 245–313.
Ouellet, M. and Jones, H. G. (1983). Palaeolimnological evidence for the long-range atmospheric transport of acidic pollutants and heavy metals into the Province of Quebec, eastern Canada. *Can. J. Earth Sci.* **20,** 23–36.
Overbeck, F. and Happach, H. (1957). Über das Wachstum und den Wasserhaushalt einiger Hochmoorsphagnen. *Flora. Jena* **144,** 335–402.
Pakarinen, P. (1978). Distribution of heavy metals in the *Sphagnum* layer of bog hummocks and hollows. *Ann. Bot. Fenn.* **15,** 287–292.
Pakarinen, P. and Gorham, E. (1983). Mineral element composition of *Sphagnum fuscum* peats collected from Minnesota, Manitoba and Ontario. In: *Proceedings of the International Symposium on Peat Utilization* (Ed. by C. H. Fuchsman and S. A. Spigarelli), Bemidji State University, Bemidji, Minnesota, pp. 417–429.
Pakarinen, P. and Tolonen, K. (1976). Studies on the heavy metals content of ombrotrophic *Sphagnum* spp. In: *Proceedings of the 5th International Peat Congress*, Poznan, Poland, pp. 264–275.
Pakarinen, P. and Tolonen, K. (1977a). Distribution of lead in *Sphagnum fuscum* profiles in Finland. *Oikos* **28,** 69–73.
Pakarinen, P. and Tolonen, K. (1977b). Pintaturpeen kasvunopeudesta ja ajoittamisesta. *Suo* **28,** 19–24.
Pakarinen, P. and Tolonen, K. (1977c). Pääravinteiden sekä sinkinja lyijyn vertikaalijakautunista rahkaturpeessa. *Suo* **28,** 95–102.
Pakarinen, P., Tolonen, K. and Soveri, J. (1980). Distribution of trace metals and sulphur in the surface peat of Finnish raised bogs. In: *Proceedings of the 6th International Peat Congress*, Duluth, U.S.A., pp. 78–91.

Pakarinen, P., Tolonen, K., Heikkinen, S. and Nurmi, A. (1983). Accumulation of metals in Finnish raised bogs. In: *Environmental Biogeochemistry* (Ed. by R. Hallberg), pp. 377–382. Stockholm: Ecological Bulletins.

Parekh, P. P. and Husain, L. (1987). Fe/Mg ratio: a signature for local coal-fired power plants. *Atmos. Environ.* **21,** 1707–1712.

Parungo, F. P. and Rhea, J. O. (1970). Lead measurement in urban air as it relates to weather modification. *J. appl. Meteorol.* **9,** 468–475.

Patchineelam, S. R. and Förstner, U. (1977). Bindungsformen von Schwermetallen in marine Sedimenten. Untersuchungen an einem Sedimentkern aus der Deutscher Bucht. *Senckenb. Marit.* **9,** 75–103.

Pattenden, N. J. (1975). Atmospheric concentrations and deposition rates of some trace elements measured in the Swansea–Neath–Port Talbot area. In: *Report of a Collaborative Study on Certain Elements in Air, Soil, Plants, Animals and Humans in the Swansea–Neath–Port Talbot Area, together with a Moss Bag Study of Atmospheric Pollution across South Wales*, pp. 11–134. Cardiff: Welsh Office.

Pattenden, N. J., Cambray, R. S. and Playford, K. (1981). Trace and major elements in the sea surface microlayer. *Geochim. cosmochim. Acta* **45,** 93–100.

Patterson, C. C. (1965). Contaminated and natural lead environments of man. *Arch. Environ. Health* **11,** 344–360.

Patterson, C. C. (1987). Global pollution measured by lead in mid-ocean sediments. *Nature, Lond.* **326,** 244–245.

Peel, D. A. and Wolff, E. W. (1982). Recent variations in heavy metal concentrations in firn and air from the Antarctic Peninsula. *Ann. Glaciol.* **3,** 255–259.

Peirson, D. H., Cawse, P. A., Salmon, L. and Cambray, R. S. (1973). Trace elements in the atmospheric environment. *Nature, Lond.* **241,** 252–256.

Pennington, W., Cambray, R. S. and Fisher, E. M. (1973). Observations on lake sediments using fallout ^{137}Cs as a tracer. *Nature, Lond.* **242,** 324–326.

Petit, D. (1974). ^{210}Pb et isotopes stables du plomb dans des sediments lacustre. *Earth Planet. Sci. Lett.* **23,** 199–205.

Petit, D., Mennessier, J. P. and Lamberts, L. (1984). Stable lead isotopes in pond sediments as tracers of past and present atmospheric lead pollution in Belgium. *Atmos. Environ.* **18,** 1189–1193.

Phillips, S. F., Wotton, D. L. and McEarchern, D. (1986). Snow chemistry in the Flin Flon area of Manitoba, 1981–1984. *Water Air Soil Pollut.* **30,** 253–261.

Picciotto, E., Crozaz, G. and de Breuck, W. (1964). Rate of accumulation of snow at the South Pole as determined by radioactive measurements. *Nature, Lond.* **203,** 393–394.

Presley, B. J., Kolodny, Y., Nissenbaum, A. and Kaplan, I. R. (1972). Early diagenesis in a reducing fjord, Saanich Inlet, British Columbia II. Trace element distribution in interstitial water and sediment. *Geochim. cosmochim. Acta* **36,** 1073–1090.

Press, M. C., Ferguson, P. and Lee, J. (1983). 200 years of acid rain. *Naturalist* **108,** 125–129.

Putnam, P. C. (1953). *Energy in the Future.* New York: Van Nostrand & Co. 556 pp.

Rahn, K. A. and McCaffrey, R. J. (1979). Compositional differences between Arctic aerosol and snow. *Nature, Lond.* **280,** 479–480.

Randhawa, N. S. and Broadbent, F. E. (1965). Soil organic matter–metal complexes: 5. Reactions of zinc with model compounds and humic acid. *Soil Sci.* **99,** 295–300.

Renberg, I. (1984). Varved sediments in chronology. In: *Proceedings of a Workshop on Palaeolimnological Studies of the History and Effects of Acidic Precipitation, sponsored by the U.S. E.P.A.* (Ed. by S. A. Norton), pp. 78–85.

Rippey, B., Murphy, R. J. and Kyle, S. W. (1982). Anthropogenically-derived changes in the sedimentary flux of Mg, Cr, Ni, Cu, Zn, Hg, Pb and P in Lough Neagh, Northern Ireland. *Environ. Sci. Technol.* **16,** 23–30.

Ritchie, J. C., McHenry, J. R. B. and Gill, A. C. (1973). Dating recent reservoir sediments. *Limnol. Oceanogr.* **18,** 254–263.

Robbins, J. A. (1978). Geochemical and geophysical applications of radioactive lead. In: *The Biogeochemistry of Lead in the Environment*, (Ed. by J. O. Nriagu), Vol. 1, pp. 285–405. New York: Elsevier.

Robbins, J. A. (1982). Stratigraphic and dynamic effects of sediment reworking by Great Lakes zoobenthos. *Hydrobiologia* **92,** 611–622.

Ross, H. B. and Granat, L. (1986). Deposition of atmospheric trace metals in northern Sweden as measured in the snowpack. *Tellus.* **38B,** 27–43.

Rühling, A. and Tyler, G. (1968). An ecological approach to the lead problem. *Bot. Notiser* **121,** 321–342.

Rühling, A. and Tyler, G. (1969). Ecology of heavy metals—a regional and historical study. *Bot. Notiser* **122,** 248–259.

Rühling, A. and Tyler, G. (1970). Sorption and retention of heavy metals by the woodland moss, *Hylocomium splendens. Oikos* **21,** 92–97.

Rühling, A. and Tyler, G. (1973). Heavy metal deposition in Scandinavia. *Water Air Soil Pollut.* **2,** 445–455.

Rühling, A. and Tyler, G. (1984). Recent changes in the deposition of heavy metals in Northern Europe. *Water Air Soil Pollut. 22,* 173–180.

Rupp, W. H. (1956). Air pollution sources and their control. In: *Air Pollution Handbook* (Ed. by P. L. Magill, F. R. Holden and C. Ackley), pp. 1.1–1.58. New York: McGraw-Hill.

Russell, R. D. and Farquar, R. M. (1960). *Lead Isotopes in Geology*. London: Interscience Publishers. 243 pp.

Rybniček, K. (1973). A comparison of the past and present mire communities of Central Europe. In: *Quaternary Plant Ecology* (Ed. by H. J. B. Birks and R. G. West), pp. 237–261. Oxford: Blackwell Scientific.

Salomons, W. and Förstner, U. (1984). *Metals in the Hydrocycle*. Berlin: Springer-Verlag. 349 pp.

Schell, W. R. (1986). Deposited and atmospheric chemicals. A mountaintop peat bog in Pennsylvania provides a record dating to 1800. *Environ. Sci. Technol.* **20,** 847–853.

Schell, W. R., Sanchez, A. L. and Granlund, C. (1986). New data from peat bogs may give a historical perspective on acid deposition. *Water Air Soil Pollut.* **30,** 393–409.

Schnitzer, M. and Khan, S. V. (1978). *Soil Organic Matter*. New York: Elsevier. 319 pp.

Shewchuk, S. R. (1985). A study of atmospheric deposition onto the snowpack in northern Saskatchewan. *Ann. Glaciol.* **7,** 191–195.

Shirahata, H., Elias, R. W., Patterson, C. C. and Koide, M. (1980). Chronological variations in concentrations and isotopic compositions of anthropogenic atmospheric lead in sediments of a remote subalpine pond. *Geochim. cosmochim. Acta* **44,** 149–162.

Sigg, L. (1985). Metal transfer mechanisms in lakes; the role of settling particles. In: *Chemical Processes in Lakes* (Ed. by W. Stumm), pp. 283–304. New York: J. Wiley & Sons.

Sigg, L., Sturm, M. and Kistler, D. (1987). Vertical transport of heavy metals by settling particles in Lake Zurich. *Limnol. Oceanogr.* **32,** 112–130.

Sillanpää, M. (1972). Distribution of trace elements in peat profiles. *Proceedings of the 4th International Peat Congress*, Helsinki, Finland, pp. 185–191.

Smith, R. A. (1872). *Air and Rain. The Beginnings of a Chemical Climatology.* London: Longman, Green & Co. 600 pp.

Stern, A. C., Boubel, R. W., Turner, D. B. and Fox, D. L. (1984). *Fundamentals of Air Pollution*, 2nd edn. New York: Academic Press. 530 pp.

Stevenson, F. J. (1977). Nature of divalent transition metal complexes of humic acids as revealed by a modified potentiometric method. *Soil Sci.* **123,** 10–17.

Sturges, W. T. and Harrison, R. M. (1986). The use of Br/Pb ratios in atmospheric particles to discriminate between vehicular and industrial lead sources in the vicinity of a lead works I. Thorpe, West Yorkshire. *Atmos. Environ.* **20,** 833–843.

Suess, H. E. (1955). Radiocarbon concentration in modern wood. *Science, N.Y.* **122,** 415–417.

Syers, J. K., Iskandar, I. K. and Keeney, D. R. (1973). Distribution and background levels of mercury in sediment cores from selected Wisconsin lakes. *Water Air Soil Pollut.* **2,** 105–118.

Tallis, J. H. (1959). Periodicity of growth in *Rhacomitrium lanuginosum. J. Linn. Soc.* **56,** 212–217.

Tallis, J. H. (1965). Studies on southern Pennine blanket peats IV. Evidence of recent erosion. *J. Ecol.* **53,** 509–520.

Tallis, J. H. (1973). Studies on southern Pennine blanket peats V. Direct observations on peat erosion and peat hydrology at Featherbed Moss (Derbys.). *J. Ecol.* **61,** 1–22.

Tallis, J. H. (1983). Changes in wetland communities. In: *Ecosystems of the World 4A. Mires: Swamp, Bog, Fen and Moor, A. General Studies* (Ed. by A. J. P. Gore), pp. 311–347. Amsterdam: Elsevier.

Tallis, J. H. and Yalden, D. W. (1984). *Moorland Restoration Project. Phase 2 Report.* Peak Park Joint Planning Board, Bakewell, Derbyshire. 95 pp.

Tamm, C. O. (1953). Growth, yield and nutrition in carpets of a forest moss (*Hylocomium splendens*). *Medd. Stat. Skogsforskningsinstitut* **43,** 1–140.

Taylor, S. R. (1964). Abundance of chemical elements in the continental crust: a new table. *Geochim. cosmochim. Acta* **28,** 1273–1285.

Ter Haar, G. L., Holtzman, R. B. and Lucas, H. F. Jr (1967). Lead and lead-210 in rainwater. *Nature, Lond.* **216,** 353–355.

Tessier, A., Campbell, P. G. C. and Bisson, M. (1979). Sequential procedure for the speciation of particulate trace metals. *Anal. Chem.* **51,** 844–851.

Thomas, M., Petit, D. and Lamberts, L. (1984). Pond sediments as historical record of heavy metals fallout. *Water Air Soil Pollut.* **23,** 51–59.

Thomas, R. L. (1972). The distribution of mercury in the sediments of Lake Ontario. *Can. J. Earth. Sci.* **9,** 636–651.

Tolonen, K. and Jaakkola, T. (1983). History of lake acidification and air pollution studies on sediments in Southern Finland. *Ann. Bot. Fenn.* **20,** 57–78.

Tracy, B. L. and Prantl, F. A. (1983). 25 years of fission product input to Lakes Superior and Huron. *Water Air Soil Pollut.* **19,** 15–27.

Tummavuori, J. and Aho, M. (1980). On the ion-exchange properties of peat I. On the adsorption of some divalent metals ions (Mn^{2+}, Co^{2+}, Ni^{2+}, Cu^{2+}, Zn^{2+}, Cd^{2+} and Pb^{2+}) on the peat. *Suo* **31,** 45–51.

Tyler, G. (1972). Heavy metals pollute nature, may reduce productivity. *Ambio* **1,** 52–59.

United Nations (1952). *World Energy Supplies in Selected Years, 1929–1950.* New York: United Nations.

United Nations (1976). *World Energy Supplies 1950–1974.* New York: United Nations.

United Nations (1981). *Yearbook of World Energy Statistics 1980*. New York: United Nations.

Valenta, P., Duursma, E. K., Merks, A. G. A., Kützer, H. and Nürnberg, H. W. (1986). Distribution of Cd, Pb and Cu between the dissolved and particulate phase in the Eastern Scheldt and Western Scheldt Estuary. *Sci. Total Environ.* **53,** 41–76.

Verdouw, H., Gons, H. and Steenbergen, C. L. M. (1987). Distribution of particulate matter in relation to the thermal cycle in Lake Vechten (Netherlands): the significance of transportation along the bottom. *Water Res.* **21,** 345–351.

Vernet, J. P. and Favarger, P.–Y. (1982). Climatic and anthropogenic effects on the sedimentation and geochemistry of Lakes Bourget, Annecy and Léman. *Hydrobiologia* **92,** 643–650.

Veron, A., Lambert, C. E., Isley, A., Linet, P. and Grousset, F. (1987). Evidence of recent lead pollution in deep North-east Atlantic sediments. *Nature, Lond.* **326,** 278–281.

Vuorinen, A., Alhonen, P. and Suksi, J. (1986). Palaeolimnological and limnogeochemical features in the sedimentary record of the polluted Lake Lippajärvi in Southern Finland. *Environ. Pollut. A* **41,** 323–362.

Wagner, G. and Müller, P. (1979). Fichten als "Bioindikation" für die Immisionsbelastung urbaner Ökosysteme unter besonderer Berücksichtigung Schwermetallen. *Verhandlungen der Gesellschaft für Ökologie* **7,** 307–314.

Walsh, T. and Barry, T. A. (1958). The chemical composition of some Irish peats. *Proc. R. Ir. Acad.* **59B,** 305–328.

Wanta, R. C. and Lowry, W. P. (1976). The meteorological setting for the dispersal of air pollutants. In: *Air Pollution*, 3rd edn (Ed. by A. C. Stern), pp. 328–400. New York: Academic Press.

Ward, N. I., Reeves, R. D. and Brooks, R. R. (1975). Lead in soils and vegetation along a New Zealand State highway with low traffic volume. *Environ. Pollut.* **9,** 243–251.

Weiss, H. V., Koide, M. and Goldberg, E. D. (1971). Mercury in the Greenland ice-sheet: evidence of recent input by man. *Science, N.Y.* **174,** 692–694.

Welsh, R. P. H. and Denny, P. (1980). The uptake of lead and copper by submerged aquatic macrophytes in two English lakes. *J. Ecol.* **68,** 443–455.

West, R. G. (1977). *Pleistocene Geology and Biology*, 2nd edn. London: Longman. 440 pp.

Williamson, H. F., Andreano, R. L., Daum, A. R. and Klose, G. C. (1963). *The American Petroleum Industry II. The Age of Energy 1899–1959*. Evanston, Illinois: Northwestern University Press. 928 pp.

Wise, S. M. (1980). Caesium-137 and lead-210: A review of the techniques and some applications in geomorphology. In: *Timescales in Geomorphology* (Ed. by R. A. Culliford, D. A. Davidson and J. Lewin), pp. 109–127. Chichester: J. Wiley & Sons.

Wolff, E. W. and Peel, D. A. (1985a). The record of global pollution in polar snow and ice. *Nature, Lond.* **313,** 535–540.

Wolff, E. W. and Peel, D. A. (1985b). Closer to a true value for heavy metal concentrations in recent Antarctic snow by improved contamination control. *Ann. Glaciol.* **7,** 61–69.

Wong, H. K. T., Nriagu, J. O. and Coker, R. D. (1984). Atmospheric input of heavy metals chronicled in lake sediments of the Algonquin Provincial Park, Ontario, Canada. *Chem. Geol.* **44,** 187–201.

Yan, N. D. and Dillon, P. J. (1984). Experimental neutralization of lakes near Sudbury, Ontario. In: *Advances in Science and Technology*, (Ed. by J. O. Nriagu), Vol. 15, pp. 417–456. New York: J. Wiley & Sons.

Yliruokanen, I. (1976). Heavy metal distributions and their significance in Finnish peat bogs. *Proceedings of the 5th International Peat Congress*, Poznan, Poland, pp. 276–283.

Young, D. R., Johnson, J. N., Soutar, A. and Isaacs, J. D. (1973). Mercury concentrations in dated varved marine sediments collected off Southern California. *Nature, Lond.* **244,** 273–275.

APPENDIX

Location of sites mentioned by name in the text. Each location is specified by latitude and longitude, respectively (degrees only for sites in Continental Europe and North America; degrees and minutes for sites in the British Isles, Greenland, Antarctica and the rest of the world.

Site	Location	Reference
A *British Isles*		
Alston Moor, England	54.45 N, 2.30 W	Hunt (1970)
Baglan, Wales	51.40 N, 3.48 W	Pattenden (1975)
Blelham Tarn, England	54.20 N, 3.00 W	Hilton *et al.* (1985)
Buxton, England	53.15 N, 1.55 W	Lee and Tallis (1973), Livett *et al.* (1979)
Collafirth, Shetland	60.30 N, 1.20 W	Cawse (1977)
Ennerdale Water, England	54.30 N, 3.30 W	Hamilton-Taylor (1983)
Esthwaite Water, England	54.22 N, 2.59 W	Mackereth (1966)
Featherbed Moss, England	53.26 N, 1.52 W	Lee and Tallis (1973), Livett (1982)
Fladdabister, Shetland	60.05 N, 1.10 W	Livett (1982)
Glenshieldaig, Scotland	50.45 N, 5.18 W	Lee and Tallis (1973), Livett *et al.* (1979)
Gordano Valley, England	51.20 N, 2.40 W	Martin *et al.* (1979)
Grassington Moor, England	54.00 N, 2.05 W	Livett *et al.* (1979)
Lake Windermere, England	54.20 N, 2.55 W	Aston *et al.* (1973), Hamilton-Taylor (1979, 1983)
Loch Lomond, Scotland	56.10 N, 4.40 W	Farmer *et al.* (1980)
Lough Neagh, N. Ireland	54.40 N, 6.25 W	Rippey *et al.* (1982)
Moor House, England	54.45 N, 2.25 W	Livett *et al.* (1979), Livett (1982)
Port Talbot, Wales	51.30 N, 3.45 W	Pattenden (1975)
Ringinglow Bog, England	53.20 N, 1.40 W	Livett *et al.* (1979), Livett (1982)

Site	Location	Reference
Trebanos, Wales	51.35 N, 3.50 W	Pattenden (1975)
Ullswater, England	54.35 N, 2.50 W	Everard and Denny (1985a,b)
Wastwater, England	54.25 N, 3.20 W	Hamilton-Taylor (1983)
Wraymires, England	54.24 N, 2.58 W	Peirson *et al.* (1973)
B *Continental Europe*		
Draved Mose, Denmark	55 N, 9 E	Aaby and Jacobsen (1979), Aaby *et al.* (1979), Madsen (1981)
Kieler Bucht, W. Baltic	54 N, 10 E	Erlenkeuser *et al.* (1974), Müller *et al.* (1980)
Lake Annecy, Lake du Bourget and Lake Geneva	46 N, 6 E	Vernet and Favarger (1982)
Lake Constance	48 N, 9 E	Müller *et al.* (1977)
Lake Öjesjön, Sweden	61 N, 14 E	Fredriksson and Qvarfort (1973)
Lake Sorvalampi, Finland	60 N, 25 E	Tolonen and Jaakkola (1983)
Mont Blanc, France	46 N, 7 E	Briat (1978), Batifol and Boutron (1984)
Myras Mire, Finland	60 N, 25 E	Pakarinen and Tolonen (1977a)
Pavin Crater Lake, France	46 N, 3 E	Martin (1985)
Southern Belgium (Lake Mirwart, Lake Vielsalm and Lake Willerzie	50 N, 5 E	Petit (1974), Oldfield *et al.* (1980a), Petit *et al.* (1984), Thomas *et al.* (1984)
Storbreen Glacier, Norway	62 N, 8 E	Jaworowski *et al.* (1975)
Store Mosse, Sweden	57 N, 20 E	Damman (1978)
Tatra Mountain Glaciers	50 N, 19 E	Jaworowski (1968)
C *North America*		
Adirondack Lakes, NY	44 N, 75 W	Galloway and Likens (1979), Heit *et al.* (1981), Norton (1986)
Algonquin Provincial Park, Ont.	46 N, 78 W	Wong *et al.* (1984)
Big Heath Bog, ME	44 N, 68 W	Norton (1987)
Clearwater Lake, Ont.	46 N, 81 W	Carignan and Tessier (1985)
Flin Flon, Manitoba	55 N, 102 W	Franzin *et al.* (1979), Shewchuk (1985), Phillips *et al.* (1986)
Found Lake, Ont.	45 N, 79 W	Evans and Dillon (1982)
Great Lakes:		
Lake Erie	42 N, 81 W	Thomas (1972), Kemp *et al.* (1974), Kemp and Thomas (1976), Nriagu *et al.* (1979)

Site	Location	Reference
Lake Huron	44 N, 82 W	Kemp *et al.* (1974, 1978)
Lake Michigan	44 N, 87 W	Edgington and Robbins (1976), Filipek and Owen (1979), Griffin and Goldberg (1979, 1983), Christensen and Chien (1981), Goldberg *et al.* (1981), Muhlbaier and Tisue (1981)
Lake Ontario	43 N, 77 W	Thomas (1972), Kemp *et al.* (1974), Kemp and Thomas (1976), Farmer (1978)
Lake Superior	47 N, 87 W	Kemp *et al.* (1978)
Jerry Lake, Ont.	45 N, 79 W	Dillon and Evans (1982)
Lake Washington, WA	48 N, 122 W	Barnes and Schell (1973), Crecelius and Piper (1973)
Lassen Volcanic Park, CA	40 N, 121 W	Chow and Johnstone (1965)
New Haven Reservoirs, CT (Lake Whitney and Lake Saltonstall)	41 N, 73 W	Bertine and Mendeck (1978)
Narragansett Bay, RI	41 N, 71 W	Goldberg *et al.* (1977)
Rocky Mountain Lakes, CO	41 N, 106 W	Heit *et al.* (1984), Baron *et al.* (1986)
Saguenay Fjord, Quebec	48 N, 70 W	Lu *et al.* (1986)
Southern California coast:		
Santa Barbara Basin, inner basins	34 N, 120 W	Chow *et al.* (1973), Young *et al.* (1973), Bruland *et al.* (1974), Bertine and Goldberg (1977), Ng and Patterson (1982)
San Clemente Basin, outer basins	33 N, 119 W	Bertine and Goldberg (1977)
Spruce Flats Bog, PA	40 N, 79 W	Schell *et al.* (1986)
Sudbury, Ont. (various lakes)	46 N, 81 W	Nriagu *et al.* (1982), Nriagu and Gaillard (1984), Yan and Dillon (1984), Nriagu and Rao (1987)
Thompson Canyon Pond, Sierra Nevada, CA	38 N, 120 W	Shirahata *et al.* (1980)
Turkey Lakes, Ont.	46 N, 82 W	Johnson *et al.* (1986)
D *Greenland*		
Camp Century and	77.10 N, 61.08 W	Murozumi *et al.* (1969)
Camp Century Virgin site	76.46 N, 58.52 W	Ng and Patterson (1981)
Camp Tuto	76.25 N, 68.20 W	Murozumi *et al.* (1969), Ng and Patterson (1981)

Site	Location	Reference
Crete	71.00 N, 37.00 W	Appelquist *et al.* (1978)
Dye-3 Camp	65.11 N, 43.50 W	Davidson *et al.* (1981)
Station Milcent	70.18 N, 44.35 W	Herron *et al.* (1977)
T-1	69.44 N, 48.03 W	Boutron (1979b)
T-46	71.10 N, 36.20 W	Boutron (1979b)
E *Antarctica*		
Adelie Land:		
Cap Prudhomme	66.40 S, 140.01 E	Boutron *et al.* (1984)
Stake D55	66.00 S, 137.46 E	Boutron and Patterson (1983)
Stake D80	70.02 S, 134.50 E	Boutron and Patterson (1987)
Antarctic Peninsula:		
Plateau Site	70.53 S, 64.57 W	Wolff and Peel (1985b)
Spaatz Island	72.53 S, 74.41 W	Landy and Peel (1981), Peel and Wolff (1982), Wolff and Peel (1985b)
Dome C	74.40 S, 124.10 E	Boutron (1979a), Boutron and Lorius (1979), Boutron *et al.* (1987)
Don Juan Pond	74.34 S, 161.11 E	Lyons *et al.* (1985)
Mirny	66.30 S, 93.00 E	Boutron *et al.* (1972)
New Byrd Station	80.01 S, 119.03 E	Murozumi *et al.* (1969), Ng and Patterson (1981)
Shea Sisters Lake	71.48 S, 162.00 E	Lyons *et al.* (1985)
South Pole (Geographic)	—	Boutron (1982)
Vostok	78.27 S, 107.00 E	Boutron *et al.* (1972), Boutron and Patterson (1986)
F *Rest of the World*		
Lake Kinneret, Israel	32.44 N, 35.34 E	Frevert (1987), Frevert and Sollmann (1987)
North-east Atlantic Ocean:		
Station 7	46.22 N, 12.23 W	Veron *et al.* (1987)
Station 8	47.30 N, 08.31 W	Veron *et al.* (1987)

Can a General Hypothesis Explain Population Cycles of Forest Lepidoptera?

JUDITH H. MYERS

I. Summary 179
II. Introduction 181
III. Evidence for Population Cycles in Forest Lepidoptera 182
IV. Characteristics of Cyclic Populations of Forest Lepidoptera . . . 188
A. Characteristics of Cyclic Species 188
B. Patterns of Population Change 189
C. The Beginning of the Decline 193
D. Insect Fecundity and Population Fluctuations 195
E. Parasitoids and Population Fluctuations 196
F. Cyclic and Non-cyclic Populations 199
G. The Impact of Forest Defoliators on the Forests 201
V. Hypotheses to Explain Population Cycles 202
A. Variation in Insect Quality 202
B. The Climatic Release Hypothesis 206
C. Variation in Plant Quality 208
D. Disease Susceptibility 212
E. Mathematical Models 216
VI. Evaluation of Hypotheses 223
VII. Population Cycles of Other Organisms 226
VIII. Conclusions and Speculations 228
Acknowledgements 231
References 232

I. SUMMARY

At least 18 species of forest Lepidoptera in North America and Europe display population cycles with average periodicities of 8 to 11 years. In most long-term studies population outbreak is indicated by defoliation, and at a particular site outbreak periodicity is often irregular. However, detailed

ADVANCES IN ECOLOGICAL RESEARCH VOL. 18
ISBN 0-12-013918-9

long-term studies which quantify insect densities show population trends of 3 to 4 years of increase, 1 to 3 years of peak density, and 1 to 3 years of decline. The increase phase of populations in a geographic area can vary in year of initiation and rate of increase, but population declines tend to be more synchronous. The decline frequently begins when fecundity is high and early larval survival is good, but late instar caterpillars survive poorly. The population decline is not dependent on defoliation. Fecundity of insects is rarely recorded, but may be reduced for several years of the decline.

Life table studies have been predominant in research on cyclic species of forest Lepidoptera. However, measurement errors, interactions between mortality agents, and heterogeneity in insect distributions prevent the emergence of clear patterns of density-dependence. The cyclic dynamics of forest Lepidoptera have stimulated a number of explanatory hypotheses, and mechanisms proposed to cause cyclic dynamics are evaluated: genetic variation, qualitative variation, climatic release, food quality deterioration, food quality improvement following plant stress, and disease susceptibility. Disease may explain several of the characteristics of population fluctuations: (1) reduced fecundity following the decline, (2) synchronous decline of populations with different histories, (3) behavioral variation of individuals at different phases, and (4) mortality of late instars in the first year of the decline and reduced survival of early instars in subsequent years. The impact of disease could be influenced by food quality, food quantity or weather, and selection for increased resistance to disease could occur during the population decline and modify the quality of the insect population. Disease need not kill a large fraction of the population to be selective.

Mathematical models direct attention to delayed, density-dependent processes as the cause of cyclical population dynamics, but have not yielded testable predictions or unique population trends that allow differentiation of causal processes from field observations. A number of different processes could be delayed, density-dependent and have their greatest impact during the decline, when accurate information is most difficult to obtain. The few experimental studies that have been done indicate that the cyclical dynamics of forest Lepidoptera are resilient to external manipulation. An exception is the early and continued suppression of outbreaks of Douglas-fir tussock moth following spraying with nuclear polyhedrosis virus.

Reduced breeding at peak and declining densities is characteristic of other animals with cyclical population dynamics and suggests a generality among different groups. The "quality" of individuals changes with density, as indicated by the propensity to breed.

Because so many biological interactions change with the density of forest Lepidoptera, determining which are sufficient and which are necessary conditions for cyclic population dynamics may be impossible. Parasites, predators, food quality and weather could interact in different ways and at different times to cause population cycles. However, deterioration in insect quality, perhaps mediated by disease, could be a general mechanism that is often overlooked in field studies, and yet theoretically could explain many of the characteristics of population cycles. Experimental manipulation of disease and quantification of the prevalence, distribution, and resistance to disease, are recommended for future studies. Because the phase of decline tends to be more synchronous among populations than the phase of increase, more general understanding of cycles may arise from detailed analysis of this period.

II. INTRODUCTION

The fluctuations of temperate, forest Lepidoptera have received an inordinately large fraction of the total hours devoted to the study of insect populations. This is so even though only 1–2% of species of forest Lepidoptera reach outbreak densities or show cyclic dynamics (Mason, 1987). Why are these fluctuations so fascinating? Trees provide monetary profits and jobs. Therefore, insects competing with man for these valuable resources are worthy of study and governments are willing to foot the bill. If the Achilles' heel can be found, lumber barons can dominate the caterpillars.

But there is more to it than that. Population cycles are repeatable phenomena that occur over large geographical areas. Increases and declines are the stuff of population ecology, the prerequisites for observing density-dependence. Large-scale patterns of population change suggest a major driving force that is waiting to be identified and quantified. A regular pattern of increase and decline allows population ecologists to compare conditions over a range of densities. An understanding of population cycles could greatly improve our general understanding of natural population regulation.

Do population cycles really exist? How consistent does the pattern have to be before we can call it a cycle? Ecologists have a tendency to look for and see patterns in nature, and therefore we might see cycles in population numbers even though the number of years between outbreaks and peak densities vary. Patterns of population fluctuations of a number of forest Lepidoptera are tantalizingly similar. These similarities have stimulated the generation of a number of explanatory hypotheses. In this paper I describe the characteristics of population cycles of forest Lepidoptera and evaluate mechanisms proposed to explain them.

III. EVIDENCE FOR POPULATION CYCLES IN FOREST LEPIDOPTERA

First we must ask if population fluctuations of forest Lepidoptera are cyclic. The pattern of population fluctuation can be shown in long-term data sets pieced together from historical observations, and in detailed studies of populations in specific areas. Both of these contribute to our understanding of population cycles and elucidate details that must be explained. Species for which population cycles have been reported and the average periodicities of outbreaks are summarized in Table 1.

An excellent example of an analysis based on historical data is the study by Tenow (1972) of the outbreaks of *Oporina* (= *Epirrita*) *autumnata* Bkh. and *Operophtera* spp. in the Scandes Mountains of Sweden and Norway. From 1862 to 1968, twelve synchronous outbreaks occurred among these species of moths with an average periodicity of 9·4 years (Table 1). Three different patterns of outbreak can be distinguished: those in which attacks occurred simultaneously over the mountain range, those beginning at one end of the mountain chain and proceeding to the other, and those in which the outbreaks were erratic among sites (Fig. 1). The duration of the outbreaks were extremely variable, ranging from 3 to 10 years over the 12 periods.

Table 1
Periodicity of outbreaks of forest Lepidoptera.

Species	Periodicity[a]			No. sites[b]	Reference
	X	*N*	*SE*		
Oporina autumnata and *Operophtera* species	8·8	12	0·5	M	Tenow (1972)
	9			M	Haukioja *et al.* (1988)
Douglas-fir tussock moth (*Orygia pseudotsugata*)	10			M	CFS[c]
Forest tent caterpillar (*Malacosoma disstria*)	8–12			M	Hodson (1941)
Western tent caterpillar (*Malacosoma californicum pluviale*)	8·2	6	0·7	1	CFS, Myers
Eastern spruce budworm (*Choristoneura fumiferana*)	~35			M	Royama (1984)

Table 1—*Continued*

Species	Periodicity[a]			No. sites[b]	Reference
	X	*N*	*SE*		
Western spruce budworm (*Choristoneura occidentalis*)	28	6	4·0	2	Harris *et al.* (1985a)
	12·8	6	3·0	M	
Larch budmoth (*Zeiraphera diniana*)	8·5	15	0·3	1	Baltensweiler and Fischlin (1988)
Pine looper (*Bupalus piniarius*)	7·9	7	1·0	1	Varley (1949)
Lymantria fumida	7·0	5	0·3	1	Entwistle and Evans (1985)
Lymantria dispar	~8			(Europe)	Keremidchiev (1972)
Black-headed budworm (*Acleris variana*)	8·0	1		1	Morris (1959)
Winter moth (*Operophtera brumata*)	8·0	1		1	Varley *et al.* (1973)
Western hemlock looper (*Lambdina fiscella lugubrosa*)	8·8	5	0·3	M	Harris *et al.* (1982)
False hemlock looper (*Nepytia freemani*)	11·3	4	1·9	M	Harris *et al.* (1985b)
Fall Webworm (*Hyphantria cunea*)	8·0	4	0·7	2	Morris (1964), Ito (1977)
Saddled prominent (*Heterocampa guttivitta*)	11·2	6	1·2	1	Martinat and Allen (1987)
Kotochalia junodi	6–8				Ossowski (1957)

[a] *X* = mean, range, or approximate periodicity; *N* = number of outbreaks counted in the determination of the mean periodicity; SE = standard error.
[b] M = more than 3.
[c] CFS indicates the data were obtained from a report of the Canadian Forestry Service.

The pattern described by Tenow for *Oporina* and *Operophtera* shows an overall cyclicity of outbreaks but considerable variation in the periodicity at any one location. This same pattern is apparent for populations of Douglas-fir tussock moth, *Orygia pseudotsugata* (McDunnough) (Fig. 2), western hemlock looper, *Lambdina fiscella lugubrosa* (Hulst), false hemlock looper, *Nepytia freemani* Munroe, and western spruce budworm, *Choristoneura occidentalis* Freeman, in British Columbia, Canada (Table 1). Outbreaks of

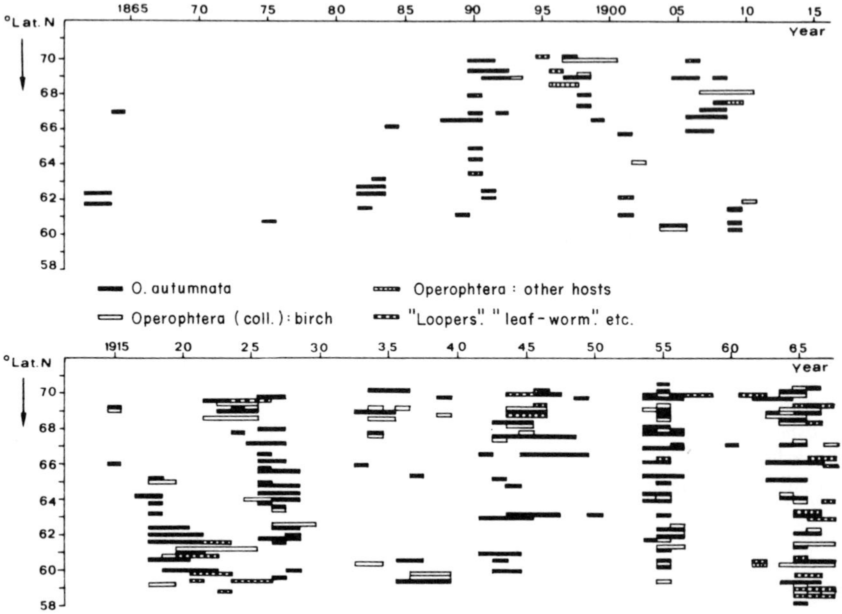

Fig. 1. Records of defoliation of birch and other host trees in the Scandinavian mountain chain and northern Finland by *Oporina* and *Operophtera*. From Tenow (1972).

these species have been recorded somewhere in British Columbia approximately every 10 years, but periodicity is variable for particular sites.

Eastern spruce budworm (*Choristoneura fumiferana* (Clem.), is well known for its population cycles (see Royama, 1984 for review; Table 1). While three outbreaks began in eastern study areas in New Brunswick, the 1955–1960 outbreak started in Quebec and the 1976 outbreak was synchronous across eastern Canada. Densities of egg masses of eastern spruce budworm were monitored at 21 sites in New Brunswick between 1952 and 1980. At some sites populations remained dense throughout this period, while others declined and increased more or less in synchrony (Royama, 1984).

Populations of forest tent caterpillars, *Malacosoma disstria*, reach high density every 8–12 years (Hodson, 1941; Varley *et al.*, 1973, p. 142) with apparent synchrony among populations in eastern North America (Massachusetts and Maine) and western North America (Minnesota and western Canada).

Long-term quantitative studies of forest Lepidoptera at specific sites show more consistent trends, perhaps because the study sites were chosen for their continuous, cyclic populations. An excellent example of a long-term study is

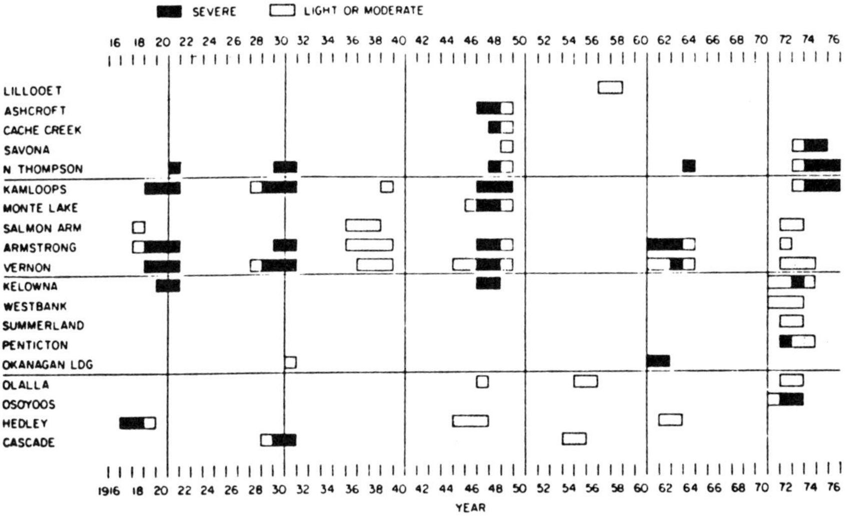

Fig. 2. Occurrence of outbreaks of Douglas-fir tussock moth in British Columbia, Canada, as indicated by defoliation of Douglas-fir. Sites in the top half of the figure represent an east–west transect of approximately 150 km, and the bottom half a north–south transect of approximately 140 km. From Canadian Forest Service (1980).

that of the larch budmoth, *Zeiraphera diniana* Gn., in the Engadine Valley of Switzerland (Fig. 3). Four outbreaks that have been monitored in detail have had a periodicity of nine or ten years between peak densities. In addition, twelve more outbreaks occurred between 1855 and 1950 (Baltensweiler and Fischlin, 1988).

Schwerdtfeger (1941) collated information collected between 1880 and 1940 on four species of Lepidoptera infesting a pine plantation in Germany. Populations of three of the four species, *Panolis flammea*, *Hyloicus pinastri*, and *Dendrolimus pini*, fluctuated erratically (Varley *et al.*, 1973). Populations of the fourth species, *Bupalus piniarius*, reached peak numbers every 6–12 years (Table 1). Klomp (1966) also studied *B. piniarius* in a pine plantation. This study covered fourteen years and the two years of highest density were eleven years apart. The data do not indicate a clear pattern of increase and decline, however.

Data on the periodicity of outbreaks of forest Lepidoptera have been summarized by McNamee (1987), who gives both the observed range of years between outbreaks and periodicity as indicated from spectral analysis of long-term data sets (Table 2). A tendency for outbreak periodicities to occur in multiples of eight years is suggested, although variation is great. A pattern

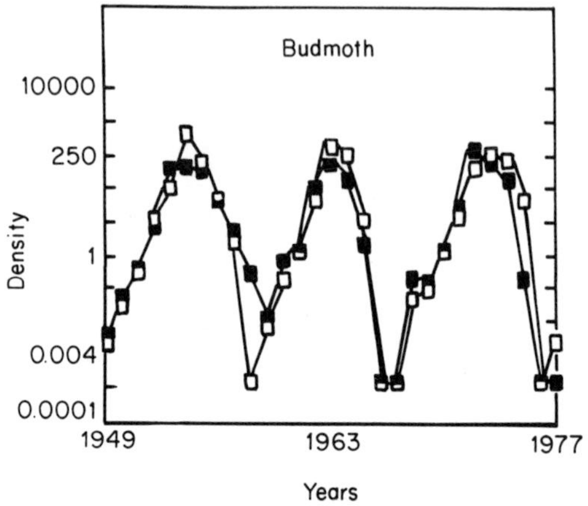

Fig. 3. Population fluctuations of larch budmoth for two sites in the Engadine Valley of Switzerland. Data from Fischlin (1982).

of outbreaks separated by eight or sixteen years would be expected if some outbreaks do not occur or are not recognized in the study.

Although only a small fraction of all forest Lepidoptera periodically attain outbreak densities, it is interesting to know if populations of outbreak species in the same geographical area fluctuate in synchrony, or if outbreaks of dominant species influence the densities of non-outbreak species. Synchronous fluctuations may indicate a common cause of the fluctuations, such as weather influences or changes in densities of shared predators and parasites. Incongruent fluctuations would be expected if dominant species depleted the resources available to other species feeding on the same trees. Miller and Epstein (1986) analysed nine sets of serial density data for sympatric populations of four to 24 species of Lepidoptera. Fluctuations in unison were weakly to strongly indicated in most data sets, while there was little evidence for incongruous synchrony. Mason (1987) gives several more examples of synchronous cycles of forest Lepidoptera. A study of western spruce budworm and associated Lepidoptera showed that non-outbreak species increased in density after the spruce budworm reached high density, increased at a faster rate, and declined at a slower rate than spruce budworm (Mason, 1987). Under crowded conditions the species of Lepidoptera that are normally rare may be better competitors, or escape some of the mortality agents that reduce the densities of the outbreak species, and therefore decline at a slower rate.

To summarize, few of the total number of species of forest Lepidoptera in the temperate region attain outbreak densities, but many of those that do show a clear pattern of periodic fluctuations. At a particular location

Table 2

Range in periodicities of outbreaks of forest lepidoptera summarized by McNamee (1987). Most of the data used in the spectral analyses of periodicity were from the Canadian Forestry Service (CFS). *N* is the number of data sets used in the spectral analyses. Only analyses based on complete and long time series are presented here. McNamee (1987) should be consulted for further details and more complete reference citations.

Species	Periodicity			Reference
	Range	As indicated by spectral analysis	*N*	
Douglas-fir tussock moth	8–12	9	1	Harris and Brown (1976)
Eastern hemlock looper	12–16	—	—	Carroll (1954)
Eastern black-headed budworm	12–16	—	—	Miller (1966)
Green striped forest looper	8–10	7–12	5	Dawson (1970)
Gypsy moth		—	—	Simionescu (1973)
Europe	6–10	3–8	8	Campbell *et al.* (1977)
Massachuesetts	6–8			
Jack Pine budworm	6–8	—	—	CFS
Pine butterfly	10–16	—	—	Cole (1971) Furniss and Carolin (1977)
Pine looper	8–16	8–12	3	Klomp (1966), Varley (1949)
Saddle backed looper	8–12	7–9	4	McNamee (1987)
Western black-headed budworm	8–16	6–16	7	McCambridge and Downing (1960)
Western false hemlock looper	10–16	8–9	9	Klein and Minnock (1971)
Western hemlock looper	6–11	5–12	6	Furniss and Carolin (1977)
Western spruce budworm	12–20	4–16	11	CFS

defoliation does not necessarily occur regularly, but quantitative studies of insect density at single sites indicate regular cycles.

Do populations of forest Lepidoptera cycle in the tropics? Tropical insect populations are as variable in density as temperate ones (Wolda, 1978). Outbreaks of *Zunacetha annulata* in the neotropics cause severe defoliation (Wolda and Foster, 1978) and occur periodically (Harrison, 1987), but do not seem to cycle. This species has seven generations a year, so over a single year this tropical species has almost as many generations as temperate Lepidoptera have between outbreaks. Wolda (personal communication) has been using light traps to monitor insects at Barro Colorado Island, Panama and has found no strong evidence for cycles. However, if species have multiple generations each year we would not necessarily expect to see 7–10 year cycles. In New Guinea, *Lymantria ninayi* in a plantation of *Pinus patula* has three generations a year. Densities were higher and more widespread in 1977 and 1983 in a study that extended from 1977–1984 (Entwistle, 1986). Entwistle (1986) mentions other tropical insects for which outbreaks are reported, but nothing is known about the periodicity of the fluctuations. Without the benefit of annual growth rings in tropical trees, long-term histories of defoliation will not be recorded. Only continuing studies of tropical insects will determine if they cycle, and if so, what the periodicity of the cycles is.

IV. CHARACTERISTICS OF CYCLIC POPULATIONS OF FOREST LEPIDOPTERA

A. Characteristics of Cyclic Species

Species of Lepidoptera that display cyclic population dynamics occur in at least seven families and feed on evergreens or deciduous trees (Table 3). A majority of the species lay their eggs in batches, which is exceptional when considering the Lepidoptera in general (Hebert, 1983; Nothnagle and Schultz, 1987). However, some cyclic species lay eggs singly or in very small groups. In many of the species that lay eggs in clusters, first instar larvae disperse by spinning down on silk threads and being blown in the wind, ballooning. Only four of the fourteen species reviewed in Table 3 are gregarious. Nothnagle and Schultz (1987) reviewed 27 species of forest Lepidoptera in North America and found that larvae of half of these were gregarious. Maximum fecundities of cyclic Lepidoptera tend to be in the 300+ range with the exception of the species that deposit eggs singly which have lower fecundities. It is typical of the Lepidoptera that species that spread their eggs have lower fecundities (Courtney, 1984; Nothnagle and Schultz, 1987). But even several species that lay eggs a few at a time (*Epirrita*, *Operophtera* and *Zeiraphera*) produce on average 120–150 eggs per female. Mason (1987) reviewed the characteristics of non-outbreak species of forest Lepidoptera and found that they tended to have lower fecundities than outbreak species.

To conclude, it is surprising that species that have similar population dynamics display a large range of fecundities typical of Lepidoptera and

Table 3

Characteristics of genera of forest Lepidoptera which demonstrate cyclic population dynamics. Numbers in parentheses are maximum fecundities.

Genus	Family	Fecundity	Egg distribution	Larval dispersal
Deciduous hosts				
Heterocampa	Notodontidae	200 (500)	30–300	Gregarious
Hyphantria	Arctiidae	400 (1000)	Masses	Gregarious
Lymantria	Lymantridae	300 (800)	Masses	Ballooning
Malacosoma	Lasiocampidae	150–225 (250)	Masses	Gregarious
Oporina (= *Epirrita*)	Geometridae	120 (250)	Spread	Ballooning
Operophtera	Geometridae	150	Spread	Ballooning
Symmerista	Notodontidae	300	50	Gregarious
Evergreen hosts				
Acleris	Tortricidae	53 (83)	Singly	Ballooning
Bupalus	Geometridae	220	2–25	Wander
Choristoneura	Tortricidae	175	5–50	Ballooning
Lambdina	Geometridae	59 (122)	2–3	Wander
Melanolophia	Geometridae	80	Singly	Solitary
Orygia	Lymantridae	110–150 (275)	Masses	Ballooning
Pieris	Pieridae	?	5–20	Local
Zeiraphera	Tortricidae	20–180 (350)	1–7	Wander

larval distributions that vary from solitary to gregarious. No simple pattern emerges.

B. Patterns of Population Change

Our image of population fluctations can be strongly influenced by how density data are presented and which sets of data are considered. Density plotted on an arithmetic scale gives the impression of sudden and rapid periods of population increase and decline separated by periods of low numbers. This elicits images of population explosion as might be expected with a release caused by improved conditions in food quality, weather, or predator escape. Royama (1984), amongst others, pointed out that density data should be presented on a logarithmic scale. This emphasizes the rates of population change.

Population trends of cyclic species plotted on a logarithmic scale show consistent patterns (Figs 3–6) of 3–4 years of increase, 1–3 years of peak

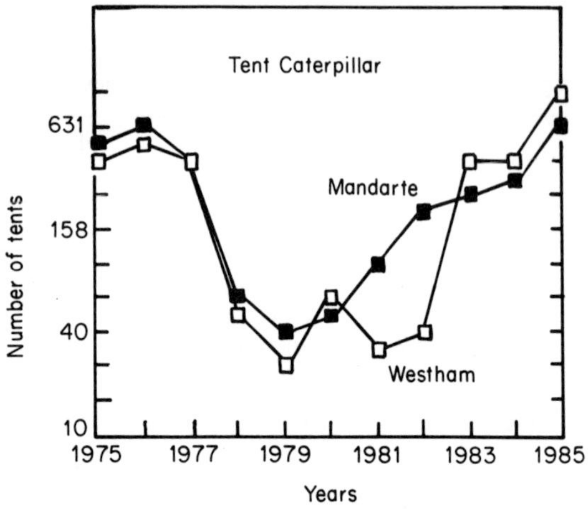

Fig. 4. Two populations of western tent caterpillar in south-western British Columbia, Canada, for which total counts of tents were made. The trees at the Westham site were cut down in winter 1985–1986. The population on Mandarte Island declined 14% between 1985 and 1986, and 78% between 1986 and 1987.

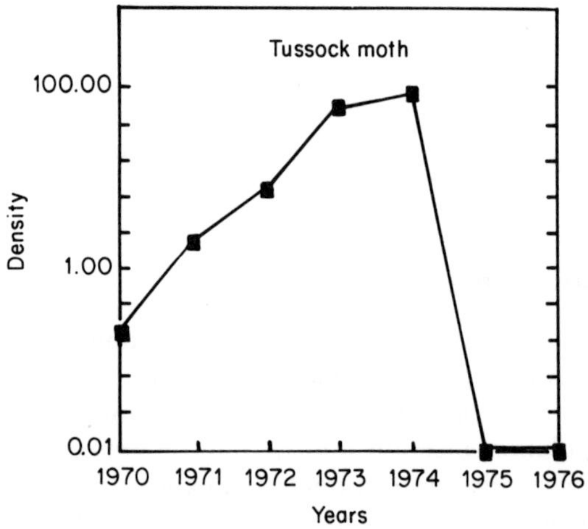

Fig. 5. Population trend for the Douglas-fir tussock moth. From Mason (1978).

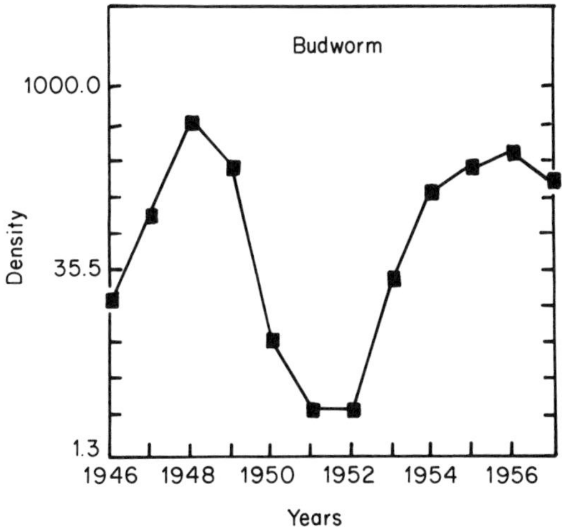

Fig. 6. Density trend for a population of black-headed budworm. From Morris (1959).

density and 1–3 years of decline. There is little evidence of prolonged periods of low numbers. Therefore, cycles that have been well studied are not characterized by sudden increases from low to high densities, but rather demonstrate gradual increase over several years. It is interesting, however, that within a particular geographic area the increase phases can vary in years of initiation, rates of increase and peak density, and still the populations decline more or less in synchrony. Examples illustrating this are presented below.

1. Tussock Moth

In four outbreaks of the Douglas-fir tussock moth monitored in British Columbia from 1917 to 1966, peaks, defined as observable defoliation, began over a period of two to five years in five or six different areas. In contrast, 20 of 21 declines (end of defoliation) occurred in the same year for these same sites (Canadian Forest Service, 1980). Mason (1974) monitored tussock moths in ten localities near Aztec Peak, Arizona, and found variation in the year of initiation, rate of population increase, and the density at the peak. However, eight of ten populations began to decline between 1969 and 1970, and by 1971 all populations had crashed regardless of their peak density (eight-fold difference) and the initial year of population increase (one year difference). Infestation at peak densities was considered by Mason to be light for all of these populations, and declines occurred even though defoliation was patchy.

Three populations of tussock moth in Oregon showed synchronous increase and decline phases, but over an order of magnitude difference in peak density (Mason, 1978). Two other populations, monitored from 1972 to 1980, showed similar patterns of outbreak and decline, but failed to reach densities sufficient to cause visible defoliation (Mason *et al.* 1983). In two other studies by Mason (1974, 1976), the rates of decline of populations of tussock moths were related to the peak density attained, such that less dense populations declined more slowly. However, all populations declined to very low densities within a year of each other, regardless of peak density or lack of defoliation.

To summarize, the onset of the outbreak phase of tussock moth varies among populations but the decline phase tends to be more synchronous. The duration of the outbreak and level of defoliation varies among populations that decline in synchrony.

2. Western Tent Caterpillar

Wellington (1960) found variations in the initiation of the increase phase and densities of subpopulations of western tent caterpillars, *Malacosoma californicum pluviale*, but almost synchronous declines (Fig. 7; Myers, 1988). Two populations of western tent caterpillar in British Columbia had quite different patterns of increase (Fig. 4), but both reached peak density in 1986. Further observations showed that a population that increased to defoliation densities between 1985 and 1986 was almost non-existent in 1987, demon-

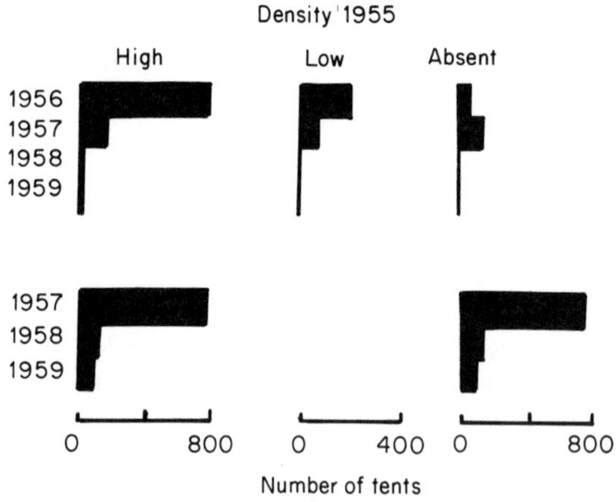

Fig. 7. Declines of populations of western tent caterpillar having different histories. Populations declined synchronously even at sites invaded late in the population fluctuation. Data from Wellington (1960).

strating that population decline does not depend on the length of the outbreak (J. H. Myers, unpublished).

A population outbreak of forest tent caterpillars, *Malacosoma disstria*, was described by Witter *et al.* (1975) to have begun in 1964 in northern Minnesota and to have spread, either by migration or a sequential wave of increase, southward and westward. The total population crashed synchronously in 1972.

3. Larch Budmoth

Larch budmoth were monitored at 20 sites in the Upper Engadine Valley of Switzerland over three outbreaks from 1949 to 1976 (Fig. 3). In 49 of 60 observations made of larch budmoth at these sites, peak densities were reached the same year, and in the other 11 cases peaks were out of phase by only one year. However, peak densities varied by an order of magnitude among populations (Fig. 3). All populations began to decline within a year of each other, regardless of the history of attack (Fischlin, 1982, pp. 88–91) or whether they reached defoliation densities (Baltensweiler *et al.*, 1977). Two additional populations of larch budmoth at different elevations cycled slightly out of phase (Baltensweiler, 1984). The lower elevation population declined without ever defoliating the larch trees, and its population build-up may have depended on immigration from the higher elevation population where defoliation occurred regularly.

4. Spruce Budworm

Spruce budworm outbreaks are more prolonged than those of other Lepidoptera (Fig. 8.). Royama (1984) found similar patterns of decline in four populations of spruce budworm in New Brunswick, even though two of the populations had reached densities sufficient to defoliate trees and two had not.

C. The Beginning of the Decline

It is important for the interpretation of population fluctuations to know at what life stage the population decline begins. For example, poor early spring weather or poor food quality might be expected to have the greatest influence on the survival of early instars. The stage of initial population decline is recorded for several studies. Wellington (1960) observed that in the year of peak density, early larval survival was good but the flight of moths following the summer of peak larval density was low (Wellington, 1957). Therefore, the initial disappearance occurred between the late larval instars and moth eclosion. A similar situation occurred in the 1976 decline of western tent caterpillar (Myers, 1981, 1988). Tussock moth has also been observed to decline initially between the late larval instars and the egg mass stage of the next generation (Mason, 1974). The decline of spruce budworm in

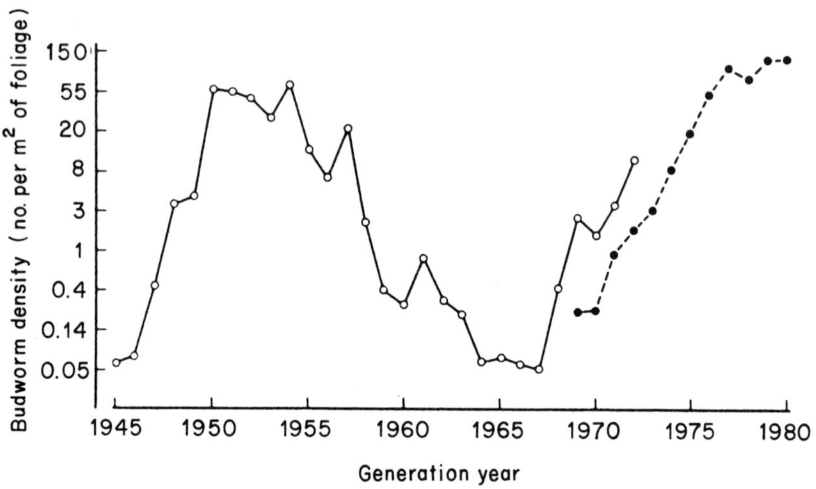

Fig. 8. Number of third to fourth instar larvae, ○, at one site in north-western New Brunswick and, ●, egg mass densities sampled in unsprayed areas covering a wider area but including the site sampled for larvae. From Royama (1984).

Newfoundland in 1981 was associated with the disappearance of late fourth and fifth instar larvae. These larvae fed little and readily dropped from trees. Thirty to 40% of the larvae died of unknown causes in this decline (Raske, 1985). In the year of peak density 5th instar larch budmoth larvae suffer high mortality from starvation and dispersal. High mortality of early instars occurs in the next several years (Baltensweiler, 1978). These observations are very interesting because they indicate that the initial decline may occur following high density and good early larval survival.

To summarize, the pattern of fluctuation of four well-studied forest insects is characterized by variable increase phases that begin in different populations or subpopulations over a span of several years. Observable or measurable tree defoliation does not always occur at peak densities. The decline phase tends to be synchronous among populations and can occur in one year or over several years. The initial decline may occur in the late developmental stages following good early larval survival and high density. For forest insects that fluctuate every eight to ten years, the duration of the outbreak, the peak density and the amount of damage to trees vary among populations that decline synchronously.

D. Insect Fecundity and Population Fluctuations

The fecundity of insects is an important factor in determining the potential of the population for increase, and variation in fecundity provides a measure of the "quality" of insects in different phases of the population fluctuation. Unfortunately, because fecundity is often more difficult to quantify than is mortality, good information on this characteristic is limited. However, some examples are presented below.

The fecundity of western tent caterpillars can easily be measured since each female lays all of her eggs in a single mass. Data for a number of populations are shown in Table 4. In one population (Mandarte), fecundity during the prolonged increase phase tended to be higher than during the decline phase. For another population (Westham), however, no consistent pattern between population increase and fecundity occurred. The first year of the decline was characterized by high fecundity. In a third population (Cambie), density

Table 4

(a) Mean number of eggs/mass (SE) at the beginning of the generation for western tent caterpillars during three population declines. Data for first decline are from Wellington (1960) and are for a population on Vancouver Island. The second (Myers, 1981) and third declines (J. H. Myers, unpublished) are of a population on Mandarte Island, approximately 15 km from the population studied by Wellington. The food plant here is wild rose, as compared to red alder on Vancouver Island. The number of eggs/mass is significantly lower at the end of the decline in both the 1976–1979 and 1985–1987 declines. Between 1985 and 1986 the population declined by 14% and between 1986 and 1987 by 78% based on total counts of late instar tents in this island population.

Year 1	Year 2	Year 3	Year 4	Area	Years
216	200	155	170	Vancouver Island	1956–1959
172 (4·4)	158 (7·2)	156 (5·1)	144 (1·7)	Mandarte	1976–1979
202 (6·9)	218 (4·7)	180 (8·4)	—	Mandarte	1985–1987

(b) Mean number of eggs/mass (SE) during the phase of population increase for three populations of western tent caterpillar. Sample sizes ranged from 13 (Westham: 1981) to 85 egg masses, and in most years 22–40 egg masses were counted. Most of the variation between years is not significant, with the exception of the increase between years 4 and 5 at Mandarte and Cambie. (Data from J. H. Myers, 1988 and unpublished.)

Year 1	Year 2	Year 3	Year 4	Year 5	Area	Years
184 (8·6)	189 (5·7)	191 (6·8)	176 (3·6)	202 (6·9)	Mandarte	1981–1985
189 (13·2)	174 (9·8)	174 (6·0)	173 (10.0)	187 (5·6)	Westham	1981–1985
—	196 (7·0)	146 (4·9)	165 (8·6)	198 (4·7)	Cambie	1982–1985

remained high over four years, during which time the mean fecundity varied. The highest fecundity preceded the summer in which the population dropped by approximately 40%, based on counts of numbers of egg masses on trees. Wellington (1960) reports synchronous declines of subpopulations of western tent caterpillar with varying fecundities. For western tent caterpillar there is no consistent association between fecundity and population increases. There is some suggestion that populations in the second and third years of the decline tend to have lower fecundities than in the phase of increase.

Other studies have shown a reduction in insect size or fecundity in declining populations. Gruys (1970) found that interactions among larvae of the pine looper, *Bupalus piniarius*, occurred more frequently at high densities and resulted in reduced adult size and fecundity in the next generation. Mason *et al.* (1977) found reduced fecundity of tussock moth in declining populations regardless of whether the population reached densities at which defoliation occurred. Declining populations of larch budmoth have reduced fecundity (Baltensweiler, 1968). Fecundity of eastern spruce budworm is directly correlated with larval density and tree defoliation (Miller, 1963).

Defoliation of oak or apple trees reduces fecundity of winter moths of the current generation, but larvae feeding on trees defoliated in the previous year develop into moths with higher fecundity than expected (Roland and Myers, 1987). Studies of tent caterpillar and fall webworm also indicate that herbivory can improve the quality of foliage in subsequent years (Williams and Myers, 1984; Myers and Williams, 1987) and could result in larger moths the next generation. On the other hand, defoliation can reduce fecundity in subsequent years (Schultz and Baldwin, 1982). These interactions between food quantity, quality, and interactions among caterpillars complicate relationships between insect density and fecundity.

To summarize, fecundity is rarely measured in population studies of forest insects. Crowding, food limitation, phenology of insects and trees, foliage quality and insect vigor can all influence fecundity. Fecundities of insects in declining populations are frequently lower than in increasing populations, but during the increase phase, fecundity can vary markedly from population to population. Fecundity can improve or decline for insects feeding on trees defoliated in the previous year, can be directly related to herbivore density, and/or can remain low several years after population density declines.

E. Parasitoids and Population Fluctuations

Most of the emphasis in the study of insect population fluctuations has been on the mortality agents, particularly parasites. Mortality caused by different agents and the progression of mortality through the life stages of the forest Lepidoptera have been summarized in many studies as life tables (Dempster, 1983; Price, 1987). Morris (1959, 1963) developed key factor analysis in an attempt to identify the mortality agents most important in determining population trends and, therefore, worthy of further study. In the Morris key factor analysis, a single stage of the life-cycle is considered in successive generations. A similar type of analysis was developed by Varley and Gradwell (1960). Using this approach, the "killing power" of successive mortality agents are compared to the total generation mortality. Price (1984) gives an excellent overview and critique of these techniques.

Parasitoids influence the density of eggs or other life stages in 75% of the studies of herbivores reviewed by Podoler and Rogers (1975), but it is not clear if they cause population changes or respond to them (Price, 1984, p. 316). For larch budmoth, mortality from parasitoids in the peak and early decline phase reaches 70–80%, but Baltensweiler and Fischlin (1988) conclude that the parasitoids are regulated by the host rather than the opposite. Similarly, Varley *et al.* (1973) proposed that parasitoids track the density of the winter moth rather than determining it. The reproductive rate of specialized parasites is less than that of the autumnal moth, allowing the host to escape from the parasites during the increase phase. However, the "killing power" of the parasites is greatest the year after the host population begins to decline (Haukioja *et al.*, 1988). This is what would be expected if the parasites were tracking their hosts. Dempster (1975) could find no key factor in his analysis of Klomp's data on *Bupalus*. Royama (1984) concluded that parasitoids partially determined the basic oscillation of eastern spruce budworm in conjunction with disease and "an intriguing complex of unknown causes" (p. 429). Harris and Dawson (1985) discounted parasitoids as being necessary for population decline of western spruce budworm since the infestation collapsed over its entire range regardless of the parasitoid populations.

The dynamics and life tables of a non-outbreak but cyclic population of the Douglas-fir tussock moth in Oregon were monitored by Mason and Torgersen (1987). Larval disappearance accounted for the highest mortality and a large unexplained "residual loss" occurred in most generations. Parasitization was delayed density-dependent, while predation was density-independent. The major mortality factors were compensatory such that their combined effects were delayed density-dependent and regulatory. However, because the cause of the major disappearance of larvae was unexplained, an element of mystery remains. It is clear in this study that defoliation is not necessary for cyclic population dynamics, but possibly because measurement errors are likely to be large when population densities are low, no obvious key factor emerged.

A more extensive review of life table studies is given by Price (1987), who points out that the same life table data sets can lead to very different conclusions depending on the method of analysis. For example, Dempster (1983) reviewed life tables of 21 species of Lepidoptera and concluded that parasitoids rarely act in a density-dependent manner. This conclusion has been criticized by Hassell (1985), whose own analyses show that temporal and spatial variation can make the demonstration of density-dependence very difficult even when it is important. A similar conclusion to Hassell's was reached by Jones *et al.* (1987) in their study of the cabbage butterfly, *Pieris rapae*, in Australia. Analysis of the mortality data for the whole juvenile cohort gave no hint of density-dependent predation. However, within plant patches predation was density-dependent and could have been regulatory. These findings call into question any conclusions from life table studies. Unless measurements are made which take account of spatial variation, and are on the appropriate spatial and temporal scale, life tables are not appropriate for the study of population regulation.

In addition to these problems of analysis, other biological complications reduce the likelihood of life table studies leading to general conclusions. For example, mortality can be compensatory. The lack of an agent that kills an early life stage may be compensated for by increased parasitization of another agent in later life stages. The most common parasite can vary among populations, as occurs with the western tent caterpillar. A tachinid fly that is a very common parasite of an island population of tent caterpillars (Myers, 1981), rarely occurs in mainland populations (J. H. Myers, personal observation). Campbell and Torgersen (1983) found that ant and bird mortality were compensatory for the western spruce budworm. Observations are even more complicated if a mortality agent acts at a different life stage in different years. For example a virus infection might kill late instars initially, but, as contamination builds in the environment, earlier life stages may succumb. Therefore, a virus could interact with different parasites as the population cycle progresses.

Disease and predators can also act in a compensatory way if predators are attracted to dead prey. For example, Roland (1986) found that experimental removal of predatory ground beetles did not influence the total mortality of pupal winter moth because beetles preferentially consumed pupae that were already dead.

Further inconsistency arises if mortality agents are not equally available for quantification. For example, caterpillars that die of nuclear polyhedrosis virus may totally disintegrate in less than a day, while the cadavers of caterpillars that have been parasitized by wasps or fungi may remain for much longer (J. H. Myers, personal observation). If mortality is scored by observed deaths, this differential persistence of dead bodies strongly biases the results.

Because stage-specific mortalities are not additive, small changes in mortality can have large effects on population density, especially in life stages in which mortality is high (Fleming, 1985). For example, a population that is stable when only 10% of later instar larvae survive, would be reduced by half if mortality of this stage were increased by an additional 5%. On the other hand, an additional 5% mortality at a life stage with high survival would have a much smaller effect on the rate of population decline. Determining a change in mortality of 5% in a field study might be very difficult because of sampling problems. The resolution of field measurements may therefore rarely be sufficient to measure biologically relevant shifts in mortality.

Experimental analysis may be useful in evaluating the impact of parasites or predators on natural populations. One type of experiment that has been done is the introduction of parasites as biological control agents for exotic pests. Winter moth in eastern Canada is frequently cited as the best example of a situation in which life tables are available for a native population and for an exotic population before and after the introduction of parasites (Varley *et al.*, 1973; Embree, 1966; Price, 1987). The introduced parasites, *Cyzenis albicans* and *Agrypon flaveolatum*, were given credit for the decline of the introduced winter moth in eastern Canada (Embree, 1971; Hassell, 1978), but a reanalysis of the data shows that native nematodes may have been responsible for the decline (Roland, 1988). Other biological control successes show that parasites can reduce host densities, but their role in regulating the densities of hosts remains controversial (Murdoch *et al.*, 1985).

Experimentally increasing the density of hole-nesting birds decreased the average density and suppressed outbreaks of pine looper (Herberg, 1960 cited in Price, 1987). The population exposed to increased bird predation fluctuated with a much reduced amplitude, but neither the control nor treatment populations in this study showed a cyclic pattern of population fluctuation. Further experiments are discussed later.

In conclusion, although many studies of forest insects have focused on mortality, they have not led to a general understanding of the driving mechanism behind population cycles. If anything, the focus on quantifying mortality may have detracted attention from the consideration of conditions of populations or individuals that modify the susceptibility to any or all mortality agents. Life table analysis, which has sometimes been seen as the epitome of a quantitative approach to population ecology, has resulted in very disappointing progress.

F. Cyclic and Non-cyclic Populations

Not all populations of species of forest Lepidoptera classified as cyclic actually cycle. Population cycles characteristically occur toward the elevational and latitudinal edge of the species' distribution. For example, larch budmoth cycles are best demonstrated in populations above 1700 m (Baltensweiler, 1984). Populations below 700 m remain at low densities, increasing only when severe defoliation occurs at higher elevations and moths immigrate to the lower elevations in search of green trees for oviposition. The sparse moth densities at lower elevations are attributed to high mortality of eggs caused by warm temperatures during diapause, early diapause development, and predation. Moths are less fecund in these areas and lay fewer eggs because trees lack lichens that stimulate oviposition (Baltensweiler, 1984).

Tenow (1983) observed dramatic demarcations between outbreak and non-outbreak areas of the autumnal moth. Birch trees surrounding lakes were undefoliated while defoliation was complete for trees on surrounding hillsides. Cold air accumulating in valleys around lakes reduces the minimum temperature below the lethal temperature for eggs, and therefore populations do not persist in these areas.

Populations of the autumnal moth in southern Finland do not reach outbreak densities, but this cannot be explained by the extreme cold temperatures. Haukioja *et al.* (1988) found differences in the foliage quality of trees originating from northern and southern areas, but the relationship of these to insect population dynamics is not known. Moths reared on transplanted trees of southern and northern biotypes at a northern site were larger than those on transplanted trees. (Haukioja and Hanhimaki, 1985).

Western spruce budmoth rarely causes defoliation in valley bottoms in British Columbia, Canada, even though many eggs can occur there laid by immigrating moths (Thompson *et al.*, 1984). Higher temperatures at lower elevations are thought to disrupt the phenology of egg hatch and bud burst in this situation, but no data are available.

Mason (1987) has speculated that the diversity of the fauna may be reduced in harsh environments, resulting in reduced predator and parasite pressure. In addition, non-outbreak species of Lepidoptera would be rare, and therefore the predators and parasites that were present would have few alternative hosts when the outbreak species declined. This could lead to the observed population instability.

The distribution of trees may influence the propensity of insects to reach outbreak densities; trees at the edge of their elevational and latitudinal distribution may form less dense stands. Our studies of western tent caterpillar show that oviposition is much more common on widely spaced trees and on the southern sides of trees (Moore *et al.*, 1988). This distribution of eggs is not obviously explained by food quality differences, and caterpillars on sections of trees with higher density suffer slightly higher mortality than those in less dense areas. We conclude that some aspect of courtship or oviposition is determined by the exposure of foliage, causing isolated trees to be more heavily attacked. Because of this, outbreaks tend to occur where trees are more widely spaced or on the edge of more densely wooded areas. We cannot generalize widely to other species from this observation, but the interactions between forest density and oviposition by Lepidoptera should be studied further. Trees growing in dense stands in areas that are central and optimal to their distribution may not provide the appropriate cues for oviposition by moths.

Differences between sparse and outbreak populations of gypsy moth in North-eastern USA have been studied by Campbell (1975, 1976). All life stages except the eggs survived better in the outbreak area. Predation of pupae in the soil, attributed to small mammals, was markedly higher in the sparse population. Sparse populations can be maintained by vertebrate predation on larvae and pupae in the soil (Campbell *et al.*, 1977), but what determines the variation in vertebrate predation? Although these studies were carried out over three to five years, the emphasis was on the differences between outbreak and sparse populations rather than on changes in predation during the population fluctuation. Conditions in the low phase of a cyclic population are not necessarily similar to those of a sparse, endemic population that never attains outbreak density.

Black headed budworm show cyclic population dynamics in the maritime provinces of Canada with outbreaks occurring every 12–16 years (Miller, 1966). However, over much of their range population densities remain low and this has not been explained.

Because sparse populations are so difficult to study, the differences between endemic and epidemic populations of forest defoliators will only be elucidated through experimental studies. It is always tempting to attribute these differences to weather, but the structure of the forest, competing insect species, variation in predators, and differences in foliage quality could all be associated (Mason, 1987; Wallner, 1986).

G. The Impact of Forest Defoliators on the Forests

There has been surprisingly little study of the impact of cyclic forest defoliators on the forests. Outbreak densities of most forest Lepidoptera reduce the growth of host trees, but widespread mortality of trees rarely occurs. Deciduous trees frequently respond to defoliation by refoliating later in the summer, which allows the trees to compensate for the damage caused by the insects (Baltensweiler, and Fischlin, 1988; Faeth, 1987; Williams and Myers, 1984). A majority, but not all species of forest Lepidoptera considered to be pests feed on new foliage early in the spring (Nothnagle and Schultz, 1987). A number of food quality characteristics could explain this phenology, but early spring feeding also permits refoliation of defoliated trees and could increase the persistence of the insect–host relationship.

Outbreaks of gypsy moth in Europe reduce the growth of oak trees, but, simultaneously, the growth of non-host trees improves (Szujecki, 1987). Gypsy moths may kill trees more readily in North America, where they have been accidentally introduced. Defoliation caused by outbreaks of forest Lepidoptera can have important impacts on the cycling of nutrients and thereby influence soil organisms as well as tree growth (Szujecki, 1987). Trees that are stunted or stressed may die following several years of defoliation, so insects can act as a selecting agent in the forest community (Szujecki, 1987). Widespread death of trees can occur and *Epirrita* outbreaks have sometimes killed birch trees over vast areas of northern Sweden and Finland. In some cases former birch woodlands have been replaced "permanently" by treeless tundra following *Epirrita* outbreaks (Tenow, 1972).

Eastern spruce budworm is one cyclic species that kills one of its host tree species, balsam fir, in pure stands. In this situation only mature trees are vulnerable to high budworm attack and support increasing budworm densities. Trees defoliated sequentially over several years die. However, other host species such as spruce and Douglas-fir are much less likely to die because only the new foliage is acceptable to spruce budworm (Mattson, 1985). Blais (1985) claims that outbreaks of eastern spruce budworm do not change forest species composition, but do change the age distribution and cause fluctuations in the volume of trees. This will vary with different site conditions.

Species of forest Lepidoptera with outbreak population dynamics appear to have evolved to have little long-term detrimental effect on the forests (Faeth, 1987). This can be contrasted to the situation with bark beetles that kill trees and therefore have a much larger impact on forest dynamics (Berryman *et al.*, 1987).

V. HYPOTHESES TO EXPLAIN POPULATION CYCLES

Population cycles elicit creative interpretation and, over the years, a number of hypotheses have been proposed. The emphasis on mortality and life table analysis suggest that, for many, the changes in population density can be understood by describing mortality. However, Chitty (1967) and Wellington (1960) focused attention on the quality of insects and recommended that the conditions that modify the susceptibility of insects to mortality agents may be more important to study than simply what kills insects.

A. Variation in Insect Quality

1. Genetic Variation

Chitty (1967) proposed that density-related selection on genetically controlled variation in behavior and physiology of animals could provide the basis for self-regulation of populations. If the type of individuals that survived and reproduced well in expanding populations was different from those that did so in declining populations, the genetic composition of the population would change through selection. If these selected traits also influenced population growth, then a mechanism for self-regulation of the population would be provided. What actually kills individuals in this situation may not be as important as their exposure and susceptibility to a variety of mortality factors. For example, if animals were polymorphic in their dispersal tendency, density-related selection could modify the proportion of the population that dispersed, and in this way modify the density of the population. Insect dispersal is related to density in some forest Lepidoptera (Gruys, 1970; Leonard, 1970; Dempster, 1975; Baltensweiler and Fischlin, 1988), but there is no strong support for dispersal being genetically based in these situations. For example, in the larch budmoth, dispersal of adult females is greatest when trees have been totally defoliated by larvae and therefore do not provide the necessary colour and quality cues for oviposition (Baltensweiler, 1984). Selection could also act on other traits that might have a genetic basis, such as mate attraction (pheromone types), oviposition behavior or disease resistance.

Perhaps the best example of a study of genetic variation and population fluctuation is that of the larch budmoth (Baltensweiler, 1984). Larch budmoth larvae can be categorized into colour morphs in the fifth instar; dark, light and intermediate. The darker form is characteristic of populations feeding primarily on larch and the lighter form occurs on *Pinus cembra*. Crosses between the two extreme types result in intermediate forms. Associated with the color types are ecologically relevant properties. Dark forms are slightly larger than light forms and the two types have different sex attractants. Eggs of the dark form have faster post-diapause development. When conditions are optimal, the offspring of the dark parents survive slightly better than do those of intermediate parents. However, nutritional stress is associated with an increase in the frequency of intermediate types in natural populations feeding on larch. At peak densities 80% of populations are dark morphotypes, while during the decline phase only 20% of the population are dark. This change in frequency of morphotypes seems to exemplify the pattern proposed by Chitty, in which one genetic form is favored during the increase phase and another during the decline.

However, is this genetically driven self-regulation of population density, or does the genetic shift result from selection caused by nutritional stress following high population density, defoliation, and an increase in the fibre content of the larch needles? Baltensweiler (1984) proposes that the cycle is driven by two processes: (1) at peak density defoliation causes a deterioration in food quality that reduces population density over several years and provides directional selection for the intermediate "fitness type"; and (2) at minimum density, assortative mating of dark and light morphs may lead to an increase in the frequency of dark morphs with their higher fecundity and better synchronization with foliage development (Baltensweiler, pers. comm.). Population density increases as optimal food quality returns. In this model the driving mechanism is the deterioration in food quality following defoliation, not the genetic shift.

To my knowledge no one has shown genetically driven self-regulation of populations. I will discuss subsequently how genetically controlled resistance to disease might change with herbivore density and select for different types of individuals.

2. Qualitative Variation

Wellington (1957, 1960, 1964) recognized behavioral and physiological variation in populations and attributed population fluctuations of the western tent caterpillar to these. Behavior types of tent caterpillars were characterized in two ways by Wellington. The ability of newly emerged larvae to orient to a light source classified larvae into active, directed larvae and sluggish, disoriented larvae. This behavioral spectrum of larvae was reflected in the shape of the tent constructed by the later instars. A group of larvae with a high proportion of active individuals constructed elongate tents. The tents of groups with a high frequency of sluggish individuals were compact and usually small. This classification of tents allowed Wellington to characterize the condition of field populations of tent caterpillars.

According to Wellington's interpretation, an infestation begins with a predominance of active colonies. As the population increases, sluggish colonies accumulate following the dispersal of active moths from the area of increasing density. The remaining sluggish colonies are susceptible to poor weather, starvation, viral disease, and parasitization. Any or all of these destroy the populations, and cause the cyclic decline. The activity phenotypes of larvae are determined initially by the distribution of maternal food reserves; eggs receiving generous amounts of yolk develop into active larvae and those with reduced yolk produce sluggish larvae. The accumulation of resources during the larval stage determines the proportion of active to sluggish larvae in the single egg mass produced by each female moth.

Wellington (1960) showed that the frequency of elongate tents, a characteristic of caterpillar groups with a high proportion of active individuals, changes with the duration of the infestation; recent infestations having 60–80% active groups and those 2–5 years old having approximately 50% elongate tents. However, subpopulations with these different compositions still declined in synchrony (Myers, 1988; Fig. 7). Population quality measured by tent type and population decline were not closely related.

During the population decline which occurred between 1956 and 1958, Wellington recorded a reduction from 13% to 10% in the average percentage of active larvae. Over this same time, the fecundity of moths declined by approximately 30%, which suggests that reduced fecundity may have had a larger impact on the population dynamics than reduced activity.

Interestingly, the initial population decline of 90% between 1956 and 1957 occurred when fecundity and proportion of active colonies was high, and early larval survival good. In the next two years the decline occurred regardless of the history of infestations or the frequency of elongate or compact tents. The deterioration in insect quality seems to follow the decline of tent caterpillar populations rather than initiating it.

From 1982 (early peak density) to 1986 (early decline) we reared tent caterpillars collected each year as early instars, from a field population. In the first three years, during which the population was at high density, larval survival through to pupation was approximately 40% on both foliage from trees in the area of high caterpillar density and from other sites with few caterpillars. In 1985 larval survival began to decline, and in 1986 few larvae survived to pupation (Fig. 9). The field population declined by approximately 40% in the summer of 1986. The "quality" of the caterpillars apparently decreased as the cycle progressed, regardless of the type of foliage the caterpillars ate. Many of the laboratory-reared caterpillars in 1986 died from virus.

Both the Chitty and Wellington hypotheses have been very valuable because they have directed attention to individual insects. But so far

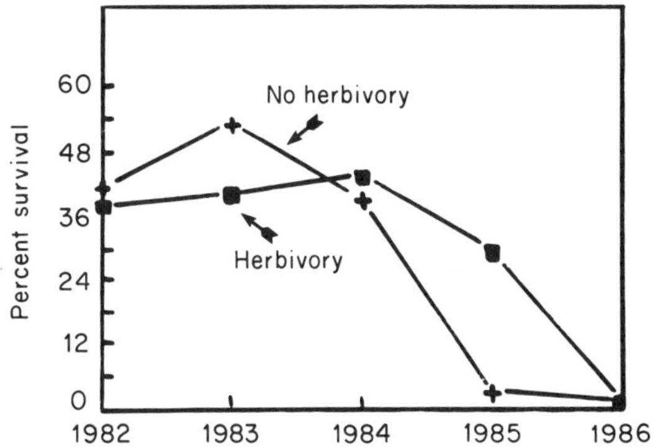

Fig. 9. Survival from early instars to pupae of western tent caterpillars collected each year from a high density field population and reared in the laboratory either on foliage from trees in an area of intense herbivory or on foliage from trees in control areas with no herbivory. The field population began to decline over the summer of 1986 and trees were cut down in winter 1986–1987.

mechanisms to explain changes in quality have not been identified. Crowded insects tend to be smaller (Barbosa and Baltensweiler, 1987; Peters and Barbosa, 1977), but we have little understanding of what component of this size reduction is caused by reduced food availability and how much might be caused by physiological changes associated with interactions among individuals (Gruys, 1970). The continued reduction in fecundity following the decline of the population observed by Wellington suggests that food limitation alone cannot explain this change. Wellington further explored how the availability of resources for yolk deposition during the maternal generation determined the activity of offspring. Maternal inheritance of condition, mediated by yolk deposition, could delay the recovery of a population following the initial decline.

B. The Climatic Release Hypothesis

Climatic release is a very popular explanation of insect outbreaks. Uvarov (1931) and Andrewartha and Birch (1954) are perhaps the best known early proponents of weather and climate as controlling factors of insect abundance—poor weather suppresses insect populations. Graham (1939), Wellington (1952) and Greenbank (1963) were impressed by the potential of good weather for improving conditions for insects and developed the idea that several years of favorable weather could increase insect survival and fecundity, leading to a population outbreak. More recently, the hypotheses of White (1974) and Rhoades (1979) implicate weather as an environmental stress on plants causing either an increase in soluble nitrogen or a decrease in "defensive" chemicals of food plants. Either of these food quality changes could allow herbivorous insects to increase.

Correlations between certain weather conditions and biologically relevant characteristics such as insect growth rate have both stimulated the climatic release hypothesis, and have been used to support it. Martinat (1987) summarizes studies attempting to relate forest insect outbreaks to climatic conditions and particularly emphasizes the spruce budworm and the forest tent caterpillar. He examines the methodological problems associated with testing the hypothesis.

An initial problem is that of relating standard meteorological data to biologically relevant conditions. Leaf surface temperatures can differ from air temperatures (Wellington, 1954) and insects can behaviorally modify their exposure to weather conditions. But statistical problems are perhaps more difficult.

How should one determine the independent weather variables to be tested and what measure of the dependent variable, density, is adequate? Without quantitative population data, one cannot identify the duration of the increase phase. As shown in Fig. 4, the time course of the increase phase can vary among populations in the same general geographical area. To determine if poor weather causes population declines, it is necessary to identify when the decline occurs. For example, the greatest density drop for the 1955–1956 outbreak of western tent caterpillar occurred in 1956 between the late larval stage and the deposition of eggs by the subsequent moth generation 4–6 weeks later (Wellington, 1962). Wellington (1962) attributed the continuing decline to poor weather, but it is interesting that although rainfall was high in April and May 1958, temperatures were above average. This raises the question of what aspect of weather is important to insects.

Martinat (1987) summarizes studies claiming an association between good weather and outbreaks of forest tent caterpillar. In the literature the following conditions are reported to be good: warm sunny spring; continuous warm, humid and cloudy spring; not unusually warm weather; warm early spring followed by cooler weather; cool fall, cold winter and warm spring. The spring of 1958 seems to fall into the category of good weather since it was warm and humid, but with this varied list of conditions that are potentially good for caterpillar growth and survival, the probability of "poor" conditions must be low.

Many studies look for associations by plotting weather variables and changes in insect density or the occurrence of outbreaks. However, there are so many possible associations to consider, that the probability of finding a significant combination becomes high. If the first guess, for example temperature at egg hatch, does not work, overwinter temperature or temperature at some other time could be tried. Average temperatures, extreme temperatures, daily temperatures or monthly temperatures are possibilities. The degrees of freedom should change with every additional relationship explored, and the normal statistical tests used are thus technically invalid.

Martinat (1987) concludes that most studies that claim support for the climatic release hypothesis "would not hold up under rigorous examination". An example is the association between temperature and larval survival found by Morris (1963) which Royama (1978) claims to be spurious. Reanalysis of life tables of spruce budworm led Royama (1984) to reject the "double equilibrium" theory with good weather being responsible for the "release" from the endemic population level. More precise long-term data will be necessary for the testing of this hypothesis, including studies of the temperature relationships between the phenologies of trees and insects, the effect of food limitation for larvae immediately following hatching, the effect of temperature on overwintering stages of caterpillars, and accurate density estimates.

It is very likely that some weather conditions, perhaps catastrophic in nature, are involved in synchronizing populations (Moran, 1953). A variety of weather conditions could have the same effect, making it very difficult to identify which conditions are necessary or sufficient to have an impact on the population cycles of forest Lepidoptera.

We must conclude that although the climatic release hypothesis could be valid, at least in some situations, it should not be assumed to be true. It is likely that a combination of insect quality and environmental conditions are required for population outbreak, and this complicates further the search for patterns. The climatic release hypothesis may be untestable.

C. Variation in Plant Quality

Chitty and Wellington called attention to changes in the quality of the insects during population fluctuations. More recently, attention has been directed to variation in the quality of foliage. Again, two different views appeared in the literature at about the same time. These can be characterized as the "food quality deterioration hypothesis" (Benz, 1974, 1977; Haukioja and Hakala, 1975) and the "food quality improvement hypothesis" (White, 1974, 1978, 1984).

1. Food Quality Deterioration

The quality of foliage may deteriorate following herbivore damage and thus act in a delayed density-dependent manner (Haukioja and Neuvonen, 1987). This deterioration in quality could be caused by increases in secondary plant compounds that are detrimental to insect growth or survival (Rhoades, 1985) or reductions in the nutrient quality, lower nitrogen and higher fiber content (Benz, 1974; Tuomi *et al.*, 1984).

Responses of plants to herbivore damage have been interpreted as induced defenses or induced resistance. They have been classified as rapidly inducible responses if the generation of herbivore causing the damage experiences the effect, and long-term inducible responses if subsequent generations of herbivores are influenced by the modification in plant quality and recovery of the foliage quality is delayed (Haukioja and Neuvonen, 1987).

The idea that plants respond to herbivore damage is very attractive, since it provides a density-dependent mechanism to reduce herbivore populations, and a delayed recovery in foliage quality could explain both the continued decline in insect numbers and quality that are characteristic of population cycles of forest Lepidoptera. It would also seem to be a testable hypothesis, since one need only monitor foliage quality through bioassay or chemical analysis over a cycle to determine if long- or short-term induced responses occur. In reality, however, testing this hypothesis has not been easy.

Many of the problems that arise in tests of the hypothesis of food plant deterioration have been discussed by Haukioja and Neuvonen (1987). A major problem has been to obtain sufficient replication or even to determine what an appropriate replicate is (Fowler and Lawton, 1985). Food quality is most readily assayed by determining development rates and final size of insects reared on foliage of different types either in the laboratory or in bagged foliage on trees in the field. For appropriate replication, each insect should be reared individually on foliage from a specific tree. Replication should include tests of foliage quality in several different areas and should be done during different stages of the population cycle. Because the quality of insects can change as well as the quality of foliage, replication through time becomes particularly important. In our studies of the western tent caterpillar, pupal weight declined for laboratory-reared insects fed foliage of trees from an area with a high density of caterpillars and from an area with a low caterpillar population (Fig. 10). However, there was no correspondence between the temporal variation in pupal weight of laboratory-reared larvae

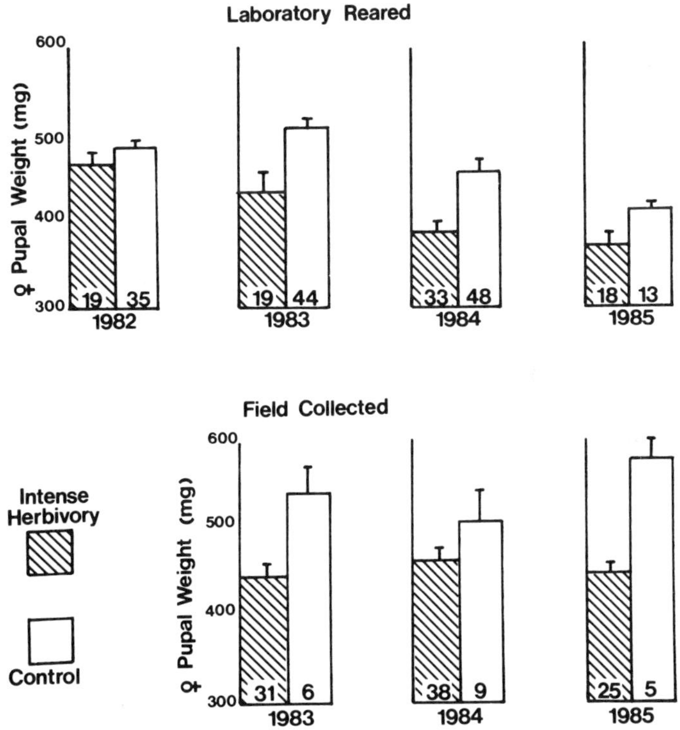

Fig. 10. Weight of female pupae of western tent caterpillar collected as early instars each year and reared in the laboratory on foliage from red alder trees in an area of intense herbivory or control areas with few caterpillars. Field-collected pupae were from the high density population which was the source of the caterpillars used for rearing in the laboratory, or from low density sites. From Myers (1988).

and that of field-collected pupae. This makes a realistic interpretation of the laboratory results difficult.

Haukioja and Neuvonen (1987) give other examples suggesting that the density of the source population of insects may influence their response to food quality, and that selection from crowding or defoliation may modify the response of the next generation of caterpillars to foliage quality. Offspring of parents reared on "induced" foliage were less influenced by foliage deterioration than offspring of parents unexposed to "induced" foliage, and crowded larvae showed less reduction in pupal weight and survival on "induced" foliage than did individually reared larvae. In addition, the responses of trees to damage can change with time (Haukioja and Neuvonen, 1987), and trees from different source populations differ in their reaction to foliage damage even when planted at the same locality (Haukioja *et al.*, 1988).

To date, a number of studies have shown that leaf damage and tree defoliation may sometimes affect insect growth rate, size, and survival (see reviews by Fowler and Lawton, 1985; Haukioja and Neuvonen, 1987; and Rhoades, 1985 for examples), but relevant associations between foliage quality and population dynamics have only been shown for the larch budmoth (Baltensweiler and Fischlin, 1988).

Raw fiber content of larch leaves varies during the population fluctuation and affects larval survival and fecundity (Benz, 1974; Baltensweiler, 1984). In the year following defoliation, needle growth and size are reduced. Raw fiber content is higher and nitrogen content lower in these leaves, and normal conditions may not return for several years after larval populations decline. In this case foliage deterioration that is biologically relevant occurs only after trees are more than 50% defoliated. Therefore, rather than acting as a defensive response, this deterioration is due to a disruption in nutrient resorption from leaves in the previous summer, and several years are necessary before the physiological condition of the tree improves.

Few field experiments have tested the relevance of foliage quality to population dynamics. However, experimental removal of larch budmoth with *Bacillus thuringiensis* did not prevent the population decline (Baltensweiler *et al.*, 1977). Similarly, protecting wild rose foliage on a small island from herbivory by western tent caterpillar did not delay the decline in the insect population, and the subsequent introduction of tent caterpillar eggs from a declining population to the area did not lead to a population increase (Myers, 1981).

The hypothesis that population cycles result from the deterioration of food plant quality due to defensive responses induced by damage to plants or nutrient stress following defoliation requires more experimental analysis. Biologically relevant tests must be done over the complete population fluctuation. In addition, more consideration should be given to the patterns of changes in fecundity and survival that must be explained. While deterioration of the quality of foliage may occur following several years of intensive herbivory, insect population declines without defoliation suggest that deterioration in foliage quality is not necessary for cyclic decline of populations.

2. Food Quality and Plant Stress

"Substantial evidence indicates that drought stress promotes outbreaks of plant-eating fungi and insects" (Mattson and Haack, 1987). Two hypotheses of insect outbreak are based on this assumption. White (1974, 1978, 1984) proposes that insect outbreaks occur when the nutritional quality of leaves improves following environmental stress from drought, root damage from waterlogging, or a variety of other stresses. Rhoades (1979) considers stress to be a mechanism for reducing the production of "defensive" chemicals of trees and therefore for improving foliage quality for herbivorous insects. However, if any pattern exists, stress seems to increase the concentration of at least some types of secondary chemicals, so this reasoning might not be consistent with data (Mattson and Haack, 1987; Myers, 1988).

First, testing the hypothesis that drought and insect outbreaks are related has the same pitfalls as testing the climatic release hypothesis; associations are based on looking for patterns among uncontrolled and poorly quantified variables. A good example of an apparent association between insect outbreaks and drought, which breaks down with appropriate analysis, is the study of outbreaks of the saddled prominent, *Heterocampa guttivitta* (Walker) in north-eastern USA by Martinat and Allen (1987). Although one outbreak was associated with drought, this was not a general pattern when more outbreaks were considered.

Secondly, it is wrong to generalize among totally different types of insects in looking for patterns. Aphids may benefit from increased concentrations of sugars and amino acids, but low water content of leaves can be detrimental to Lepidoptera growth (Scriber and Feeny, 1979). Some grasshoppers may prefer to feed on stressed, yellowish plants (Lewis, 1979), but cabbage white butterflies preferentially oviposit on greener plants (Myers, 1985). It is interesting that almost all of the examples cited in White (1984) to show an association between plant stress and insect outbreak involve sucking insects or grasshoppers.

It is unlikely that the association between stress and any measurable physiological characteristic of plants is linear, but rather most are probably curvilinear. Without quantifying plant stress and its influence on growth and survival of particular insects, no general relationship can be expected or should be inferred.

The hypothesis that plant stress causes insect outbreaks may be tested experimentally. While it is not a perfect technique, simulating drought stress by cutting roots and observing the impact on insects is a good start. McCullough and Wagner (1987) studied the influence of watering and trenching ponderosa pine on the pine sawfly (*Neodiprion autumnalis*). In the first year of the study there were no differences among treatments, but in the second year larvae on stressed trees produced smaller pupae, grew more slowly, had lower survival and rejected more foliage. Similarly, I cut roots of wild rose to study simulated drought stress on western tent caterpillar and found no effect (Myers, 1981). Louda (1986) observed that cutting the roots of a montane crucifer increased herbivore damage. However, no measurements were made on the numbers or growth of insects. Insects may eat or damage more leaf material of stressed plants without benefitting in growth and survival.

Tussock moth outbreaks occur on sites on ridges where tree growth is reduced and drought stress is higher than in adjacent drainages, and this observation is often used as evidence for the plant stress hypothesis. Mason (1981) compared the foliage quality of white fir in paired sites by putting out tussock moth larvae in cages on tree branches. Larval survival was the same in both types of sites, but larvae consumed more foliage when feeding on the tops of trees in outbreak sites. Increased foliage consumption is usually interpreted as an indication of poor food quality. The fecundity of moths reared on trees in the outbreak sites was significantly higher than in non-outbreak sites, but Mason (1981) did not think this was sufficient to explain the difference in population dynamics of tussock moth in the two types of locations.

Mattson and Haack (1987) nicely review the conditions associated with drought that could cause insect outbreaks, including better microclimate (warmer and drier), better nutrient balance, and beneficial associations with micro-organisms. But the fact remains that the statistical problems associated with correlational data, and the negative results of experimental tests, suggest that the question of whether drought stress mediated through food quality causes outbreaks of forest Lepidoptera is still very much open to debate (Baltensweiler, 1985).

D. Disease Susceptibility

Disease, particularly baculoviral disease, is frequently reported in declining populations of forest Lepidoptera (Cunningham, 1982; Entwistle and Evans, 1985; Entwistle, 1986; Tanada and Fuxa, 1987; and Tanada, 1976 provide reviews of an extensive literature). The sequence of events between disease and forest Lepidoptera might be as follows. As the population of an insect herbivore increases, individuals interact more frequently, thus allowing the transmission of disease. Stress associated with food limitation or poor weather could further accentuate the susceptibility to disease which results in an epizootic. High mortality from disease selects for resistant individuals and the epizootic ends as the host density declines. Sublethal effects of disease may reduce vigor and fecundity for several subsequent generations. We can look at the components of this hypothesized sequence to evaluate its feasibility.

First is the most important question of where the disease comes from. Most of the discussion here will pertain to nuclear polyhedrosis virus (NPV), since this has received the most attention in the past, but similar relationships could exist with other disease organisms. Viral polyhedra can remain active in the environment for long periods of time (see Hostetter and Bell, 1985; Kaupp and Sohi, 1985 for reviews). High density populations may be more likely to contact refuges of viral polyhedra. On the other hand, over the years there has been considerable speculation on whether latent forms of virus or other diseases might exist (Entwistle, 1983; Entwistle and Evans, 1985; Kaupp and Sohi, 1985). Positive identification of latent virus has been difficult because contamination with active virus is hard to rule out. Proof will only come when it is shown that the pathogen exists in a non-infective and non-replicative state and is transformed to an infective and replicative state when the insect is stressed (Tanada and Fuxa, 1987). The latent virus would have to be transmitted among generations in order to be maintained in populations. If latent viruses do exist, they could explain sudden epizootics in natural and introduced populations of insects.

Transmission of virus is most likely at high host density, but virus can occur at low host density (see Benz, 1987, for a review). Benz points out that the proportion of larvae infected in a particular tree is not always directly associated with the density of larvae on the tree. Therefore, while the relationship between virus and host tends to be density dependent, this is not always the case.

How does starvation or food limitation influence the susceptibility of Lepidoptera to virus? Benz (1987) reviewed studies that show that starvation may have no effect on the prevalence of virus, or may even suppress the development of NPV. However, in several interesting studies, food quality (leaf age, toughness, species of food plant) has influenced the susceptibility of insects to virus; virus infection is more likely when leaf quality is poor. The presence of sugar in the food at the time of viral infection reduced infectivity of a granulosis virus to *Pieris brassicae* (David and Taylor, 1977). Reduced sugar in the diet possibly increases the permeability of the insect gut to virus and, therefore, infectivity is higher when virus particles occur on food low in sugar. Similarly, Keating and Yendol (1987) have shown variation in the mortality of gypsy moth larvae associated with the tree host species on which NPV is presented. White (1974) proposed that the sudden outbreak of disease in high density insect populations could be associated with reduced nitrogen content of food following defoliation. Food quantity and quality are likely to be associated with the susceptibility of forest Lepidoptera to disease, particularly NPV.

An epizootic of a granulosis virus occurred in the 1955 outbreak of larch budmoth in the Engadine Valley, Switzerland. Martignioni (1957) observed that budmoth were more susceptible to virus before the population decline than after. Several years later Martignioni and Schmid (1961) further developed the hypothesis that mortality from virus epizootics at high density selects for increased resistance to viral disease in subsequent generations. This hypothesis was supported by Wellington's (1962) observation that in the second year of population decline of western tent caterpillar, virus could be induced by starvation. However, in the next two years starvation did not "induce" viral disease. Greater resistance to virus could result from high larval mortality during the decline.

A number of studies show a genetic as well as an environmental component to disease resistance of insects. Recent reviews of the genetics of disease resistance have been prepared by Briese and Podgwaite (1985) and Watanabe (1987). Polygenic inheritance is more common than single gene control. Geographically distinct populations of Lepidoptera can vary in their susceptibility to disease, and selection for increased disease resistance has been successful in some laboratory populations. Therefore, variation in disease resistance during population cycles is possible.

The potential association between disease and population cycles received support from the simulation models of Anderson and May (1980) and Anderson (1982). In these models, which assume continuous host generations, disease transmission occurs at a threshold density. Population cycles are one of the possible outcomes of these models, and values determined from early studies of larch budmoth were used to show how granulosis virus could cause population cycles. However, because the granulosis virus was only prevalent in one cycle of larch budmoth, this example has been criticized as not representing a general phenomenon (Baltensweiler and Fischlin, 1988). In addition, Getz and Pickering (1983) showed that the dynamics predicted by the Anderson and May models are very sensitive to the way the transmission of virus is modelled. Adaptations of the Anderson and May models to the tussock moth did not result in appropriate periodicities or amplitudes of outbreaks (Vezina and Peterman, 1985), and these authors conclude that other components of the system are also important. Regniere (1984) used models to explore the potential of diseases transmitted vertically (from parent to offspring) on the population dynamics of insects with discrete generations. His results show that diseases with low virulence could play an important role in population cycles. Until we understand how disease is maintained in populations at low density and transmitted among individuals at increasing densities, and how genetic or environmental modifications in the susceptibility of insects to disease might occur, epidemiological models cannot be very realistic (see next section).

Disease is only one of many causes of mortality in outbreak populations, and in most cases, only a small portion of the total mortality is attributed to disease. These observations could be biased by the rapid disappearance of the cadavers, particularly of young instars dying of virus. Diseased pupae could be preferentially eaten by predators (Roland, 1986) and the cause of mortality misidentified. Therefore, quantification of disease may be very difficult in field populations. However, even if disease represented a small portion of total mortality, selection for disease resistance could be strong, particularly if mortality by other agents was high and non-selective in the early instars. In addition, even a small increase in mortality from disease could cause the population to decline.

A fascinating dimension of viral infection is the effect it has on behavior. Viral infections are sometimes called "tree top" disease because the infected larvae move to the tops of trees (Evans and Entwistle, 1987). Evans and Allaway (1983) observed that NPV-infected *Mamestra brassicae* were twice as active as non-infected larvae. Diurnal behavior and increased activity have been observed in several species of Lepidoptera when densities are high (Knapp and Casey, 1986).

With these observations in mind, it is interesting to re-evaluate the observations of Wellington on western tent caterpillar behavior and population fluctuations. In the summer of 1956, larval density was very high but the number of moths flying in July was low, and the numbers of egg masses declined between 1956 and 1957. During 1956, Wellington observed that many of the tents were large, elongate, and situated on the ends of branches and tops of trees. In subsequent years most tents were smaller, compact, and more often formed in the crotches of trees. Wellington also observed extensive wandering by caterpillars at peak density, and I too found this to be characteristic of peak density tent caterpillars. These behavioral changes might be associated with viral infection.

Wellington (1962) described the distribution of viral disease in the declining populations of tent caterpillars. Sluggish colonies, i.e. those forming small compact tents, were frequently destroyed by disease. Disease persisted in a few "very active colonies." These active colonies may have been composed of individuals that were more resistant to virus, causing the virus infection to develop more slowly. Rather than the larvae dying of virus at an early stage, behavioral modification, particularly in later instars, might occur.

Another characteristic of declining populations that might be explained by non-lethal viral infections is the reduced size and fecundity of moths. Moths surviving exposure to virus are frequently smaller and have reduced fecundities (reviews in: Briese and Mende, 1983; Entwistle and Evans, 1985; Vargas-Osuna and Santiago-Alvarez, 1986), but the relationship can vary depending on the dose and the time of infection (Perelle and Harper, 1986). An alternative explanation is that genetic resistance to disease and reduced fecundity are correlated characters.

To conclude, disease has been rejected as a causal mechanism of population cycles because the proportion of the population observed dying of disease is frequently low. As shown earlier, however, even small increases in mortality at stages for which mortality is already high can have large effects on population density. Sublethal effects of disease can explain some of the behavioral and life-history characteristics of populations undergoing population cycles. The influence of vertically transmitted diseases will vary with the reproductive rate of the host population and may cause more frequent oscillations with greater amplitude in areas with lower reproductive rates (Regniere, 1984). The susceptibility to disease of populations in different phases of the cycle should be monitored to determine if increased resistance occurs as the population declines. We must understand how disease is maintained in low density populations to develop realistic epidemiological models.

E. Mathematical Models

Populations do not continue to increase indefinitely and the relationship between density, reproductive rate, population limit or carrying capacity, and time can be described by the logistic equation:

$$N_{t+1} = \frac{N_t R}{(1 + aN_t)}$$

where N_t = density at time t, R = the potential reproductive rate, and $a = (R-1)/K$, where K = carrying capacity.

This equation describes a population increasing in a sigmoid manner to the carrying capacity (see Begon *et al.*, 1986; Krebs, 1985; or other introductory ecology texts for further explanation). As the population increases, the actual reproductive rate ($R/(1+aN_t)$) declines.

A more general model of this sort allows for variation in the type of density dependence (Maynard Smith and Slatkin, 1973).

$$N_{t+1} = \frac{N_t R}{(1 + aN_t)^b}$$

where $b < 1$ gives undercompensation, $b = 1$ gives perfect compensation, $b > 1$

gives overcompensation, and $b=0$ gives density independence. The influences of b and R in these models on the dynamics of populations were explored by May (1975) and Bellows (1981), and the variety of population patterns is shown in Fig. 11.

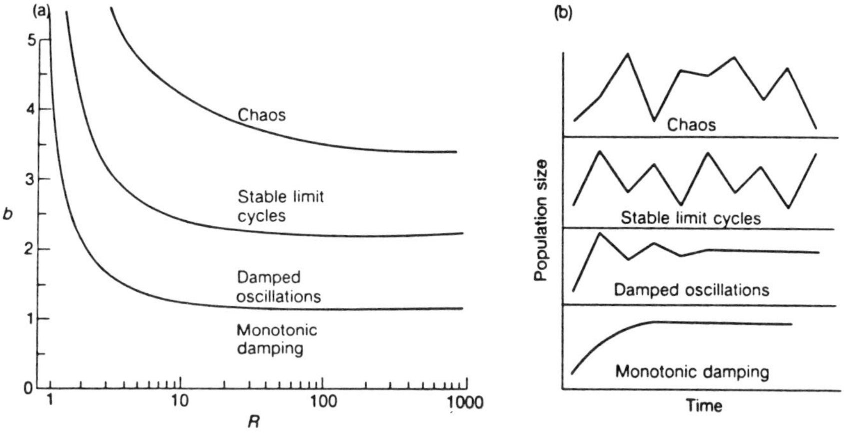

Fig. 11. The range of population behaviors (a) and their patterns (b) generated by models incorporating R=reproductive rate and b=density dependence (see text). After May (1975) and Bellows (1981) from Begon *et al.* (1986).

Time-lags in the response of populations to density can also vary the population dynamics (Cunningham, 1954; see examples in Krebs, 1985, p. 227). Longer time-lags in general increase population instability. Mathematical models show that time-lags, high reproductive rates and over-compensating density-dependence are all capable of causing population fluctuations, and these relationships have been widely explored (Anderson and May, 1978; Berryman, 1987; Hassell *et al.*, 1976; Hutchinson, 1948; May, 1974a, b; May *et al.*, 1974).

Population regulation through negative feedback occurs when forces act on natality and mortality in direct response to the density of the population. These forces can be fast acting, such as vertebrate predators that will eat larger numbers of readily available prey, or slow acting, such as plant regrowth after defoliation. The biotic variables and their potential impacts on high and low density populations of insect herbivores are summarized in Table 5.

In the single species models of Berryman *et al.* (1987, and other references therein), the combined feedback of all density-dependent variables are incorporated in a single function referred to as an R-function or replacement

Table 5

Possible responses of host plants and natural enemies to increasing density of forest Lepidoptera, and their potential impact as density regulators. Modified from Berryman *et al.* (1987)

Variable	Response characteristic	Effectiveness in population regulation	
		Low herbivore density	High herbivore density
Plants	(a) Rapid induced defense	Stabilizing	May suppress outbreaks
	(b) Slow induced defense	Rapid response and slow recovery keeps density low	Foliage deterioration at high density may cause cycles
	(c) Foliage depletion	Slow recovery keeps density low	Starvation, exposure to other mortality and slow recovery may cause cycles
	(d) Tree death	Keeps density low	Slow recovery—long cycle period
Vertebrate predators	(a) Rapid sigmoidal response	Can potentially keep density low	Ineffective because of weak reproductive response
	(b) Immigration (c) Prey switching (d) Large appetites		May create outbreak threshold
Invertebrate parasitoids and predators	(a) Short generation (b) High fecundity (c) Numerical response	Effective if adapted to find hosts and reproductive responses rapid	Can terminate outbreaks Can generate cycles by delayed response
Pathogens	(a) Rapid numerical response at high density	Usually ineffective because of transmission thresholds	Often terminates outbreaks, possibly prolongs reduction of host fecundity and vigor
	(b) Infection and virulence may depend on herbivore density (c) Other stresses may induce infection		May synchronize cycles

curve (Fig. 12(b)). Replacement curves are similar to the reproduction curves of Ricker (1954; see also Morris, 1963 and Holling, 1973), which are plots of density at t_1 on density at t. R-functions are identified by either observing $R_t = N_{t+1}/N_t$ or by adding the per capita density-dependent death rates and subtracting the density-dependent birth rates (Fig. 12(a)). This second method is essentially the same as that used in the multispecies models discussed below, since the relationship between density and the probability of death from vertebrate predators, invertebrates and microparasites and food shortage must be considered.

Berryman *et al.* (1987) show that in the simplest case, fast acting processes stabilize the populations. However, second-order processes characterized by time-lags arising from the impact of one generation on the reproduction and survival of the next or later generations result in cyclic behavior (Royama, 1977). In this case the plot of the replacement rate on density has an orbital trajectory. Time delays in negative feedback processes destabilize populations. Bimodal R-functions result when a variety of processes influence

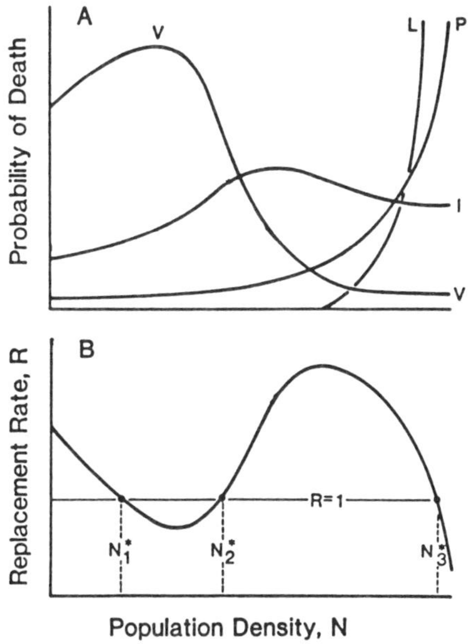

Fig. 12. (a) Hypothetical curves for the probability of death of forest Lepidoptera from vertebrate predators (V), insect parasitoids (I), food shortage (L) and viral pathogens (P). (b) The replacement function resulting from synthesis of the mortality functions and assuming average female fecundity and a reduction in the proportion of females at high density. N_1* and N_3* are stable equilibria and N_2* is an unstable equilibrium. After Campbell (1975) from Berryman *et al.* (1987).

reproduction and survival at different rates and different densities (Fig. 12). This can result in several equilibria for populations and may lead to eruptive behavior of populations characterized by long periods of low density and an occasional explosion to high density (but see below). The population release can be triggered by a density-independent variable such as improved weather (see Section V.B) or a positive feedback process. Examples of the latter might be the escape from predators or parasites (Southwood and Comins, 1976) or improved food quality with intermediate levels of herbivory (Myers and Williams, 1984; Roland and Myers, 1987).

Multispecies models are similar to the single species models, although they consider separate processes for interacting trophic levels. Rosenzweig and MacArthur (1963) used graphical representations to analyze the stability of predator–prey cycles, and May (1971) analyzed the stability properties of multispecies models. The implications of these to population stability was further considered by Holling (1973).

An example in which generalized functional and numerical responses were used to represent the qualitative properties of defoliating insect systems is the study of McNamee *et al.* (1981). These models use functions for intraspecific competition, density-dependent foliage growth, defoliation, and density-dependent forest growth. Depending on the insect system being considered, effects due to ground-searching predators, crown-searching predators, parasitism or disease, tree mortality and stochastic weather effects can be included. By categorizing the forest insect as being subjected to ground predators (pupates in the soil), or bird predators (lacks predator defenses), whether it has parasites or diseases that act on the same time-scale or lacks these (the latter applies primarily to introduced insects) and whether the defoliators kill the trees, the dynamics or the number of equilibria can be determined. Because little can be predicted about the impact of weather on particular populations dynamics of insects, this is included as a stochastic effect. From these analyses four classes of population behavior can be generated: (1) chronic endemic, (2) defoliator/parasitoid disease cycles, (3) defoliator/foliage cycles, and (4) defoliator/forest cycles (Fig. 13).

An example of three modelling approaches to the same data set concerns population cycles of the black-headed budworm. Morris (1959) concluded from his long-term study of this species that population cycles were due to insect parasitoids. Parasitism and larval density were highly correlated and populations declined before food became limiting. McNamee (1979) modelled the dynamics of this population and included effects of parasites, food depletion, weather and bird predation. The addition of bird predation causes the relationship between the rate of population change and density to be bimodal, which results in two stable equilibria (see Fig. 12(b)); an equilibrium at low density determined by bird predation, and one at high

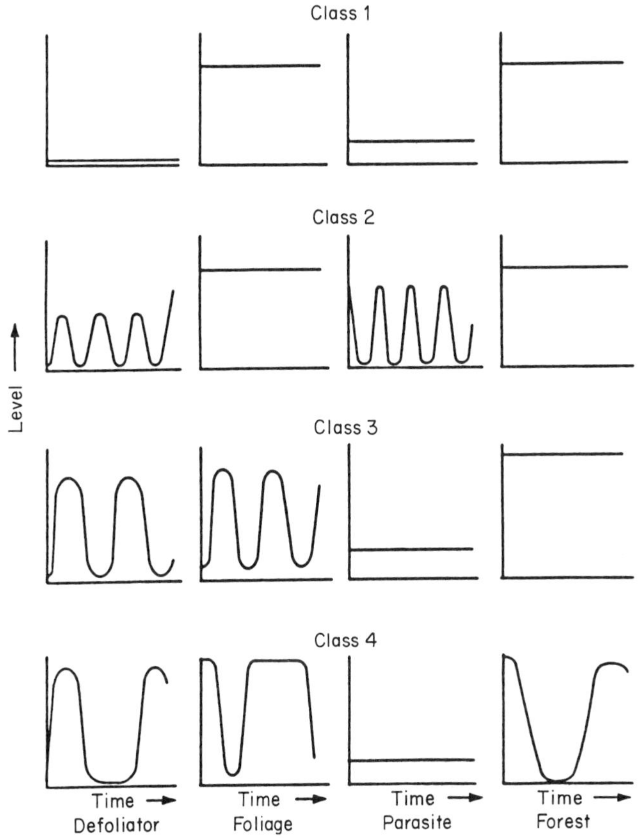

Fig. 13. Four classes of population behavior resulting from possible interactions between defoliators, foliage, parasites and forests. From McNamee *et al.* (1981).

density determined by food limitation. Including weather effects in the model did not change the qualitative behavior. Berryman (1986) also re-analyzed the Morris data. He plotted the relationship between the natural logarithm of the rate of population change and density and concluded that there was no evidence for the multimodal relationship used by McNamee (1979). Berryman fitted a time-delayed logistic equation to the data and he concluded that a one-generation delayed feedback process, such as numerical response by a parasitoid, adequately explains the relationship between density and the rate of population increase. Because this solution leads to damped cycles, a random weather variable was added which was sufficient to maintain cyclic behavior. Thus one does not know which is more realistic, a model that produces cyclic dynamics with the addition of a function for bird predation even though this is not apparent from the data, or one that produces cyclic behavior only if a random weather variable is included. Both models could produce cyclic behavior for the wrong reason.

One should be able to determine experimentally which processes to include in models of particular lepidopterous species. Perhaps the most interesting result is the similarity in dynamics that can be achieved in several ways. The multispecies models show that cycles caused by parasites (Class 2) may be expected to have slightly shorter periodicities than those caused by depletion of food (Class 3) (McNamee *et al.*, 1981), but these differences are not apparent in the data summarized in Tables 1 and 2. This is perhaps because for any given species, some outbreaks are terminated by parasites and others by food depletion, and the average periodicity includes longer and shorter outbreaks. As described previously, populations of tussock moth in the same general geographical area have been observed to decline synchronously, even though some populations defoliated trees and others did not. Thus in this situation Class 2 and Class 3 behaviors are indistinguishable. If the classes of behavior cannot be identified from density data, they will only be determined by the observation of the prevalence of parasites or the occurrence of foliage depletion. The outcome of the models is determined by the processes they incorporate; they do not produce testable predictions.

While models show what might happen to populations, they cannot show what does happen. Models will be most useful if they are used as part of an experimental program. For example, spraying with *Bacillus thuringiensis*, a bioinsecticide that does not kill predaceous and parasitic insects, could be used to determine if density suppression can push populations into a lower equilibrium or if protecting foliage from herbivore damage can prolong the increase in the density of the insect herbivores toward an upper equilibrium.

Finally, of interest to the field ecologist is whether populations having multiple equilibria could be recognized from field data. Populations that fluctuate every 8–10 generations, with outbreaks lasting 2–3 generations and increase and decline transitions lasting 2–4 generations each, cannot remain at a low density equilibrium for long. Ludwig (personal communication) has shown that a model of spruce budworm dynamics that includes a lower density equilibrium determined by bird predation still produces a continously changing population trend when viewed as a plot of log (density) against time (Fig. 14). There is not a rapid transition or release from a lower threshold to an upper one, but rather a slight shoulder on the increase trend. Other processes that allow low density populations to increase override the theoretical equilibrium caused by bird predation, and both the lower equilibrium and the unstable outbreak equilibrium only slightly below the population 'eruption'. This trend should be compared with Fig. 8, which plots density data for spruce budworm. The phase of peak density is much

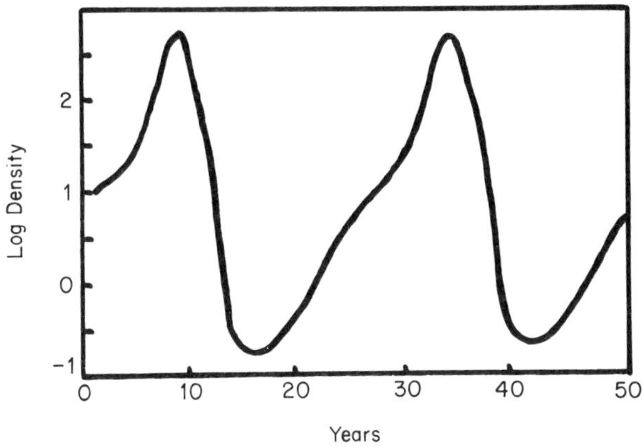

Fig. 14. Population trace based on a model of eastern spruce budworm and incorporating a function for bird predation that causes an unstable equilibrium or outbreak threshold at intermediate densities (Ludwig, personal communication).

longer in the population data than in the population trend arising from the model, which suggests that some aspect of the dynamics has not been captured by the model.

To conclude, even the simplest models of population change produce cyclic dynamics when time-lags are incorporated. A variety of biologically relevant processes can produce such time-lags and are possible causes of cycles of forest Lepidoptera. The multiple equilibrium view of population cycles, while theoretically valid, has little relevance to field populations. To my knowledge, no experimental tests of predictions of population models of forest Lepidoptera have been done, and testable predictions are rarely made from models, even though this should be their primary function.

VI. EVALUATION OF HYPOTHESES

A dichotomy has been made historically between self-regulation of animal populations and regulation by extrinsic causes such as predation, food limitation or weather. Self-regulation is the reduction of numbers prior to over-exploitation of the habitat through genetic or physiological change of individuals in the population. It is difficult to imagine how density alone would select among individual insects in which social behaviour is not well developed. Attributes associated with high density, such as food limitation or disease transmission, could select for a change in the "quality" of individuals composing populations. If the genetic or physiological response is to an extrinsic component of high density, self-regulation *per se* is not in operation (Haukioja *et al.*, 1983).

High parasitization is frequently recorded at high host densities, but because parasitoids do not track host densities during the decline, they do not explain the continued population reduction. Improved food quality might be loosely associated with population increase, but variation in the initiation and rate of population increase among areas and the relative synchrony of peak populations in areas with different qualities of food plant (Myers and Williams, 1987) suggest this is not a general explanation of population outbreaks. Reduced plant quality occurs following defoliation by larch budmoth in areas of Switzerland, but more appropriate field studies are necessary to determine the generality of this phenomenon to other areas and other species. Populations of forest Lepidoptera frequently do not defoliate trees prior to decline and populations decline from different peak densities. Therefore, a variable response between insect damage and reduction in food quality is necessary for this mechanism to have general applicability.

The disease susceptibility hypothesis has been previously rejected because few individuals are recorded as having died of disease in declining populations. However, while selection for resistance could occur even when few individuals in total die of disease, little contamination of the environment would occur and transmission of virus would be slow. Latent viruses might be relevant to the maintenance of disease at low density, but this requires further investigation. Sublethal disease could explain observed behavioral changes in populations and reduced fecundity during the population decline. Disease could account for the more or less synchronous declines of populations established out of phase, particularly in new areas invaded by immigration from adjacent high density populations. Interactions between different types of disease such as NPV and small RNA viruses might be important to the dynamics of populations, but this has never been studied.

What is most fascinating is the similar periodicity of outbreaks shown by a number of species of forest Lepidoptera. Surely this is a general phenomenon, although variation in peak densities, amounts of defoliation, species of parasites and patterns of spread make one wonder how generality can exist. Models show that different processes may produce very similar dynamics if they prolong the decline phase by one or two generations. Therefore, periodicities might be expected to be similar for these species of forest Lepidoptera which have a single generation a year. It is interesting that in Japan, where fall webworm has two generations a year, cycles maintain the eight generation length (Ito, 1977).

There is an unfortunate lack of experimental investigation of population cycles of forest Lepidoptera. It is regrettable that spray programs have not yielded more information. In the initial studies of eastern spruce budworm, populations sprayed with DDT declined at the same time as unsprayed populations (Morris, 1963). Seven years of spraying with DDT during the 1958 outbreak of spruce budworm in northern New Brunswick prevented extensive tree mortality, but budworm populations declined at the same time in sprayed and unsprayed areas (Webb *et al.*, 1961). Tree mortality is not necessary for decline of spruce budworm. Although extensive insecticide spraying occurred in many, but not all, areas of eastern Canada in the late 1970s and early 1980s, populations of spruce budworm have declined in all provinces from Ontario to New Foundland in the last few years (Sanders *et al.*, 1985; and personal communications from a number of forest entomologists).

Spraying larch budmoth populations with *Bacillus thuringiensis* at peak density and low density did not substantially modify the population fluctuation (Auer *et al.*, 1981; Baltensweiler *et al.*, 1977). Introducing 35 000 larch budmoth pupae to an area two years after population decline initiated the phase of increase one year prior to that of control populations, but densities peaked on both the experimental and two control areas at the same time. This shows that the delayed recovery of control populations cannot be completely explained by poor food quality; the forest could support an increasing population in the third year after peak defoliation. Cropping populations of tent caterpillars and introducing egg masses during the population decline did not modify the dynamics of two island populations (Myers, 1981).

Two excellent experimental studies have been done using sprays of NPV on tussock moth populations in British Columbia, Canada. Spraying populations at peak densities caused an earlier epizootic and protected foliage of the attacked Douglas-fir trees (Otvos *et al.*, 1987a,b). Viral deaths were significantly more common on sprayed plots early in the subsequent summer, but by late summer viral deaths were equally common on both control and sprayed plots. Populations in both sprayed and unsprayed areas declined over the next two years. In another study (Shepherd *et al.*, 1984), tussock moths were treated early in the outbreak (1981) with aerial and ground spraying of NPV. Measured densities declined to zero by the next year in six experimental plots and two of three control plots to which virus may have spread. Mean densities at ten other sites peaked in 1982 and declined through 1983, associated with increased incidence of naturally occurring disease. Resurgence of the populations did not occur on treated plots even though foliage was protected.

These experimental manipulations show that it is difficult to perturb the basic dynamics of population fluctuations, although spraying with virus early in the outbreak terminated that phase in tussock moth populations. The resilience of population dynamics to manipulation may be because immigration to manipulated areas swamps experimental effects. But spraying, cropping and introduction experiments show that preventing defoliation or protecting foliage do not affect the population dynamics as would be expected if deterioration of food quality induced by herbivore damage were the driving mechanism behind fluctuations. Disease epizootics have been initiated through spraying early in the peak of population density of tussock moth; therefore, populations which have recently increased are susceptible to virus.

Predator exclusion experiments have been carried out on eastern spruce budworm, western spruce budworm (reviewed in Sanders *et al.*, 1985) and tussock moth. Torgersen (1985) enclosed whole trees in four areas to exclude bird predators and prevented ants from climbing trees with sticky barriers that had little effect on caterpillar dispersal. Birds and/or ants reduced adult densities by ten or fifteen times in low density populations, but had much less effect on high density populations. Predator exclusion experiments showed that invertebrate and vertebrate predators cause major mortality of tussock moth as well (Mason *et al.*, 1983; Mason and Torgersen, 1983). Predator exclusions have not been done over a sufficiently long time to measure the impact of predators on cyclic dynamics. While predators can reduce low density populations, it is not known if this prolongs the decline and slows the initial increase. It is interesting that some species, like the western tent caterpillar, are largely resistant to bird predators. While ant predation of eastern tent caterpillars has been observed (Tilman, 1978), it is not a major component of the mortality of western tent caterpillars (J. H. Myers, personal observation). However, population fluctuations of western tent caterpillar are very similar to those of tussock moth, where bird and ant predation can reduce density.

To conclude, experiments show that bird, spider, and insect predation cause high mortality to some populations of Lepidoptera with cyclic population dynamics. However, there is no evidence that predation is an integral part of the mechanisms causing cyclicity. The most important conclusion from the experimental manipulations of populations of forest Lepidoptera is that they are very resilient. In some cases outbreaks have been suppressed by spraying once with NPV, or continued spraying with insecticides, but experimental manipulations have not changed cyclic populations into stable populations. More large-scale and long-term manipulations should replace long-term quantification of mortality that has dominated forest entomology for so long.

VII. POPULATION CYCLES OF OTHER ORGANISMS

Population cycles range from the single generation cycles of *Daphnia* (McCauley and Murdoch, 1987) to 10-year cycles of snowshoe hares, grouse and associated predators (Keith, 1963). It is interesting that population fluctuations of plankton, small mammals and birds are associated with changes in reproductive attributes of individuals at high density, and reduced reproduction could provide the necessary delayed density-dependent factor to cause cyclic population dynamics. For example, suppression of reproduction at high density of *Daphnia* can occur in both laboratory populations with controlled food sources and in field populations with variable food sources. Suppressed reproduction results in the dominance of one cohort, which continues until the population declines, after which another cohort is produced. Stable populations also have been observed, and these have constant age distributions. The mechanisms causing reproductive suppression are not understood (McCauley and Murdoch, 1987).

Self-regulation through behavioral interactions has received more support from studies of small mammals for which dominance and spacing behavior can determine the impact of extrinsic factors such as predation and food availability, as well as intrinsic factors such as growth, reproduction and dispersal (Taitt and Krebs, 1985). Peak populations are characterized by large individuals, and the age at first reproduction is delayed in the smaller individuals characteristic of declining populations (Krebs and Myers, 1974). Whether these changes are genetically or environmentally induced is still not known (Taitt and Krebs, 1985), but maternal effects are important (Boonstra and Boag, 1987). A relationship between microparasites and population fluctuations of *Microtus* was hypothesized in one recent study (Mihok *et al.*, 1985).

Reproductive periods of peak and declining populations of snowshoe hares are shorter because females produce fewer litters (review in Krebs *et al.*, 1986). This does not seem to be influenced by winter food availability, but the role of summer food has not been studied (C. J. Krebs, personal communication). In addition, continued pressure from predators as the population declines is thought to influence the amplitude and periodicity of the snowshoe hare cycles (Sinclair *et al.*, 1988).

Red grouse populations in north-east Scotland fluctuate with a periodicity of about six years (Moss and Watson, 1985). Emigration of female grouse in the summer occurs in declining populations and greatly reduces the number of hens. Low recruitment for several years following peak density occurs from poor egg and chick survival, the high sex ratio resulting in unmated males and summer emigration of females. These characteristics could provide the necessary delayed density-dependent factor to cause cyclic dynamics, but it is not understood what modifies the migration behavior of the hens. The burden of nematode worms carried by grouse varies with density with a lag of one to two years in some fluctuating populations, particularly in northern England (Hudson *et al.*, 1985), but this is not a general characteristic (Moss and Watson, 1985). However, parasites could influence behavior.

Dragonfly migrations occurring several years in sequence have been observed in Europe to have a periodicity of ten years (Dumont and Hinnekint, 1973). These authors speculate that migrations are associated with a build-up of metacercariae of trematode worms in the dragonflies. As dragonfly density builds, so do infection levels. At intermediate levels of parasite infection, irritation could result in migration. Higher levels of parasite infection cause lethargy and death. Decline in host density causes a collapse in the parasite population, with a lag of one to two years. This is an example of a parasite having both lethal and non-lethal effects on the host that modify behavior and reproduction.

This overview of cyclic population dynamics of several species of vertebrates and invertebrates is superficial by necessity. However, it shows a common feature of population cycles; reproduction is reduced at peak density and into the population decline. Predation and parasitization related to density may account for high mortality during peak and declining populations, but they cannot be directly responsible for the suppressed, delayed or shortened breeding season observed in these studies. It is tempting to compare the occurence of large voles that are characteristic of high density populations to the large egg masses found in tent caterpillar populations at peak density. Disease could be associated with changes in reproduction and behavior observed in cyclic populations of other organisms besides Lepidoptera.

VIII. CONCLUSIONS AND SPECULATIONS

The understanding of population cycles has been hindered by a number of factors. Examples are the focus on observation rather than on experimentation, on single populations rather than on comparative data, on single mechanisms, on the collection of data for life table analysis, and on the failure to outline the details and patterns of the changes in population density and characteristics of individuals to be explained. Insects with high fertility (200+ eggs per female) will have high mortality even in increasing populations, and any study of just this aspect is doomed to failure. Declines of populations might be caused by a variety of different agents. If we are to find generality, it will be in identifying whether some agent reduces populations when everything else fails.

Predators appear to have their greatest impact on Lepidoptera when populations are at low density. Parasitoids vary from population to population, but tend to increase with host density. However, parasitization is frequently discounted as causing cyclic dynamics of their hosts because they rapidly disappear as host density declines; they cannot be responsible for continuing the decline. In many life table studies, unexplained losses are strongly correlated to changes in population density and account for most of the total mortality. Therefore parasitoids and predators kill many hosts, but they appear to be insufficient to explain cycles of forest Lepidoptera. A prolonged reduction in fecundity and/or poor survival or vigor for several years following peak density appears to be necessary to explain cyclic dynamics.

Disease is usually rejected as a cause of insect cycles because it is only observed in a small proportion of the population if at all. This may not be surprising. If individuals dying of disease rapidly disintegrate, they would not be readily observed in field studies. However, I think disease is a likely explanation for continued declines in population density over several generations. In Table 6 I have listed characteristics that have been found in some studies to be associated with population cycles of forest Lepidoptera. I certainly cannot claim that these characteristics have always been observed, but then, depending on the methods used and the characteristics of the insects and trees involved, I would not expect them to have always been observed. Rather, they are proposed as guidelines for characteristics to look

Table 6

Characteristics of population cycles of forest Lepidoptera that have been found in many studies. These should be investigated in future analyses in an attempt to determine which are consistent attributes of population cycles.

(1) Increase phase varies in year of initiation and rate of increase among subpopulations.

(2) Fecundity may be highest in the generation prior to that in which the decline begins.

(3) The initial decline may start in late life-history stages, with mortality increasing in the early instars of the next generation.

(4) Survival may be poorer for larvae collected from field populations at the beginning of the decline phase, and reared in the laboratory, than for larvae collected during the increase phase.

(5) Fecundity may be reduced for several generations during which the population continues to decline.

(6) Dispersal and high larval activity and migration may characterize populations at peak density.

(7) Disease or microparasites may be widespread among subpopulations, although not necessarily observed to kill a high proportion of individuals.

(8) Populations may decline without reaching densities sufficient to defoliate their host trees.

for in future analyses. For example, in conversations one discovers that often laboratory experiments planned to be done using caterpillars from high density populations fail because the insects die of disease even though the previous year survival was good. This information is rarely recorded because the outbreak of disease is blamed on the laboratory technician rather than the population phase of the insect (or snowshoe hare).

I speculate that the 8–10 year population cycles of forest Lepidoptera are caused by prolonged effects of disease or microparasites reducing the vigor and fecundity of populations for several generations after the beginning of the population decline. I propose that disease builds following interactions among individuals in localities where population increases begin, and spreads with immigration to other areas so that populations slightly out of phase have similar compositions in terms of the health and quality of individuals. Transition from latent to active disease forms may be associated with the deterioration in the health of individuals, and the effects of the environment, weather and food, and behavior of insects might be modified by the presence of disease.

Testing the proposed relationship between disease and population cycles will not be easy. In addition to determining how robust the characteristics outlined in Table 6 are, other testable predictions can be made: (1) larvae collected from field populations and reared in the laboratory should have poor survival when collected from declining populations, (2) spraying populations with virus early in the outbreak phase should terminate the outbreak while spraying with insecticide should not, (3) introduction of individuals from declining populations to areas of suitable habitat that has been free of larvae should not cause immediate outbreak, and (4) protecting trees from insect herbivores should not prolong the outbreak at a particular site once this manipulation is stopped.

An advantage of this proposed interaction between insects and disease is that it helps to explain the resilience of populations to perturbations. It may explain why protecting foliage does not prevent population decline and why introducing insects from a declining population does not immediately result in population increase. A variety of different microparasites such as microsporidia, viruses or bacteria could have similar influences on insects. What is necessary is a lag in the return to normal vigor and fecundity that lasts several generations. Spraying early outbreak populations of tussock moth with virus is the only example I know of in which the dynamics of a population cycle have been experimentally manipulated, resulting in a premature and permanent decline of the population. This suggests that changing the quality of the insects through disease may be a prerequisite for continued population suppression. This area deserves more intensive research.

The take-home message from this analysis is that quantifying mortality will be insufficient for understanding population cycles because measurement errors are too large. Previous hypotheses are not consistent with observed characteristics of population cycles. Disease has been frequently overlooked and resistance to disease has only rarely been studied. I may simply be emphasizing the importance of disease because it is the least known factor in studies of insect cycles. But better studied interactions fail to explain the phenomenon, and I think disease may provide a missing link. I also recommend that work should emphasize the phase of population decline, since the increase phase is variable among populations and may be initiated by a number of different conditions.

Even with a complete understanding of population cycles of forest Lepidoptera, we will not necessarily be able to control populations to economically acceptable levels. Rather, modifications in the forest industry, which may include different patterns of land use to allow the most susceptible areas to be logged first, changes in tree species used in reforestation, and reductions in our economic expectations, will have to occur. After all, most of these species are well adapted to periodically reach high densities without destroying their food plant species. The Achilles heel is a human trait—not an insect one.

ACKNOWLEDGEMENTS

I would like to dedicate this paper to my two colleagues, Dennis Chitty and William Wellington, who have been very influential to my studies and interpretations of animal population cycles. I appreciate the discussions, reprints and preprints, and hospitality provided to me by Werner Baltensweiler and Erkki Haukioja. Their work has made very important contributions to the interpretation of insect population cycles. Charles Krebs contributed through discussions and in reading the manuscript, and Robert Mason provided reprints and commented on the manuscript when it was in an embryonic stage, for which I am grateful. This manuscript was begun when I was on sabbatical leave at the Institute of Virology, NERC, Oxford and I appreciate the discussions and opportunity to be surrounded by insect virology there. In particular I wish to thank Philip Entwistle and Jenny Cory. In addition, I thank Chris Perrins of the Edward Grey Institute, Oxford University, for providing office space and facilities. Alan Berryman and Peter Martinat kindly provided me with preprints and many others provided reprints, for which I am grateful. Jamie Smith helped through division of labor on parental duties, but unfortunately this caused time contraints that precluded his red pen approach to this paper. My research has been supported by the Natural Science and Engineering Research Council of Canada, to which I am also grateful.

REFERENCES

Anderson, R. M. (1982). Theoretical basis for the use of pathogens as biological control agents of pest species. *Parasitology* **84**, 3–33.

Anderson, R. M. and May, R. M. (1978). Regulation and stability of host–parasite population interactions. I. Regulatory processes. *J. Anim. Ecol.* **47**, 219–249

Anderson, R. M. and May, R. M. (1980). Infectious diseases and population cycles of forest insects. *Science* **210**, 658–661.

Andrewartha, H. G. and Birch, L. C. (1954). *The Distribution and Abundance of Animals*. Chicago, Illinois: University of Chicago Press.

Auer, C., Roques, A., Goussard, F. and Charles, P. J. 1981. Effets de l'accroissement provoque du niveau de population de la tordeuse du meleze *Zeiraphera diniana* guenee (Lep., Tortricidae) au cours de la phase de regression dans un massif forestier du Brianconnais. *Z. angew. Entomol.* **92**, 286–302.

Baltensweiler, W. (1968). The cyclic population dynamics of the grey larch tortrix, *Zeiraphera griseana* Hubner (= *Semasia diniana* Guenee) (Lepidoptera, Tortricidae). In: *Insect Abundance* (Ed. by T. R. E. Southwood), *Symp. Roy. Entomol. Soc. London* **4**, 88–98.

Baltensweiler, W. (1978). Insect population dynamics. *Proc. XVI. IUFRO World Congress, Norway 1976*, Div. 2.

Baltensweiler, W. (1984). The role of environment and reproduction in the population dynamics of the larch bud moth, *Zeiraphera diniana* Gn. (Lepidoptera, Tortricidae). *Adv. Invert. Reprod.* **3**, 291–302.

Baltensweiler, W. (1985). "Waldsterben": forest pests and air pollution. *Z. angew. Entomol.* **99**, 77–85.

Baltensweiler, W. and Fischlin, A. (1988). The Larch bud moth in the European Alps. In: *Dynamics of Forest Insect Populations: Patterns, Causes and Management Strategies* (Ed. by A. Berryman) pp. 331–351. New York: Academic Press.

Baltensweiler, W., Benz, G., Bovey, P. and Deluchi, V. (1977). Dynamics of larch bud moth populations. *Ann. Rev. Entomol.* **22**, 79–100.

Barbosa, P. and Baltensweiler, W. (1987). Phenotypic plasticity and herbivore outbreaks. In: *Insect Outbreaks* (Ed. by P. Barbosa and J. Schultz), pp. 469–504. New York: Academic Press.

Bellows, T. S. (1981). The descriptive properties of some models for density dependence. *J. Anim. Ecol.* **50**, 139–156.

Begon, M., Harper, J. L. and Townsend, C. R. (1986). *Ecology: Individuals, Populations, and Communities*. Sunderland, MA: Sinauer Associates, 876 pp.

Benz, G. (1974). Negative Ruckkoppelung durch Raum- und Nahrungskonkurrenz sowie syklische Veranderung der Nahrungsgrundlage als Regelprinzip in der Populationsdynamik des Grauen Larchensicklers, *Zeiraphera diniana* (Guenee) (Lep. Tortricidae). *Z. angew. Entomol.* **76**, 31–49.

Benz, G. (1977). Insect reduced resistance as a means of self defence of plants. Report of the 1st Meeting of the EUCARPIA/IOBC Working Group Breeding for Resistance to Insects and Mites. *WPRS Bull.* **3**, 155–159.

Benz, G. (1987). Environment. In: *Epizootiology of Insect Diseases* (Ed. by J. R. Fuxa and Y. Tanada), pp. 177–214. New York: John Wiley.

Berryman, A. A. (1986). On the dynamics of blackheaded budworm populations. *Can. Ent.* **118**, 775–779.

Berryman, A. A. (1987). The theory and classification of outbreaks. In: *Insect Outbreaks* (Ed. by P. Barbosa and J. C. Schultz), pp. 3–30. New York: Academic Press.

Berryman, A. A., Stenseth, N. C. and Isaev, A. S. (1987). Natural regulation of herbivorous forest insect populations. *Oecologia (Berlin)* **71**, 174–184.

Blais, J. R. (1985). The spruce budworm and the forest. In: *Recent Advances in Spruce Budworms Research* (Ed. by C. J. Sanders, R. W. Stark, E. J. Mullins and J. Murphy), pp. 135–136. Ottawa: Canadian Forest Service.

Briese, D. and Mende, H. A. (1983). Selection for increased resistance to a granulosis virus in the potato moth, *Plethorimaea operculella* Zeller (Lepidoptera: Gelechiidae). *Bull. ent. Res.* **71**, 1–9.

Briese, D. T. and Podgwaite, J. D. (1985). Development of viral resistance in insect populations. In: *Viral Insecticides for Biological Control* (Ed. by K. Maramorosch and K. Sherman), pp. 81–120. New York: Academic Press.

Boonstra, R. and Boag, D. (1987). A test of the Chitty hypothesis; inheritance of life history traits in meadow voles *microtus pennsylvanicus*. *Evolution* **41**, 929–947.

Campbell, R. W. (1975). The gypsy moth and its natural enemies. *U.S.D.A. For. Serv. Agric. Info. Bull.* **381** (cited in Berryman *et al.*, 1987).

Campbell, R. W. (1976). Comparative analysis of both numerically stable and violently fluctuating gypsy moth populations. *Environ. Entomol.* **5**, 1218–1224.

Campbell, R. W. and Torgersen, T. R. (1983). Compensatory mortality in defoliator population dynamics. *Environ. Entomol.* **12**, 630–632.

Campbell, R. W., Hubbard, D. L. and Sloan, R. J. (1975). Location of gypsy moth pupae and subsequent pupal survival in sparse, stable populations. *Environ. Entomol.* **4**, 597–600.

Campbell, R. W., Sloan, R. J. and Biazak, C. E. (1977). Sources of mortality among late instar gypsy moth larvae in sparse populations. *Environ. Entomol.* **6**, 865–871.

Canadian Forest Service (1980). Operational field trials against the Douglas-fir tussock moth with chemical and biological insecticides BC-X-201, Victoria, B.C.

Carroll, W. (1954). History of the hemlock looper, *Lambdina fiscellaria*, in Newfoundland and notes on its biology. *Can. Entomol.* **88**, 587–599.

Chitty, D. (1967). The natural selection of self-regulatory behaviour in animal populations. *Proc. ecol. Soc. Aust.* **2**, 51–78.

Cole, W. E. (1971). Pine butterfly. *U.S. Dept. Agr. For. Serv. For. Pest Leafl.* No. 66, 3 pp.

Courtney, S. P. (1984). The evolution of egg clustering by butterflies and other insects. *Am. Nat.* **123**, 276–281.

Cunningham, J. (1982). Field trials with baculoviruses: control of forest insect pests. In: *Microbial and Viral Pesticides* (Ed. by E. Kurstak), pp. 335–386. New York: Marcel Dekker.

Cunningham, W. J. (1954). A nonlinear differential-difference equation of growth. *Proc. Nat. Acad. Sci. USA* **40**, 708–713.

David, W. A. L. and Taylor, C. E. (1977). The effect of sucrose content of diets on susceptibility to granulosis virus disease in *Pieris brassicae*. *J. Invert. Pathol.* **30**, 117–118.

Dawson, A. F. (1970). Green striped forest looper in British Columbia. *Can. For. Serv. For. Ins. Dis. Surv. For. Pest Leafl.* No. 22, 5 pp (cited in McNamee, 1987).

Dempster, J. P. (1975). *Animal Population Ecology*. New York: Academic Press.

Dempster, J. P. (1983). The natural control of populations of butterflies and moths. *Biol. Rev.* **58**, 461–481.

Dumont, H. J. and Hinnekint, B. O. N. (1973). Mass migration in dragonflies, especially in *Libellula quadrimaculata* L: A review, a new ecological approach and a new hypothesis. *Odonatologica* **2**, 1–20.

Embree, D. G. (1966). The role of introduced parasites in the control of the winter moth in Nova Scotia. *Can. Entomol.* **102**, 759–768.

Embree, D. G. (1971). The biological control of winter moth in eastern Canada by introduced parasites. In: *Biological Control* (Ed. by C. B. Huffaker), pp. 217–226. New York: Plenum Press.

Entwistle, P. F. (1983). Control of insects by virus diseases. *Biocontrol News Inf.* **4**, 202–228. Commonwealth Institute of Biological Control.

Entwistle, P. F. (1986). Epizootiology and strategies of microbial control. In: *Biological Plant and Health Protection*, pp. 257–278. Stuttgart: G. Fischer Verlag.

Entwistle, P. F. and Evans, H. F. (1985). Viral Control. In: *Comprehensive Insect Physiology, Biochemistry, and Pharmacology* (Ed. by L. I. Gilbert and G. A. Kerkut), pp. 347–412. Oxford: Pergamon.

Evans, H. F. and Allaway, G. P. (1983). The influence of larval maturation on responses of *Mamestra brassicae* L. (Lepidoptera: Noctuidae) larval populations introduced into small cabbage plots. *Appl. environ. Microbiol.* **45**, 493–500.

Evans, H. F. and Entwistle, P. F. (1987). Viral diseases. In: *Epizootiology of Insect Diseases* (Ed. by J. Fuxa and Y. Tanada), pp. 257–322. New York: John Wiley.

Faeth, S. (1987). Community structure and folivorous insect outbreaks: The roles of vertical and horizontal interactions. In: *Insect Outbreaks* (Ed. by P. Barbosa and J. Schultz), pp. 135–172. New York: Academic Press.

Fischlin, A. (1982). Analyse eines Wald-Insekten-Systems. Der subalpine Larchen-Arvenwald und der graue Larchenwickler *Zeiraphera diniana* Gn. (Lep.,Tortricidae). Diss. ETH Nr. 6977. ADAG Administration & Druck AG, Zurich.

Fleming, R. (1985). Infectious diseases as part of the group of mortality factors affecting eastern spruce budworm population dynamics. In: *Recent Advances in Spruce Budworms Research* (Ed. by C. J. Sanders, R. W. Stark, E. J. Mullins and J. Murphy), pp. 110–111. Ottawa: Canadian Forestry Service.

Fowler, S. V. and Lawton, J. H. (1985). Rapidly induced defences and talking trees: the devil's advocate position. *Am. Nat.* **126**, 181–195.

Furniss, R. L. and Carolin, V. M. (1977). Western Forest Insects. *U.S. Dept. Agr. For. Serv. Misc. Publ. No. 1339*, 654 pp. (cited in McNamee, 1987).

Getz, W. M. and Pickering, J. (1983). Epidemic models: thresholds and population regulation. *Am. Nat.* **121**, 893–898.

Graham, S. A. (1939). Forest insect populations. *Ecol. Mongr.* **9**, 301–310.

Greenbank, D. O. (1963). Climate and the spruce budworm. *Mem. entomol. Soc. Can.* **31**, 174–180.

Gruys, P. (1970). Growth in *Bupalus piniarius* (Lepidoptera, Geometridae) in relation to larval population density. *Verh. Rijksinst. Natuurbeheer* **1**, 1–127.

Harris, J. W. E. and Dawson, A. F. (1985). Parasitoids of the western spruce budworm in British Columbia. In: *Recent Advances in Spruce Budworms Research* (Ed. by C. J. Sanders, R. W. Stark, E. J. Mullins, and J. Murphy), pp. 102–103. Ottawa: Canadian Forestry Service.

Harris, J. W. E., Dawson, A. F. and Brown, R. G. (1982). *The Western Hemlock Looper in British Columbia 1911–1980*. Information Report BC-X-234, Canadian Forestry Service, Pacific Forest Research Centre, Victoria, B.C. Canada. 18 pp.

Harris, J. W. E., Alfaro, R. I., Dawson, A. F. and Brown, R. G. (1985a). *The Western Spruce Budworm in British Columbia 1909–1983*. Information Report BC-X-257, Canadian Forestry Service, Pacific Forest Research Centre, Victoria, B.C., Canada. 32 pp.

Harris, J. W. E., Dawson, A. F. and Brown, R. G. (1985b). *The Western False Hemlock Looper in British Columbia 1942–1984*. Information Report BC-X-269, Canadian Forestry Service, Pacific Forest Research Centre, Victoria, B.C., Canada. 11 pp.

Harris, J. W. E. and Brown, C. E. (1976). *Data Recording and Retrieval Manual*. Canadian Forestry Service, Pacific Forest Research Centre, Victoria, B.C., Canada (cited in McNamee, 1987).

Harrison, S. (1987). Treefall gaps versus forest understory as environments for a defoliating moth on a tropical forest shrub. *Oecologia* **72**, 65–68.

Hassell, M. (1978). *The Dynamics of Arthropod Predator–Prey Systems*. New Jersey: Princeton University Press. 237 pp.

Hassell, M. (1985). Insect natural enemies as regulating factors. *J. Anim. Ecol.* **54**, 323–334.

Hassell, M., Lawton, J. H. and May, R. M. (1976). Patterns of dynamical behaviour in single-species populations. *J. Anim. Ecol.* **45**, 471–488.

Haukioja, E. (1980). On the role of plant defences in the fluctuations of herbivore populations. *Oikos* **35**, 202–213.

Haukioja, E. and Hakala, T. (1975). Herbivore cycles and periodic outbreaks. Formulation of a general hypothesis. *Rep. Kevo Subarctic Res. Stat.* **12**, 1–9.

Haukioja, E. and Hanhimaki, S. (1985). Rapid wound-induced resistance in white birch (*Betula pubescens*) foliage to the geometrid *Epirrita autumnata*, a comparison of trees and moths within and outside the outbreak range of the moths. *Oecologia (Berlin)* **65**, 223–228.

Haukioja, E. and Neuvonen, S. (1987). Insect population dynamics and induction of plant resistance: The testing of hypotheses. In: *Insect Outbreaks* (Ed. by P. Barbosa and J. Schultz), pp. 411–432. New York: Academic Press.

Haukioja, E., Kapiainen, K., Niemela, P. and Tuomi, J. (1983). Plant availability hypothesis and other explanations of herbivore cycles: complementary or exclusive alternatives? *Oikos* **40**, 419–432.

Haukioja, E., Neuvonen, S., Hanhimaki, S. and Niemela, P. (1988). The autumnal moth in Fennoscandia. In: *Dynamics of Forest Insect Populations: Patterns, Causes, and Management Strategies* (Ed. by A. A. Berryman), pp. 163–177. New York: Plenum Press.

Hebert, P. D. N. (1983). Egg dispersal patterns and adult feeding behaviour in the Lepidoptera. *Can. Entomol.* **115**, 1477–1481.

Herberg, M. (1960). Drei Jahrzehnte Vogelhege zur Niederhaltung waldschadlicher Insekten durch die Ansiedlung von Hohlenbrutern. *Archiv Forstwesen* **9**, 1015–1048 (cited in Price, 1987).

Hodson, A. C. (1941). An ecological study of the forest tent caterpillar, *Malacosoma disstria* Hbn., in northern Minnesota. *Tech. Bull. Minn. agric. Exp. Sta.* **148**, 1–55.

Holling, C. S. (1973). Resilience and stability of ecological systems. *Ann. Rev. ecol. Syst.* **4**, 1–24.

Hostetter, D. L. and Bell, M. R. (1985). Natural dispersal of baculoviruses in the environment. In: *Viral Insecticides for Biological Control* (Ed. by K. Maramorosh and K. E. Sherman), pp. 249–284. New York: Academic Press.

Hudson, P. J., Dobson, A. P. and Newborn, D. (1985). Cyclic and non-cyclic populations of red grouse, a role for parasitism? In: *Ecology and Genetics of Host–Parasite Interactions* (Ed. by D. Rollinson and R. M. Anderson). New York: Academic Press.

Hutchinson, G. E. (1948). Circular causal systems in ecology. *Ann. N. Y. Acad. Sci.* **50**, 221–246.

Ito, Y. (1977). Birth and death. In: *Adaptation and Speciation in the Fall Webworm* (Ed. by T. Hidaka), pp. 101–128. Tokyo: Kodansha Ltd.

Jones, R. E., Nealis, V. G., Ives, P. M. and Scheermeyer, E. (1987). Seasonal and spatial variation in juvenile survival of the cabbage butterfly *Pieris rapae*: evidence for patchy density-dependence. *J. Anim. Ecol.* **56**, 723–737.

Katagiri, K. (1981). Pest control by cytoplasmic polyhedrosis viruses. In: *Microbial Control of Pests and Plant Diseases 1970–1980* (Ed. by H. D. Burges), pp. 433–440. London: Academic Press.

Kaupp, W. J. and Sohi, S. S. (1985). The role of viruses in the ecosystem. In: *Viral Insecticides for Biological Control* (Ed. by K. Maramorosh and K. E. Sherman), pp. 441–465. New York: Academic Press.

Keating, S. T. and Yendol, W. G. (1987). Influence of selected host plants on gypsy moth (Lepidoptera: Lymantriidae) larval mortality caused by baculovirus. *Environ. Entomol.* **16**, 459–462.

Keith, L. B. (1963). *Wildlife's Ten-Year Cycle.* Madison: University of Wisconsin Press.

Keremidchiev, M. T. (1972). Dynamics of outbreaks of the Gypsy moth (*Lymantria dispar* L.) in the People's Republic of Bulgaria. *Proc. 13th Int. Congr. Entomol. 1968* **3**, 51–54. Nauk. Leningrad (cited in Entwistle and Evans, 1985).

Klein, W. H. and Minnock, M. W. (1971). On the occurrence and biology of *Nepytia freemani* (Lepidoptera: Geometridae) in Utah. *Can. Entomol.* **103**, 119–124.

Klomp, H. (1966). The dynamics of a field population of the pine looper, *Bupalus piniarius* L. (Lep. Geom.). *Adv. ecol. Res.* **3**, 207–305.

Knapp, R. and Casey, T. M. (1986). Thermal ecology, behavior, and growth of gypsy moth and eastern tent caterpillars. *Ecology* **67**, 598–608.

Krebs, C. J. (1985). *Ecology, The Experimental Analysis of Distribution and Abundance.* New York: Harper and Row. 800 pp.

Krebs, C. J. and Myers, J. H. (1974). Population cycles in small mammals. *Adv. ecol. Res.* **8**, 267–399.

Krebs, C. J., Gilbert, B. S., Boutin, S., Sinclair, A. R. E. and Smith, J. N. M. (1986). Population biology of snowshoe hares. 1. Demography of food-supplemented populations in southern Yukon, 1976–84. *J. Anim. Ecol.* **55**, 963–982.

Leonard, D. E. (1970). Intrinsic factors causing qualitative changes in populations of *Porthertia dispar* (Lepidoptera: Lymantridae). *Can. Entomol.* **102**, 239–249.

Lewis, A. C. (1979). Feeding preference for diseased and wilted sunflower in the grasshopper *Melanoplus differentialis. Ent. exp. appl.* **26**, 202–207.

Louda, S. M. (1986). Insect herbivory in response to root-cutting and flooding stress on a native crucifer under field conditions. *Acta Ecologia.* **7**, 37–53.

McCambridge, W.F. and Downing, G. L. (1960). Blackheaded budworm. *U.S. Dept. Agr. For. Ser. For. Pest Leaf.*. No. 45. (cited in McNamee, 1987).

McCauley, E. and Murdoch, W. W. (1987). Cyclic and stable populations: plankton as paradigm. *Am. Nat.* **129**, 97–121.

McCullough, D. G. and Wagner, M. R. (1987). Influence of watering and trenching ponderosa pine on a pine sawfly. *Oecologia* (*Berlin*) **71**, 382–387.

McNamee, P. J. (1979). A process model for eastern black-headed budworm. *Can. Entomol.* **111**, 55–66.

McNamee, P. J. (1987). *The Equilibrium Structure and Behavior of Defoliating Insect Systems*. Ph.D. Thesis, Dept. Zoology, Univ. British Columbia, Vancouver, Canada. 268 pp.

McNamee, P. J., McLeod, J. M. and Holling, C.S. (1981). The structure and behavior of defoliating insect/forest systems. *Res. Pop. Ecol.* **23**, 280–298.

Martignoni, M. E. (1957). Contributo all conoscenza di una granulosi de *Eucosma griseana* (Hubner) quale fattore limitante il pullulamento dell'insetto nella Engadine alta. *Mitt. schweiz. Anst. forsth. Verswes.* **32**, 371–418.

Martignoni, M. E. and Schmid, P. (1961). Studies on the resistance to virus infection in natural populations of Lepidoptera. *J. Invert. Pathol.* **3**, 62–74.

Martinat, P. J. (1987). The role of climatic variation and weather in forest insect outbreaks. In: *Insect Outbreaks* (Ed. by P. Barbosa and J. Schultz), pp. 241–268. New York: Academic Press.

Martinat, P. J. and Allen, D.C. (1987). Relationship between outbreaks of saddled prominent, *Heterocampa guttivitta* (Lepidoptera: Notodontidae), and drought. *Environ. Entomol.* **16**, 246–249.

Mason, R. R. (1974). Population change in an outbreak of the Douglas-fir tussock moth, *Orgyia pseudotsugata* (Lepidoptera: Lymantriidae), in Central Arizona. *Can. Entomol.* **106**, 1171–1174.

Mason, R. R. (1976). Life tables for a declining population of the Douglas-fir tussock moth in Northeastern Oregon. *Ann. Entomol. Soc. Am.* **69**, 948–958.

Mason, R. R. (1978). Synchronous patterns in an outbreak of the Douglas-fir tussock moth. *Environ. Entomol.* **7**, 672–675.

Mason, R. R. (1981). Host foliage in the susceptibility of forest sites in central California to outbreaks of the Douglas-fir tussock moth, *Orgyia pseudotsugata* (Lepidoptera: Lymantriidae). *Can. Entomol.* **113**, 325–332.

Mason, R. R. (1987). Nonoutbreak species of forest Lepidoptera. In: *Insect Outbreaks* (Ed. by P. Barbosa and J. Schultz), pp. 31–58. New York: Academic Press.

Mason, R. R. and Torgersen, T. R. (1983). Mortality of larvae in stocked cohorts of the Douglas-fir tussock moth, *Orgyia pseudotsugata* (Lepidoptera: Lymantriidae). *Can. Entomol.* **115**, 1119–1127.

Mason, R. R. and Torgersen, T. R. (1987). Dynamics of a nonoutbreak population of the Douglas-fir tussock moth (Lepidoptera: Lymantriidae) in southern Oregon. *Environ. Entomol.* **16**, 1217–1227.

Mason, R. R., Beckwith, R. C. and Paul, H. G. (1977). Fecundity reduction during collapse of a Douglas-fir tussock moth outbreak in Northeast Oregon. *Environ. Entomol.* **6**, 623–626.

Mason, R. R., Torgersen, T. R., Wickman, B. E. and Paul, H. G. (1983). Natural regulation of a Douglas-fir tussock moth (Lepidoptera: Lymantriidae) population in the Sierra Nevada. *Environ. Entomol.* **12**, 587–594.

Mattson, W. J. (1985). The role of host plants in the population dynamics of the spruce budworm. In: *Recent Advances in Spruce Budworms Research* (Ed. by C. J. Sanders, R. W. Stark, E. J. Mullins and J. Murphy), pp. 124–125. Ottawa: Canadian Forest Service.

Mattson, W. J. and Haack, R. A. (1987). The role of drought in outbreaks of plant-eating insects. *Bioscience* **37**, 110–118.

May, R. M. (1971). Stability in multi-species community models. *Math. Biosci.* **12**, 59–79.

May, R. M. (1974a). *Stability and Complexity in Model Ecosystems*. New Jersey: Princeton University Press.

May, R. M. (1974b). Biological populations with nonoverlapping generations: stable points, stable cycles, and chaos. *Science* (*Wash. D.C.*) **186**, 645–647.

May, R. M. (1975). Biological populations obeying difference equations: stable points, stable cycles and chaos. *J. theor. Biol.* **49**, 511–524.

May, R. M., Conway, G. R., Hassell, M. P. and Southwood, T. R. E. (1974). Time-delays, density-dependence and single-species oscillations. *J. Anim. Ecol.* **43**, 747–770.

Maynard Smith, J. and Slatkin, M. (1973). The stability of predator–prey systems. *Ecology* **54**, 384–391.

Mihok, S., Turner, B. N. and Iverson, S. L. (1985). The characterization of vole population dynamics. *Ecol. Monogr.* **55**, 399–420.

Miller, C. A. (1963). The analysis of fecundity proportion in the unsprayed area. In: *The Dynamics of Epidemic Spruce Budworm Populations* (Ed. by R. F. Morris), pp. 75–86. *Mem. ent. Soc. Can.* **31**.

Miller, C. A. (1966). The blackheaded budworm in eastern Canada. *Can. Entomol.* **98**, 592–613.

Miller, W. F. and Epstein, M.E. (1986). Synchronous population fluctuations among moth species (Lepidoptera). *Environ. Entomol.* **15**, 443–447.

Moore, L. V., Myers, J. H. and Eng, R. (1988). Western tent caterpillars prefer the sunny side of the tree, but why? *Oikos* **51**, pp. 321–326.

Moran, P. A. P. (1953). The statistical analysis of the Canadian lynx cycle. II. Synchronization and meteorology. *Aust. J. Zool.* **1**, 291–298.

Morris, R. F. (1959). Single-factor analysis in population dynamics. *Ecology* **40**, 580–587.

Morris, R. F. (1963). The dynamics of epidemic spruce budworm populations. *Mem. ent. Soc. Can.* **31**, 1–332.

Morris, R. F. (1964). The value of historical data in population research, with particular reference to *Hyphantria cunea* Drury. *Can. Entomol.* **96**, 356–368.

Moss, R. and Watson, A. (1985). Adaptive value of spacing behaviour in population cycles of red grouse and other animals. In: *Behavioural Ecology* (Ed. by R. M. Sibly and R. H. Smith), pp. 275–294. Oxford: Blackwell Scientific.

Murdoch, W. W., Chesson, J. and Chesson, P. L. (1985). Biological control in theory and practice. *Am. Nat.* **125**, 344–366.

Myers, J. H. (1978). A search for behavioural variation in first and last laid eggs of western tent caterpillar and an attempt to prevent a population decline. *Can. J. Zool.* **56**, 2359–2363.

Myers, J. H. (1981). Interactions between western tent caterpillars and wild rose: A test of some general plant herbivore hypotheses. *J. Anim. Ecol.* **50**, 11–25.

Myers, J. H. (1985). Effect of physiological condition of the host plant on the ovipositional choice of the cabbage white butterfly, *Pieris rapae*. *J. Anim. Ecol.* **54**, 193–204.

Myers, J. H. (1988). The induced defense hypothesis, Does it apply to the population dynamics of insects? In: *Chemical Mediation of Coevolution* (Ed. by K. Spencer), pp. 530–557. New York: Academic Press.

Myers, J. H. and Williams, K. S. (1984). Does tent caterpillar attack reduce the food quality of red alder foliage? *Oecologia* (*Berlin*) **62**, 74–79.

Myers, J. H. and Williams, K. S. (1987). Lack of short or long term inducible defenses in the red alder-western tent caterpillar system. *Oikos* **48**, 73–78.

Nothnagle, P. J. and Schultz, J. (1987). What is a forest pest? In: *Insect Outbreaks* (Ed. by P. Barbosa and J. Schultz), pp. 59–79. New York: Academic Press.

Ossowski, L. L. J. (1957). The biological control of the wattle bagworm (*Kotochalia junodi* Heyl.) by a virus disease. I. Small-scale pilot experiments. *Ann. appl. Biol.* **45**, 81–89.

Otvos, I. S., Cunningham, J. C. and Friskie, L. M. (1978a). Aerial application of nuclear polyhedrosis virus against Douglas-fir tussock moth, *Orgyia pseudotsugata* (McDunnough) (Lepidoptera: Lymantriidae): I. Impact in the year of application. *Can. Entomol.* **119**, 697–706.

Otvos, I. S., Cunningham, J. C. and Alfaro, R. I. (1987b). Aerial application of nuclear polyhedrosis virus against Douglas-fir tussock moth, *Orgyia pseudotsugata* (McDunnough) (Lepidoptera: Lymantriidae): II. Impact 1 and 2 years after application. *Can. Entomol.* **119**, 707–715.

Perelle, A. D. and Harper, J. D. (1986). An evaluation of the impact of sublethal dosages of nuclear polyhedrosis virus in larvae on pupae, adults, and adult progeny of the fall armyworm, *Spodoptera frugiperda*. *J. Invert. Pathol.* **47**, 42–47.

Peters, T. M. and Barbosa, P. (1977). Influence of population density on size, fecundity, and developmental rate of insects in culture. *Ann. Rev. Entomol.* **22**, 431–450.

Podoler, H. and Rogers, D. (1975). A new method for the identification of key factors from life-table data. *J. Anim. Ecol.* **44**, 85–114.

Price, P. (1984). *Insect Ecology*. Chichester: John Wiley, 607 pp.

Price, P. (1987). The role of natural enemies in insect population dynamics. In: *Insect Outbreaks* (Ed. by P. Barbosa and J. C. Shultz), pp. 287–313. New York: Academic Press.

Raske, A. G. (1985). Collapsing budworm populations. In: *Recent Advances in Spruce Budworms Research* (Ed. by C. J. Sanders, R. W. Stark, E. J. Mullins and J. Murphy), pp. 141–142. Ottawa: Canadian Forestry Service.

Regniere, J. (1984). Vertical transmission of diseases and population dynamics of insects with discrete populations: A model. *J. theor. Biol.* **107**, 287–301.

Rhoades, D. F. (1979). Evolution of plant chemical defense against herbivores. In: *Herbivores: Their Interaction with Secondary Plant Metabolites* (Ed. by G. A. Rosenthal and D. H. Janzen), pp. 3–54. New York: Academic Press.

Rhoades, D. F. (1983). Herbivore population dynamics and plant chemistry. In: *Variable Plants and Herbivores in Natural and Managed Systems* (Ed. by R. F. Denno and M. S. McClure). New York: Academic Press.

Rhoades, D. F. (1985). Offensive–defensive interactions between herbivores and plants, their relevance in herbivore population dynamics and ecological theory. *Am. Nat.* **125**, 205–238.

Ricker, W. E. (1954). Stock and recruitment. *J. Fish. Res. Bd. Can.* **20**, 257–284.

Roland, J. (1986). *Success and Failure of* Cyzenis albicans *in Controlling its Host the Winter Moth.* Ph.D. Thesis, Dept. Zoology, University of British Columbia, Vancouver, Canada.

Roland, J. (1988). Decline of winter moth populations in North America: direct *v.* indirect effect of introduced parasites. *J. Anim. Ecol.*, in press.

Roland, J. and Myers, J. H. (1987). Improved insect performance from host-plant defoliation: winter moth on oak and apple. *Ecol. Entomol.* **12**, 409–414.

Rosenzweig, M. L. and MacArthur, R. H. (1963). Graphical representation and stability condition of predator–prey interactions. *Am. Nat.* **171**, 385–387.

Royama, T. (1977). Population persistence and density dependence. *Ecol. Monogr.* **51**, 473–493.

Royama, T. (1978). Do weather factors influence the dynamics of spruce budworm populations? *Can. For. Serv. bimonth. Res. Notes* **34**, 9–10 (cited in Royama, 1984).

Royama, T. (1984). Population dynamics of the spruce budworm *Choristoneura fumiferana*. *Ecol. Monogr.* **54**, 429–462.

Sanders, C. J., Stark, R. W., Mullins, E. J. and Murphy, J. (Eds.) (1985). *Recent Advances in Spruce Budworms Research*. Ottawa: Canadian Forestry Service. 527 pp.

Schultz, J. C. and Baldwin, I. T. (1982). Oak leaf quality declines in response to defoliation by gypsy moth larvae. *Science* **217**, 149–151.

Schwerdtfeger, F. (1941). Uber die Ursachen des Massenwechsels der Insekten. *Z. angew. Entomol.* **28**, 254–303.

Scriber, J. M. and Feeny, P. (1979). Growth of herbivorous caterpillars in relation to feeding specialization and to growth form of their food plants. *Ecology* **60**, 829–850.

Shepherd, R. F., Otvos, I. S., Chorney, R. J. and Cunningham, J. C. (1984). Pest management of Douglas-fir tussock moth (Lepidoptera: Lymantriidae): prevention of an outbreak through early treatment with a nuclear polyhedrosis virus by ground and serial applications. *Can. Entomol.* **116**, 1533–1542.

Simionescu, A. (1973). Development of gradations of *Lymantria dispar* L. in Rumania, and control measures (in French, English summary). *Zast. Bilja.* **24**, 275–284 (cited in McNamee, 1987).

Sinclair, A. R. E., Krebs, C. J., Smith, J. N. M. and Boutin, S. (1988). Population biology of snowshoe hares. III. Nutrition, plant secondary compounds and food limitation. *J. Anim. Ecol.*, in press.

Southwood, T. R. E. and Comins, H. N. (1976). A synoptic population model. *J. Anim. Ecol.* **45**, 949–965.

Szujecki, A. (1987). *Ecology of Forest Insects*. Boston: Junk. 600 pp.

Taitt, M. J. and Krebs, C. J. (1985). Population dynamics and cycles. In: *The New World Microtus* (Ed. by R. H. Tamarin), pp. 567–620. American Society of Mammalogists.

Tanada, Y. (1976). Ecology of insect viruses. In: *Perspectives in Forest Entomology* (Ed. by J. F. Anderson and H. K. Kaya), pp. 265–283. New York: Academic Press.

Tanada, Y. and Fuxa, J. R. (1987). The pathogen population. In: *Epizootiology of Insect Diseases* (Ed. by J. R. Fuxa and Y. Tanada), pp. 113–158. New York: John Wiley.

Tenow, O. (1972). The outbreaks of *Oporina autumnata* Bkh. and *Operophtera* spp. (Lep., Geometridae) in the Scandinavian mountain chain and northern Finland 1862–1968. *Zool. Bidrag Uppsala, Suppl.* **2**, 1–107.

Tenow, O. (1983). Topoclimatic limitations to the outbreaks of *Epirrita* (= *Oporina*) *autumnata* (Bkh.) (Lepidoptera: Geometridae) near the forest limit of the mountain birch in Fennoscandia. In: *Tree-line Ecology* (Ed. by P. Morisset and S. Payette), pp. 159–164. Quebec: Univ. Laval.

Thompson, A. J., Shepherd, R.F., Harris, J. W. E. and Silversides, R. H. (1984). Relating weather to outbreaks of western spruce budworm, *Choristoneura occidentalis* (Lepidoptera: Tortricidae), in British Columbia. *Can. Entomol.* **116**, 375–381.

Tilman, D. (1978). Cherries, ants, and tent caterpillars: timing of nectar production in relation to susceptibility of caterpillars to ant predation. *Ecology* **59**, 686–692.

Torgersen, T. R. (1985). Role of birds and ants in western spruce budworm dynamics.

In: *Recent Advances in Spruce Budworms Research* (Ed. by C. J. Sanders, R. W. Stark, E. J. Mullins and J. Murphy), pp. 97–98. Ottawa: Canadian Forestry Service.

Torgersen, T. R., Campbell, R. W., Srivastava, N. and Beckwith, R. C. (1984). Role of parasites in the population dynamics of the Western spruce budworm (Lepidoptera: Tortricidae) in the northwest. *Environ. Entomol.* **13**, 568–573.

Tuomi, J., Niemela, P., Haukioja, E., and Siren, S. (1984). Nutrient stress: an explanation for plant anti-herbivore responses to defoliation. *Oecologia (Berlin)* **57**, 298–302.

Uvarov, B. P. (1931). Insects and climate. *Trans. entomol. Soc. Lond.* **79**, 1–247.

Varley, G. C. (1949). Population changes in German forest pests. *J. Anim. Ecol.* **18**, 117–122.

Varley, G. C. and Gradwell, G. R. (1960). Key factors in population studies. *J. Anim. Ecol.* **29**, 399–401.

Varley, G. C., Gradwell, G. R. and Hassell, M. P. (1973). *Insect Population Ecology*. Oxford: Blackwell Scientific. 212 pp.

Vargas-Osuna, E. and Santiago-Alvarez, C. (1986). Differential response of male and female *Spodoptera littoralis* (Boisduval) individuals to baculovirus infections. In: *Fundamental and Applied Aspects of Invertebrate Pathology*. (Ed. by R. A. Sampson, J. M. Vlak, and D. Peters). IV. Int. Coll. Invert. Pathol., Wageningen, The Netherlands (Abstract).

Vezina, A. and Peterman, R. (1985). Tests of the role of nuclear polyhedrosis virus in the population dynamics of its host, Douglas-fir tussock moth, *Orgyia pseudotsugata* (Lepidoptera: Lymantridae). *Oecologia (Berlin)* **67**, 260–266.

Wallner, W. E. (1986). Factors affecting insect population dynamics: differences between outbreak and non-outbreak species. *Ann. Rev. Entomol.* **32**, 317–340.

Watanabe, H. (1971). Resistance of the silkworm to cytoplasmic-polyhedrosis virus. In: *The Cytoplasmic-Polyhedrosis Virus of the Silkworm* (Ed. by H. Aruga and Y. Tanada), pp. 169–184. Tokyo: University of Tokyo Press, (cited in Briese and Podgwaite, 1985).

Watanabe, H. (1987). The host population. In: *Epizootiology of Insect Diseases*, (Ed. by J. R. Fuxa and Y. Tanada) pp. 71–112. New York: John Wiley.

Webb, F. E., Blais, J. R. and Nash, R. W. (1961). A cartographic history of spruce budworm outbreaks and aerial forest spraying in the Atlantic region of North America, 1949–1959. *Can. Entomol.* **93**, 360–379.

Wellington, W. G. (1952). Air-mass climatology in Ontario north of Lake Huron and Lake Superior before outbreaks of the spruce budworm, *Choristoneura fumiferana* (Clem.) and the forest tent caterpillar, *Malacosoma disstria* Hbn. *Can. J. Zool.* **30**, 114–127.

Wellington, W. G. (1954). Atmospheric circulation processes and insect ecology. *Can. Entomol.* **86**, 312–333.

Wellington, W. G. (1957). Individual differences as a factor in population dynamics; the development of a problem. *Can. J. Zool.* **35**, 293–323.

Wellington, W. G. (1960). Qualitative changes in natural populations during changes in abundance. *Can. J. Zool.* **38**, 290–314.

Wellington, W. G. (1962). Population quality and the maintenance of nuclear polyhedrosis between outbreaks of *Malacosoma pluviale* (Dyar). *J. Insect Pathol.* **4**, 285–305.

Wellington, W. G. (1964). Qualitative changes in populations in unstable environments. *Can. Entomol.* **96**, 436–451.

White, T. C. R. (1974). A hypothesis to explain outbreaks of looper caterpillars, with special reference to populations of *Selidosema suavis* in a plantation of *Pinus radiata* in New Zealand. *Oecologia* (*Berlin*) **16**, 279–301.

White, T. C. R. (1978). The importance of a relative shortage of food in animal ecology. *Oecologia* (*Berlin*) **33**, 71–86.

White, T. C. R. (1984). The abundance of invertebrate herbivores in relation to the availability of nitrogen in stressed food plants. *Oecologia* (*Berlin*) **63**, 90–105.

Williams, K. S. and Myers, J. H. (1984). Previous herbivore attack of red alder may improve food quality for fall webworm larvae. *Oecologia* (*Berlin*) **63**, 166–170.

Witter, J. A., Mattson, W. J. and Kulman H. K. (1975). Numerical analysis of a forest tent caterpillar (Lepidoptera: Lasiocampidae) outbreak in northern Minnesota. *Can. Entomol.* **107**, 837–854.

Wolda, H. (1978). Fluctuations in abundance of tropical insects. *Am. Nat.* **112**, 1017–1045.

Wolda, H. and R. Foster (1978). *Zunacetha annulata* (Lepidoptera: Dioptidae), an outbreak insect in a neotropical forest. *Geo-Eco-Trop* **2**, 443–454.

Mycorrhizal Links Between Plants: Their Functioning and Ecological Significance

E. I. NEWMAN

I. Introduction 243
II. Evidence that Links Between Plants Occur 245
III. Possible Functioning of Mycorrhizal Links 249
A. More Rapid or Greater Mycorrhizal Infection 249
B. Transport of Substances Between Plants 250
IV. Possible Roles of Mycorrhizal Links in Ecosystems 261
V. Conclusions 265
Acknowledgements 266
References 266

I. INTRODUCTION

A mycorrhiza is a symbiotic, non-pathogenic association between a plant root and a fungus. Several thousand vascular plant species have now been examined for mycorrhizal infection, and if they are representative then mycorrhizas occur on the vast majority of vascular plants (Newman and Reddell, 1987). There are, however, a few families, including the Cruciferae, Cyperaceae and Proteaceae, many of whose members apparently do not form mycorrhizas; such species will here be called "non-mycotrophic". Mycorrhizal plants have been found in every continent and in every major vegetation type. Non-mycorrhizal plants are quite commonly found in disturbed habitats and also in sites where the soil is extremely wet, but even in such conditions some species can be mycorrhizal (Pendleton and Smith, 1983; Medve, 1984; Clayton and Bagyaraj, 1984).

Mycorrhizas are conveniently divided into a few major types. Most herbaceous species and some woody species have vesicular-arbuscular

ADVANCES IN ECOLOGICAL RESEARCH Vol. 18
ISBN 0-12-013918-9

mycorrhizas (VAM); ectomycorrhizas (ECM) are confined almost entirely to woody species; the remaining types are much less widespread. Descriptions of the different types, together with an extensive review of their structure and functioning, are given by Harley and Smith (1983).

Mycorrhizas involve an intimate association of the host vascular plant's root tissue and the fungus, but there is also fungal tissue that extends into the soil as individual hyphae and in some species as more complex strands. These external hyphae can take up mineral nutrients from the soil and transport them into the host root. The promoting effect of mycorrhizas on growth of the host has most often been attributed to increased nutrient uptake, although there is also evidence that they can affect water relations and pathogen resistance (Parke *et al.*, 1983; Nelsen and Safir, 1982; Davis and Menge, 1981). Mycorrhizas are not universally beneficial to the host: on occasions growth may be unaffected or even reduced (Bethlenfalvay *et al.*, 1983; Koide, 1985), and this seems to be more common in field than in pot experiments (Fitter, 1985).

Plants, even when they grow close together, are normally assumed to be physiologically separate from each other. When the roots of trees come into contact natural grafts can sometimes form, but this has usually been observed only between members of the same species (Graham and Bormann, 1966). Many of the fungi that form mycorrhizas have low host specificity, i.e. they can infect many plant species, and this has led to the suggestion that the fungi may link plants. The mycelial network of a particular fungus in the soil could be connected directly to the fungal structures within the roots of two or more plants, thus forming hyphal links between their mycorrhizal roots. If this phenomenon is widespread, it could have profound implications for the functioning of ecosystems. Among the possibilities raised are:

(1) seedlings may quickly become linked into a large hyphal network and begin to benefit from it at an early stage;
(2) one plant may receive organic materials from another via hyphal links, perhaps sufficient to increase the "receiver's" growth and chance of survival;
(3) the balance of competition between plants may be altered if they are obtaining mineral nutrients from a common mycelial network, rather than each separately taking them up from the soil;
(4) mineral nutrients may pass from one plant to another, thus perhaps reducing competitive dominance;
(5) nutrients released from dying roots may pass directly via hyphal links to living roots, without ever entering the soil solution.

If any of these possibilities operate, it would in turn suggest that relationships between plants that cannot form links, because they have different mycorrhizal fungi or because one of them cannot form mycorrhizas, would be different from relationships between species that can form links.

This paper first summarizes evidence that mycorrhizal links between plants do occur. It then reviews evidence on various possible functions of these links, and finally considers whether there is any evidence that the links influence the species composition and structure of plant communities and ecosystem processes such as nutrient cycling.

II. EVIDENCE THAT LINKS BETWEEN PLANTS OCCUR

Among fungi that form vesicular-arbuscular mycorrhizas there appears to be very little host specificity. For example, in temperate grasslands, where almost all the plant species form VAM, as far as we know all these plants can be infected by the same fungal species. Since the roots of the plants intermingle closely, it is difficult to see how hyphal links between plants can fail to form. Living roots of some herbaceous plants impose dormancy on VAM spores (Tommerup, 1985), making it all the more likely that infection of new seedlings must be from existing mycelium. Among ectomycorrhizal fungi, some appear to infect only one or a few plant species, while others have a wide host range (Molina and Trappe, 1982); thus even among ECM trees, hyphal links seem likely.

Direct observation of hyphal links in the field is very difficult, primarily because they are so easily damaged and require microscopic examination. Determining the plant species to which fine roots belong can also be difficult. So far direct observation of hyphal links has been accomplished only with plants grown in containers. Roots were either grown down microscope slides buried in the soil, the slides and roots being removed for examination, or specially designed containers with a transparent wall allowed roots and hyphae at the soil–container interface to be examined. Figure 1 shows VAM and ECM links between roots observed in such containers. Table 1 lists pairs of species between which mycorrhizal links have been observed. Although the number of species so far studied is small, the results do show that links can occur between annuals, herbaceous perennials or tree species, and that the linked species need not belong to the same family or even closely related families. Photographs by Ritz (Fig. 1(b); Ritz, 1984) show hyphal links between *Plantago lanceolata* roots with frequencies up to several links per mm of root. Read *et al.* (1985) show photographs of VA hyphae which are fatter and less branched than normal, linking roots several centimetres apart; these they termed "arterial hyphae". Some ECM fungi can form strands many cells thick that can extend several centimetres from the host root, and

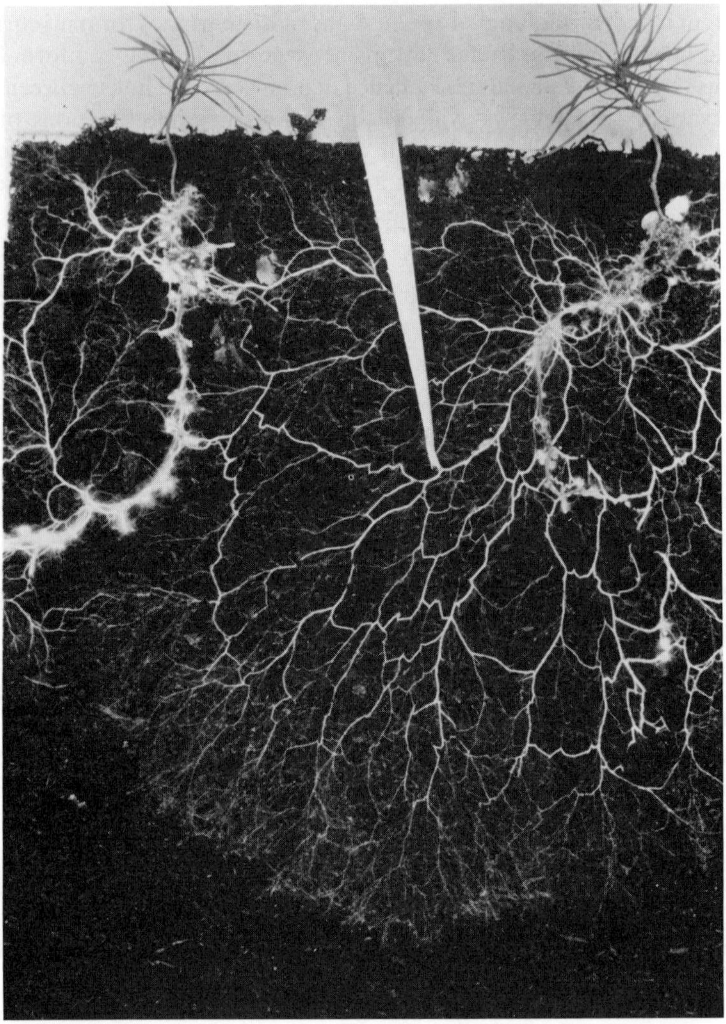

Fig. 1. Mycorrhizal fungi forming links between roots. (a) Links formed by ectomycorrhizal fungus *Suillus bovinus* between seedlings of *Pinus sylvestris* (right) and *P. contorta* (left). Pointer indicates a main hyphal strand. Magnification ×0·7. Photograph by D. J. Read.

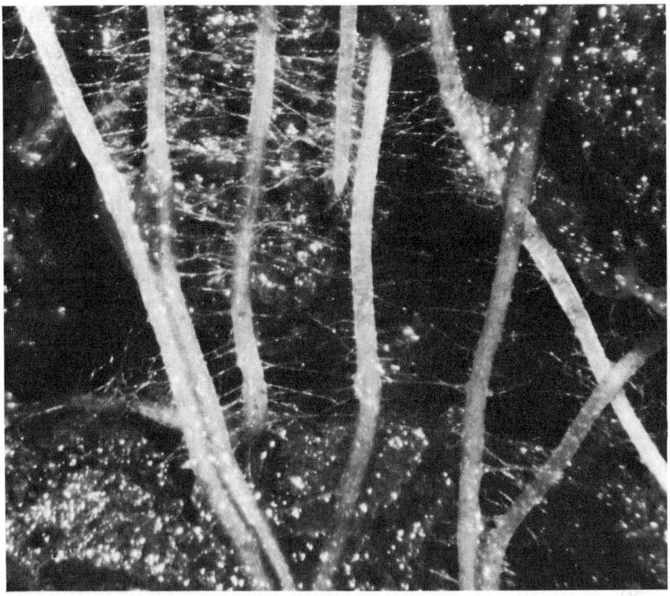

Fig. 1 (cont.). (b) *Plantago lanceolata* infected with vesicular-arbuscular mycorrhiza. Magnification × 15. It is not possible to be sure that all the hyphae in this photograph are of mycorrhizal fungi, but other observations have shown that *P. lanceolata* does form mycorrhizal links in this soil. Photograph by K. Ritz.

Table 1
Species between which mycorrhizal links have been observed.

Mycorrhizal type		Reference
(a) Vesicular-arbuscular mycorrhiza		
Lolium perenne	*Plantago lanceolata*	Heap and Newman (1980a)
Festuca ovina	*Plantago lanceolata*	Read *et al.* (1985)
Clarkia rubicunda	*Plantago erecta*	Chiariello *et al.* (1982)
(b) Ectomycorrhiza		
Pinus sylvestris	*Pinus contorta*	Read *et al.* (1985)
	Picea abies	Read *et al.* (1985)
	Picea sitchensis	Read *et al.* (1985)
	Betula pubescens	Read *et al.* (1985)
Pinus contorta	*Picea abies*	Read *et al.* (1985)
	Picea sitchensis	Read *et al.* (1985)
	Betula pubescens	Read *et al.* (1985)

such strands have been observed to link the roots of small tree seedlings (Fig. 1(b); Read *et al.*, 1985).

There is some indirect evidence that links do form in the field. In most research on mycorrhizas, plants are sown into bare soil, and infection must be initiated by spores or from fungi surviving in dead roots. If, in contrast, seedlings establish close to existing plants with the same type of mycorrhizal fungus, their roots may grow into contact with an existing mycelium and so form links. This might well result in more rapid infection than from spores or dead roots. Such spread of the fungus from plant to plant has been directly observed with ECM *Pinus* species in transparent-sided containers (Brownlee *et al.*, 1983; Finlay and Read, 1986a). Read *et al.* (1976) and Birch (1986) found that seedlings of various herbaceous species in calcareous grassland became infected at a very early stage of development. The four species studied by Birch (1986) became infected within 3–6 days and showed substantial infection within about 12 days. It seems likely that infection would have been slower in the absence of existing, living mycorrhizal roots, but this was not checked. Miller *et al.* (1983) found that *Atriplex confertifolia* was slow to develop VAM when colonizing disturbed soil following mining in Wyoming, even though in nearby undisturbed vegetation the species was well infected. In both disturbed and undisturbed sites *A. confertifolia* was only mycorrhizal when growing close to another species that was VAM-infected; the percentage of *A. confertifolia* root length infected was then similar to that of whichever species it grew next to. This suggests that *A. confertifolia* could become infected only by direct hyphal links from another plant.

There is also indirect field evidence, from work on birch (*Betula pubescens* and *B. pendula*) in Scotland, that some ectomycorrhizal fungi can form links. In the even-aged stand studied there was a succession of mycorrhizal fungi, "early-stage" species giving way to "late-stage" species (Mason *et al.*, 1983). Birch seedlings in pots of soil readily developed mycorrhizas of early-stage fungi, but failed to develop mycorrhizas of late-stage species even when spores of the fungus were known to be present (Fox, 1983). In contrast, when birch seedlings were planted close to established birch trees, within a few months the seedlings were heavily infected by late-stage fungi. Seedlings at the same site but isolated from living roots of the trees by trenching developed similar numbers of mycorrhizal rootlets, but mainly of early-stage fungi (Fleming, 1983, 1984). These results strongly suggest that the seedlings growing close to trees became infected by hyphae growing from the trees. Some of the late-stage fungi, e.g. *Leccinium* spp. and *Lactarius pubescens*, form hyphal strands which may well be involved in forming links between the plants. Similar but more limited observations on birch seedlings in oak–chestnut (*Quercus*–*Castanea*) stands in England (Fleming, 1983) suggest that there, too, the seedlings became infected by hyphal links from the trees. This provides an example where tree and seedling were unrelated species.

This section has shown that at present the direct evidence for mycorrhizal links is confined to a few plant species grown in containers, and the evidence from the field is limited. It seems likely that mycorrhizal links between plants are common, but much more evidence is needed.

III. POSSIBLE FUNCTIONING OF MYCORRHIZAL LINKS

A. More Rapid or Greater Mycorrhizal Infection

By growing into contact with existing mycorrhizal mycelium seedlings might (a) become infected more rapidly than if the only source of infection was spores or dead roots, (b) become infected by different species of mycorrhizal fungus, and/or (c) become connected with a mycelial system much larger than they could quickly develop on their own. The evidence relating to these possibilities was summarized in the previous section. Effects on mycorrhizal infection might continue beyond the seedling stage. Plant species can differ markedly in the amount of mycorrhizal infection they carry, even when growing under identical conditions. The possibility should be examined that a highly mycorrhizal plant could, via hyphal links, support substantial mycorrhizal infection on the roots of plants of another species, which would otherwise be much less infected or even non-mycorrhizal. One would then observe the second species being more highly infected when growing with the first than when on its own. In fact most of the available evidence is against this.

Ocampo *et al.* (1980) grew pairs of crop species in pots of soil both together and separately. The species that had the lower amount of VAM infection when alone did not usually show an increased infection when in two-species mixture: either the infection was not significantly altered or it was reduced in the mixture. Christie *et al.* (1978) and Lawley *et al.* (1982) carried out similar experiments with two-species mixtures of British grassland plants. Again most of the statistically significant differences between mixture and monoculture showed species having higher VAM infection when growing on their own. So none of these results supports the hypothesis that one plant promoted increased internal mycorrhizal infection in another. Fitter (1977), however, did find increased infection in *Lolium perenne* in mixtures with the more heavily infected *Holcus lanatus* than in monoculture. Hirrel *et al.* (1978) and Ocampo *et al.* (1980) grew species of Cruciferae and Chenopodiaceae, normally non-mycotrophic, with mycorrhizal species.

Some of the non-mycotrophs developed some internal infection, but it was always confined to a small proportion of the root length, and arbuscules were never formed; so it is unlikely to be of functional significance to the host.

B. Transport of Substances Between Plants

Attention has been focused on whether mycorrhizal links provide a pathway for transfer of organic materials and of mineral nutrients between plants. Whether other materials, such as toxic chemicals or inoculum of pathogens, also pass through such links has not been investigated as far as I know. In the following section the plant from which material may be moving is called the "donor", and the plant into which material may be moving the "receiver".

Much of the relevant evidence is from movement of isotopes, especially ^{14}C and ^{32}P. It is important to make clear that movement of an isotope does not of itself prove net movement of the element. For example, suppose two (non-living) compartments contain solutions of $H_2PO_4^-$ at equal concentration. If the compartments are separated by a barrier permeable to $H_2PO_4^-$, individual ions will move across it in each direction but in equal numbers. So no net movement will occur, i.e. the concentration in each compartment will not change; but if carrier-free $H_2{}^{32}PO_4^-$ is added to one compartment some ^{32}P will soon be present in the other. To take a mycorrhizal example, net carbon movement is normally from host to fungus; amino-acids can move from fungus to host (Smith, 1980), but the carbon in them is normally more than equalled by carbon in other organic compounds passing from host to fungus. So ^{14}C movement from fungus to host would not demonstrate net carbon gain by the host.

1. Transfer of Carbon-containing Substances

It is generally accepted that mycorrhizal fungi receive most of their organic carbon from the host root (Harley and Smith, 1983). A key question here is whether significant amounts can also pass from the fungus into another plant; and if so, whether enough organic matter is ever transported from one green plant to another to enhance significantly the survival or growth of the receiver.

Carbon can pass into mycorrhizal orchid seedlings from the infecting fungus (Smith, 1966, 1967; Alexander and Hadley, 1985). However, the carbon is in these cases obtained by the fungus from dead organic matter. Orchidaceous mycorrhizal fungi do not normally form associations with members of other plant families, so their ability to form links will be limited.

Björkman (1960) investigated the organic carbon supply to the chlorophyll-less angiosperm *Monotropa hypopitys*, which grows in association with trees in Europe. His evidence that *Monotropa* obtains its carbon from trees via mycorrhizal links, rather than from dead organic matter, was as follows:

(1) *Monotropa* can be found in sandy soils almost devoid of organic matter, but close to trees.
(2) A mycorrhizal fungus isolated from *Monotropa* roots could form a mycorrhizal association with pine roots.
(3) In Sweden Björkman injected ^{14}C-glucose into the bark of stems of spruce and pine (presumably *Picea abies* and *Pinus sylvestris*) and later found ^{14}C in nearby *Monotropa*.
This was the earliest published evidence of organic matter transfer between plants by mycorrhiza. Clearly *Monotropa*, lacking chlorophyll, has unusual carbon relations.

Reid and Woods (1969) provided the first clear evidence of carbon transport from an ectomycorrhizal fungus to a green plant host. They fed ^{14}C-glucose to hyphal strands of the ECM fungus *Thelephora terrestris* and later detected ^{14}C in leaves of the pine with which the fungus was mycorrhizal. However, when the fungus formed links between two pine seedlings, and ^{14}C was fed to one of them, no ^{14}C transport to the other was detected. The first attempt to demonstrate that VA mycorrhizas promote carbon transfer between plants (Hirrel and Gerdemann, 1979) also was inconclusive.

Much information on ^{14}C transport between plants has come from the work of D. J. Read and co-workers at the University of Sheffield (Francis and Read, 1984; Read *et al.*, 1985; Finlay and Read, 1986a; Grime *et al.*, 1987; Duddridge *et al.*, 1988). The experiments involved feeding $^{14}CO_2$ to leaves of one plant (the donor) and later assessing the distribution of ^{14}C in another plant (the receiver), by autoradiography or quantitative analysis. In a field experiment $^{14}CO_2$ was fed to crowns of *Pinus contorta* trees (which were ECM). Eight weeks later, ^{14}C was detected in roots of other *P. contorta* trees and of several other species; *Chamaecyparis lawsoniana*, which is VAM, contained no detectable $^{14}CO_2$, which supports the hypothesis that mycorrhizal links were involved in ^{14}C transfer. However, direct evidence on the role of mycorrhizal links has come only from experiments with small plants in containers. These allowed mycorrhizal and non-mycorrhizal plants to be compared, and the formation of mycorrhizal links between plants to be confirmed. Measurements of ^{14}C were confined to the first two or three days after the start of $^{14}CO_2$ feeding. Some experiments involved VAM species, others ECM species, and the conclusions for both were broadly similar:

(1) ^{14}C was found by autoradiography to be located not only within external mycorrhizal hyphae attached to the donor, but also in regions of these fungi within the receiver's roots.

(2) Although there was detectable ^{14}C in the roots of the receiver plant in non-mycorrhizal containers, mycorrhizal infection greatly increased this (Table 2). Whole-root analyses, as in Table 2, include ^{14}C in the fungus within the root; but autoradiography showed that some of the ^{14}C was transferred into the tissue of the root itself (Read *et al.*, 1985; Duddridge *et al.*, 1988).

(3) The concentration of ^{14}C in the receiver's shoot was sometimes increased by mycorrhizal infection, though much less than in the root (Table 2).

Thus there is strong evidence that ^{14}C can be transported from one plant to another by mycorrhizal links, either ECM or VAM. The key question arising from this is whether the receiver plant ever benefits significantly from transport of carbon compounds through mycorrhizal links. At present it is not possible to answer this question. The benefit could happen in two ways: (1) the receiver might obtain mineral nutrients from a mycorrhizal fungus, part or all of whose organic carbon was supplied by the donor, thus reducing the carbon drain on the receiver; or (2) the receiver might receive organic carbon via mycorrhizal links, supplementing its own photosynthetic carbon

Table 2

^{14}C transfer between plants in experiments of Read *et al.* (1985). Figures are radioactivity (dpm per μg dry weight) in receiver plants, measured 72 h (part (a)) or 48 h (part (b)) after start of feeding of $^{14}CO_2$ to leaves of donor plant.

(a) Transfer between *Pinus contorta* plants that were ectomycorrhizal (ECM) with one of two fungus species or non-mycorrhizal (NM).

	Part of receiver plant assayed	
	Root	Shoot
NM	0·30	0·10
ECM (*Suillus bovinus*)	3·6**	0·30*
ECM (*Suillus granulatus*)	1·6**	0·15 (ns)

(b) Transfer from *Plantago lanceolata* to *Festuca ovina*. The receiver plants were either fully illuminated or shaded during the transfer period.

	Full light		Shade	
	Root	Shoot	Root	Shoot
NM	0·08	0·10	0·03	0·04
VAM	2·8***	0·44**	36·4***	0·32***

Statistical significance of differences between mycorrhizal and non-mycorrhizal: *** $P<0{\cdot}001$; ** $P<0{\cdot}01$; * $P<0{\cdot}05$; (ns), not significant. Figures in part (b) calculated from raw data in Table 4 of Read *et al.* (1985), statistical significance by Mann–Whitney U-test.

gain. A severely shaded plant might be expected to provide the greatest "demand" for organic carbon. Shading the receiver can increase the amount of ^{14}C transferred into its roots (Table 2(b)), but whether all this increase remains within the fungal tissue is not known. Shading has not been found to increase the amount transferred to the shoot of the receiver (Table 2(b)). So this does not provide support for net carbon transfer between plants. If net carbon transfer does occur, it would be important to know how large it is compared with the receiver's carbon gain by photosynthesis. No data which allow calculation of this are available. However, in a few experiments the ^{14}C assimilated by photosynthesis by the donor is given (Read *et al.*, 1985); for every million units of ^{14}C assimilated by the donor over 2–3 days, about 1–10 units were transferred to the shoot of the receiver in VAM experiments, and about 100 units in ECM experiments. This suggests that if net carbon transfer between plants occurs, it is a small amount.

Thus, while there is strong evidence that ^{14}C can be transported between plants by mycorrhizal links, there is no clear indication whether net transfer of carbon between linked plants ever occurs, and if so whether the amount is large enough to benefit significantly the receiver plant.

2. *Transfer of Mineral Nutrients Between Living Plants*

It is generally assumed that when interactions between plants involving nutrients occur they are competitive: the plants are withdrawing nutrients from the same soil pool. Cycling of nutrients from one plant to another is normally considered to happen after death of the plants or parts of them. There has been interest in whether nitrogen is transferred from living roots of legumes to other plants growing with them, but older work mostly indicated that little nitrogen is released from the legume until its roots or shoots die (Whitehead, 1970). However, several field experiments have shown that when a radioisotope of a mineral nutrient is applied to one plant the isotope can soon afterwards be detected in some (though not necessarily all) of the neighbouring plants. Table 3 shows that transfer can occur between unrelated species and even from trees to herbs, so it cannot be solely via root grafts. There is no proof that mycorrhizal links were involved; in some transfers they cannot have been, since the two plants form different types of mycorrhiza. Ledgard *et al.* (1985) found no significant transfer of ^{15}N between plants in the field, although they did find it in a pot experiment. Transfer of ^{15}N (which is not radioactive) is less readily detected than transfer of radioisotopes.

Several experiments have been conducted in which ^{15}N or ^{32}P transfer between two plants in a container has been measured, comparing containers where both plants were mycorrhizal with those where neither was mycorrhizal. Transfer of ^{15}N from *Trifolium subterraneum* to *Lolium rigidum* (Hay-

Table 3

Transfer of radioisotopes of mineral nutrients between plants in the field. The radioisotope was fed to the shoot of the donor plant and later detected in shoot material of the receiver plant. ECM, VAM denote normal mycorrhizal status of the species, recorded at other sites; ECM + VAM indicates that the species can have either mycorrhizal type.

Location	Isotopes	Donor species	Receiver species	Reference
Byelorussia, USSR	^{32}P	ECM trees	ECM, ECM + VAM, VAM trees VAM grasses	Rakhteyenko (1958)
		ECM + VAM trees VAM trees	ECM, ECM + VAM trees VAM trees	
South-eastern USA (2 sites)	^{32}P, ^{45}Ca	VAM tree: cut stumps	ECM, VAM trees	Woods and Brock (1964)
	^{32}P, ^{45}Ca, ^{86}Rb, ^{35}S	ECM + VAM tree, ?ECM tree: cut stumps and intact trees	Trees and shrubs	Woods (1970)
California, USA	^{32}P	VAM annual herb	VAM and ?VAM annual herbs	Chiariello *et al.* (1982)

stead, 1983) and from *Glycine max* to *Zea mays* (Kessel *et al.*, 1985) was significantly increased when the plants were VA mycorrhizal. The same was true for transfer of ^{32}P between two *Plantago lanceolata* or between two *Festuca ovina* (Whittingham and Read, 1982) and transfer from *Lolium perenne* to *Plantago lanceolata* (Newman and Ritz, 1986). In contrast, when Finlay and Read (1986b) fed ^{32}P to non-mycorrhizal roots of *Pinus* spp., although the isotope moved to other parts of the fed plant none could be detected in its ECM fungus nor in another pine to which the fungus linked.

These tracer experiments do not prove net transfer of nitrogen or phosphorus. Ritz and Newman (1984) grew two mycorrhizal *Lolium perenne* per pot, of the same age and about equal size, and showed that when ^{32}P was fed to one, some of it was soon transferred to the other. Since the ^{32}P fed was carrier-free, it would have increased the phosphorus content of the fed plant by a negligible amount, so in this case presumably no net transfer of phosphorus was occurring; the presence of the isotope indicates that exchange of phosphorus atoms between the plants was going on. Net transfer would be expected only if the plants differ in some way sufficient to set up a source–sink relationship. It is not clear that a difference of size will of itself promote net transfer between plants. A large difference in tissue nutrient concentration might do so.

Francis *et al.* (1986) grew *Festuca ovina* or *Plantago lanceolata* (donor plant) with the root system split between two small pots of infertile dune sand. One of the pots could be used for feeding nutrients to the donor plant; in the other pot were receivers, two other plants of *F. ovina* or *P. lanceolata* plus two plants of *Arabis hirsuta* (which does not form mycorrhizas). There were three treatments involving supplying either nutrient solution or water alone to the donor plant, and with or without mycorrhizal inoculation of the donors and receivers (see Table 4). In general, the receiver *Festuca* and *Plantago* had greatest dry weight and nutrient content if nutrients had been supplied to the donor and the plants were mycorrhizal, whereas the non-mycotrophic *Arabis* showed no response to the donor's nutrient supply. Similar but more limited results were obtained by Whittingham and Read (1982). The authors considered this as clear evidence of nutrient transfer between mycorrhizal plants. These two species can form mycorrhizal links when growing in similar soil to that used in these experiments (Read *et al.*, 1985). However, one should consider whether the results may be explained on the more conventional assumption that the plants were competing for nutrients. If the "donor" had a higher tissue nutrient concentration as a result of being fed extra nutrients, it might well have taken up less nutrients from the pot it shared with the "receivers", hence leaving more for them. The failure of *Arabis* to respond could simply reflect the poor ability of non-mycorrhizal plants to acquire nutrients from this very nutrient-poor soil.

Table 4

Results from experiment of Francis *et al.* (1986) in which nutrients were fed to part of the root system of a "donor" *Plantago lanceolata* plant. The weight and nutrient content of shoots of smaller "receiver" plants of *Arabis hirsuta* and *Festuca ovina* were later determined, and are shown in the table.

	Arabis	*Festuca*		
	Dry wt (mg)	Dry wt (mg)	Nutrient content (μg) N	P
M-nut	1·4a	9·7c	32b	18b
NM-nut	1·5a	3·8a	10a	7a
M-dist	1·5a	6·5b	6a	10a

Treatments: M = *Plantago* and *Festuca* VA-mycorrhizal, NM = non-mycorrhizal; nut = complete nutrient solution supplied to donor plant, dist = distilled water only. Within any column, figures followed by different letters are significantly different ($P < 0{\cdot}05$).

Unfortunately there was no control treatment without donor plants to check for a competitive effect. Ocampo (1986) carried out an experiment with *Sorghum vulgare* similar to the M-nut and NM-nut treatments of Francis *et al.* (1986), but including other treatments where the "donor" and "receiver" roots were separated by a root-proof barrier. Table 5 shows that whether the plants were mycorrhizal or not, the small plants did not benefit from their roots intermingling with the large plants, but on the contrary were smaller and had lower nutrient concentration than if their roots were separated. Evidently even when mycorrhizal links could form, competition was the dominating influence, not net gain of nutrients by the "receiver" from the "donor".

Thus it remains to be demonstrated that net nutrient transfer from one plant to another occurs. Isotope transfer must be considered as showing the upper limit for net transfer, and one may still usefully ask whether this upper limit is fast enough to be ecologically significant. Ritz and Newman (1984) showed that ^{32}P transfer between mycorrhizal *Lolium perenne* and *Plantago lanceolata* proceeded at a steady rate after the first few days. The amount arriving in the receiver's shoot per day was 0·1–0·3% of the amount in the donor's roots. We cited examples of rates of phosphorus uptake from soil or solutions which were mostly 10 or more times faster than that, suggesting that phosphorus exchange between plants is sometimes, though not necessarily always, slow compared with uptake from soil. An experiment (Ritz and Newman, 1986) in which non-radioactive nutrients were fed to one *L. perenne* for six months and the final nutrient content of its neighbouring *L. perenne* determined, led to a similar conclusion.

Table 5

Results of an experiment by Ocampo (1986) in which nutrients were fed to a large *Sorghum vulgare* plant, whose roots were either intermingling with those of a smaller *S. vulgare* or were separated from them. The plants were either VA-mycorrhizal (M) or non-mycorrhizal (NM).

	Dry wt (g)		Nutrient concentration (mg g^{-1})			
			Nitrogen		Phosphorus	
State of roots	M	NM	M	NM	M	NM
			Small plant (receiver)			
Intermingling	0·74	0·63	9·0	7·0	0·8	0·6
	*	*	*	ns	*	*
Separate	1·16	0·86	13·5	8·0	1·6	0·7
			Large plant (donor)			
Intermingling	2·36	1·60	27·4	12·6	2·3	1·1
	ns	ns	ns	ns	ns	ns
Separate	2·40	1·70	27·5	13·6	2·3	1·2

Data refer to shoots only. Statistical significance of difference between roots intermingling and separate: * $P<0{\cdot}05$; ns, not significant.

Another question is whether nutrient movement between plants occurs via direct hyphal links. This requires that nutrient elements pass directly from the host root tissue to its internal fungus, the opposite direction to normal. Since some loss of ^{32}P and ^{15}N to the external medium does occur from labelled roots, even while net uptake of phosphorus and nitrogen is going on (Elliott *et al.*, 1984, Morgan *et al.*, 1973), the increase in isotope transfer between plants caused by mycorrhizas could alternatively be nutrients passing out of the donor's roots into soil and there being captured by hyphae attached to the receiver plant. Newman and Ritz (1986) attempted to determine which of these pathways was the major one in transfer of ^{32}P between *Lolium perenne* and *Plantago lanceolata*. From experiments on the two species separately we predicted the amount and time-course of ^{32}P transfer by each of the two pathways. The observed amounts and time-courses of transfer between these two species agreed better with the predictions for ^{32}P passing from the donor's roots into the soil before being taken up again by the receiver's mycorrhizal fungi. In contrast, Francis *et al.* (1986) argued for transport through mycorrhizal links, on the grounds that if in their experiment nutrients had first to pass out of the donor's roots, the non-mycotrophic *Arabis* should have shown some growth response, i.e. been larger in the M-nut treatment than M-dist (Table 4). In fact *Arabis* consistently failed to respond, whereas the species that could form mycorrhizal links often did.

In summary, VA mycorrhizas increase ^{15}N and ^{32}P transfer between living plants, but it has not been clearly demonstrated that net transfer of N or P occurs, nor that it occurs through direct hyphal links. If such transfer does occur its influence on the nutrient content of the plants may well be fairly small, relative to uptake from soil sources.

3. Transfer of Mineral Nutrients from Dying Roots

In perennial vegetation types such as grassland and forest a substantial proportion of the biomass and annual tissue death is below ground (Sims and Singh, 1978; Vogt *et al.*, 1986), so the recycling of nutrients from dying roots is important. The mycorrhizal fungus within a dying root may well remain alive longer than the root itself, especially if the fungus is linked to living roots nearby. Hence it seems quite possible that mineral nutrients released as the root dies could be taken up by mycorrhizal fungus within the dying root and transported to living roots. This could reduce the amount of nutrients captured by saprophytic micro-organisms proliferating in or near the dying root, or lost by leaching and fixation in the soil; hence it could increase the amount cycled rapidly into other plants. This possibility has been investigated in a series of experiments at Bristol University, mostly involving *Lolium perenne* and *Plantago lanceolata*, which are known to form VAM links. ^{32}P was fed to the shoot of one plant; after 4 or 7 days to allow some ^{32}P to be translocated to its roots, the fed shoot was cut off and removed and the roots left to die in the soil. Another plant in the pot was left intact and ^{32}P arrival in its shoot measured. If the plants were mycorrhizal ^{32}P transfer was almost always significantly greater, usually 2- to 4-fold, than between non-mycorrhizal plants. This was found for several combinations of donor and receiver species (Table 6). Heap and Newman (1980b) measured ^{32}P in the receiver plant's shoot 7 days after the donor was detopped. About 2–3% of the ^{32}P in the roots of *Lolium* donors had been transferred to *Plantago* if the plants were mycorrhizal, though in one *Lolium*-to-*Lolium* experiment 9% was transferred. Later Ritz and Newman (1985) showed that most of the transfer from dying *Lolium* roots occurred in a surge during the second to third week after detachment of the donor shoot, so the measurements by Heap and Newman had missed most of it. In an experiment with mycorrhizal plants Ritz and Newman (1985) found that of the ^{32}P present in the *Lolium* roots at the time of detachment, 31% was transferred to the *Plantago* shoot within 18 days. This high value merits confirmation. Since only 29% of the *Lolium* root length was mycorrhizal in this experiment it seems unlikely that all of this transfer was via direct hyphal links. Even if the 10% transfer between *Plantago* plants found by Eason (1987; see Table 6) is a more

Table 6

Summary of experiments in which plants were labelled via the shoot with ^{32}P, then after 4 or 7 days the labelled shoot was removed. After the detached roots had been left for some days (column 1) the amount of ^{32}P in the shoot of other living plants was measured. A comparison was made of pots in which plants were VA-mycorrhizal with pots in which plants were non-mycorrhizal.

Time[a] (days)	Amount[b] of transfer when plants VAM	Transfer[c] +VAM / Transfer −VAM	Reference
Lolium perenne to *Plantago lanceolata*			
7	2·9	4·2***	Heap and Newman (1980b)
7	2·2	2·2*	Heap and Newman (1980b)
7	2·1	3·2*	Heap and Newman (1980b)
29	[d]	3·1 (ns)	Ritz (1984)
L. perenne to *L. perenne*			
7	8·6	3·7*	Heap and Newman (1980b)
L. perenne to *Trifolium repens*			
7	0·5	7·9***	Heap and Newman (1980b)
P. lanceolata to *P. lanceolata*			
7	4·1	2·8*	Heap and Newman (1980b)
21	[d]	1·3 (ns)	Ritz (1984)
30	9·5	2·5*	Eason (1987)

[a] Time between detaching shoot of donor plant and measuring ^{32}P in shoot of receiver plant.
[b] ((^{32}P in receiver shoot)/(^{32}P in donor root)) × 100.
[c] Ratio of transfer between mycorrhizal plants to transfer between non-mycorrhizal plants. Transfer calculated as in [b], except when indicated by [d]. Ratio significantly greater than 1 if indicated by *** $P<0{\cdot}001$, * $P<0{\cdot}05$. ns = not significantly different from 1.
[d] Transfer calculated by (^{32}P in receiver shoot)/(^{32}P in donor shoot).

representative value than 31%, the amounts are large enough to be of potential ecological importance.

One can ask whether the ^{32}P transfer represents net transfer of phosphorus; an alternative is that net immobilization of phosphorus by micro-organisms is occurring within the dying roots, and the ^{32}P movement merely represents exchange. Clear evidence of net phosphorus loss was obtained by Eason (1987), who placed detached roots of *Lolium*, which had similar phosphorus concentration to those in the transfer experiments, in nylon litter bags in soil. Within three weeks they had lost 32% of their initial phosphorus.

Another question is whether there is phosphorus transfer directly via hyphal links from dying to living roots. It seems quite possible that when roots are dying phosphorus can pass from host to fungus, the reverse of the normal direction, thus allowing direct transfer to occur. Alternatively, phosphorus might be passing out of dying roots into the soil, and there be taken up by mycorrhizal fungi attached to the living plant. To decide between these two alternatives, one needs to compare ^{32}P transfer between plants under three circumstances:

(1) the dying and living roots are linked by mycorrhizas;
(2) the donor plant is not mycorrhizal, but the receiver is so that its associated hyphae can take up ^{32}P leaking from the dying roots;
(3) neither plant is mycorrhizal.

No experiment has so far compared all three of these at one time, but evidence comes from experiments where only the receiver was mycorrhizal, so that links could not form. Cabbage *Brassica oleracea*, which does not form mycorrhizas, was used as donor with *Plantago lanceolata* as receiver (Eason, 1987). In an experiment of Heap and Newman (1980b), *Lolium* remained non-mycorrhizal, while *Plantago* in the same pot averaged 15% mycorrhizal. (This may represent more rapid infection of *Plantago*.) In none of these experiments, where links were impossible, did mycorrhizal infection of *Plantago* increase ^{32}P transfer significantly compared with the non-mycorrhizal state (Table 7). This contrasts with the almost invariable increase of transfer when links can form (Table 6), suggesting that mycorrhizas are not acting simply by increasing uptake by the receiver of ^{32}P lost from the donor to the soil.

Contrary evidence, against a major involvement of direct links, comes from an experiment by Eason (1987), who found that if glucose was added to the soil at the time the donor was detopped, ^{32}P transfer to the receiver was reduced by about two-thirds, though not eliminated. This reduction was presumably due to increased microbial immobilization of phosphorus in the

Table 7

Experiments similar to those in Table 6, but comparing pots where receiver (but not donor) was mycorrhizal with pots where both were non-mycorrhizal.

Time[a] (days)	Amount[b] of transfer to VAM receiver	Transfer[c] +VAM / Transfer −VAM	Reference
Lolium perenne to *Plantago lanceolata*			
7	2·3	0·6 (ns)	Heap and Newman (1980b)
Brassica oleracea to *Plantago lanceolata*			
28	[d]	0·8 (ns)	Eason (1987)
28	17·2	0·6 (ns)	Eason (1987)

ns = not significantly different from 1.
[a, b, c, d] See Table 6.

soil, which direct transfer of phosphorus via hyphal links would be expected to bypass. However, this does not preclude direct links being responsible for some of the transfer.

All these experiments involving roots suddenly detached from their shoots simulate only a few and perhaps uncommon causes of root death. They are open to the criticism that if a root senesces while attached to the shoot, nutrients might be withdrawn from it into living parts of the plant. At present evidence on this is lacking.

In summary, the evidence clearly indicates that VA mycorrhiza can promote rapid cycling of phosphorus from dying roots to living plants, but the extent to which the transfer occurs via direct hyphal links is still uncertain.

IV. POSSIBLE ROLES OF MYCORRHIZAL LINKS IN ECOSYSTEMS

The previous section described experiments, mainly carried out under artificial conditions, aimed at investigating some possible functions of mycorrhizal links between plants. In this final section I consider whether there is any evidence that mycorrhizal links influence the structure and dynamics of ecosystems, using information from the field where possible.

Although much of the evidence so far reviewed has been inconclusive, it suggests some possible influences of mycorrhizal links.

(1) They may assist seedling establishment, by allowing seedlings to become infected with mycorrhizas more rapidly and perhaps to become connected to a large mycorrhizal mycelium supported by carbon from other plants. In this way the seedlings could rapidly begin to acquire nutrients. They may also receive organic carbon from another plant.
(2) If carbon can pass from well-illuminated to shaded plants, and mineral nutrients from higher-nutrient to lower-nutrient plants, this could tend to reduce competitive dominance by some plants and hence promote coexistence and species diversity.
(3) If mycorrhizal links promote nutrient cycling they may keep more nutrients in the biomass, and hence increase ecosystem biomass and productivity.

There are great difficulties in experimentally altering mycorrhizas of established vegetation in the field, which necessarily limits the evidence available. Nevertheless, these suggestions lead to some testable hypotheses:

(i) that seedlings grow or survive better if close to established plants of the same mycorrhizal type;

(ii) that plants growing together are more equal in size and growth rate if they have the same type of mycorrhiza than if they have different types of mycorrhiza or are non-mycorrhizal;
(iii) that species with the same type of mycorrhiza cycle nutrients preferentially among themselves, and less to species with a different type of mycorrhiza.

Experiments on root interactions between seedlings and established plants show these interactions to be deleterious, not beneficial, to the seedlings. Cook and Ratcliff (1984) in Australia and Snaydon and Howe (1986) in England sowed grass seedlings into established turf, some of them within tubes inserted into the soil to prevent root interactions between seedling and turf species. The tubes markedly increased seedling growth, except when soil nutrient status was very high or sward density very low. Cutting the soil around tree seedlings to separate them from tree roots also increases their growth (Jarvis, 1964; Fleming, 1983). In contrast, the chlorophyll-less plant *Monotropa hypopitys* grew very poorly when its roots were separated from those of neighbouring trees (Björkman, 1960).

One can next ask whether competitive relations are less deleterious to small plants if they can form mycorrhizal links with the larger plants than if they cannot. Grime *et al.* (1987) grew the grass *Festuca ovina* in trays of nutrient-poor sand, either VA-mycorrhizal or non-mycorrhizal. Seeds of 20 grassland species were sown into each tray. In the mycorrhizal trays *F. ovina* grew less well, but seedlings of many of the other species grew substantially better (Table 8). This illustrates, in a "microcosm", mycorrhizas promoting species coexistence and diversity. When $^{14}CO_2$ was fed to the *F. ovina* the

Table 8

Results from experiment of Grime *et al.* (1987) in which trays of sand were mycorrhiza-free (−VAM) or inoculated with VA mycorrhiza (+ VAM). The influence of mycorrhiza on initially-sown *Festuca ovina* and later-sown species is expressed as the ratio:

$$\frac{\text{dry weight per plant in +VAM treatment}}{\text{dry weight per plant in −VAM treatment}}$$

	Mean ratio
Initially-sown *Festuca ovina*	0·62
Seedlings	
10 dicotyledonous species which can form VAM	2·3–7·5 [−145]
2 non-mycotrophic dicotyledonous species	[0·8], 1·2
Festuca ovina	0·73
5 other grass species, which can form VAM	0·8–2·4

[ ] = few plants survived.

amount of ^{14}C reaching the seedlings was far greater in mycorrhizal trays than in non-mycorrhizal. This is evidence that links between species were formed, but does not demonstrate that the increased seedling growth was due to carbon received from the larger plants (see Section III. B.1). An important question is whether this experiment illustrates a general phenomenon that growth of small plants in the presence of large plants is promoted by mycorrhizal links between the plants; or does it show rather that mycorrhizal infection promotes growth of certain species but not others (at least when they are in competition)? The evidence favours the second alternative. Mycorrhizas reduced the seedlings and larger plants of *Festuca ovina* to about the same extent (Table 8), showing no relative benefit to the smaller plant within the species. The experiment of Ocampo (1986) also investigated interactions between large and small plants of the same species (Table 5). When the roots were intermingling the ratio (weight of large plant/weight of small plant) was if anything greater when the plants were mycorrhizal than when non-mycorrhizal, and the same was true for the nutrient concentrations in the plants. So these results do not show mycorrhizal links favouring the smaller plant. The results of Grime *et al.* (1987) could alternatively be interpreted as showing mycorrhizas altering the balance in favour of the dicotyledonous species and against the grasses. Of the twelve dicotyledonous species sown, all grew better in the mycorrhizal trays except for two that do not form mycorrhizas (Table 8). In contrast, most of the grass species were not benefited by the mycorrhizal treatment, although all of them can form mycorrhizas and at least three of them did so. There have been several reports of mycorrhizal infection reducing grass growth or else having little effect on it (e.g. Sparling and Tinker, 1978). When the dicotyledons *Trifolium repens* or *Plantago lanceolata* were grown with *Lolium perenne* (ryegrass) in nutrient-poor soil, under non-mycorrhizal conditions the grass grew well but the dicotyledon remained very small; VAM markedly increased the growth of the dicotyledonous species (Hall, 1978; Buwalda, 1980; Newman and Ritz, 1986). In contrast, when the two grasses *Holcus lanatus* and *Lolium perenne* were grown together (Fitter, 1977) the two were most equal in size when mycorrhizal infection was low, and the size ratio of *Holcus* to *Lolium* progressively increased the greater the amount of infection.

The available evidence supports the hypothesis that where mycorrhizal links can form between dicotyledonous herbs and grasses, growth and survival of the dicotyledons, and hence species diversity, is promoted. However, it is not clear whether links between the dicotyledons and the dominant grasses are essential for this, or whether mycorrhizal infection of the dicotyledons is sufficent.

It has frequently been suggested that species incapable of forming mycorrhizas have poor ability to compete in well established vegetation. Non-mycotrophic species are common at some disturbed sites but uncommon in many stable vegetation types. Gay *et al.* (1982) noted that some "turf-incompatible" species in English chalk grassland, i.e. small plants found only in locally disturbed sites, were non-mycotrophic, whereas "turf-compatible" annuals and biennials, which could grow in dense turf, were mycorrhizal.

There are many vegetation types where plants that normally have different mycorrhizal types coexist. Examples are in tundra, temperate deciduous forest, heathland and savanna. If mycorrhizal links play an important part in seedling establishment, one would expect seedlings to establish better near plants of the same rather than different mycorrhizal type. This has not yet been critically investigated. Tobiessen and Werner (1980) found that the VAM species *Acer rubrum* and *Fraxinus americana* established poorly under the ECM species *Pinus resinosa* in New York state, apparently due to poor VAM infection. However, in nearby stands of *P. sylvestris* there were many saplings of these species which had substantial VAM infection, so there appears to have been a more specific relationship between VAM infection and *P. resinosa* rather than a general influence of ECM. Under ECM oak (*Quercus* spp.) in Britain the commonest saplings are often *Fraxinus excelsior* and *Acer pseudoplatanus*, which are usually VAM-formers (e.g. Linhart and Whelan, 1980). The reverse interaction, grasses reducing ECM infection in *Pinus radiata*, has also been reported (Theodorou and Bowen, 1971).

Herbaceous undergrowth plants in forests are often VAM, and their survival might be influenced by whether they can form mycorrhizal links with the trees. Kovacic *et al.* (1984) reported that in two stands of *Pinus ponderosa* (ECM) in Colorado most of the sparse undergrowth was of non-mycorrhizal plants; but in two neighbouring *P. ponderosa* stands that had been killed by beetle attack there had been a large invasion of VAM undergrowth species. In the Great Smoky Mountains in eastern USA there is a wide variety of forest vegetation. The descriptions published by Whittaker have recently been re-examined (Newman and Reddell, 1988), scoring each tree species for its normal mycorrhiza type. Multiple regression showed no significant relation of number of herb species to altitude or to Whittaker's xeric-mesic scale, but a highly significant positive relation to the proportion of trees that were VAM-formers. In forests dominated by ECM trees most sites had 10 herb species or fewer, whereas forests with 50% or more of VAM-forming trees had 15–30 herb species. This is in agreement with the hypothesis that ability of herbs to form mycorrhizal links with canopy trees promotes survival of at least some of the herb species; but other explanations of these results are possible.

It appears that at some of the sites where plants with different mycorrhizal fungi coexist they are exploiting different soil layers. Högberg and Piearce (1986) observed that in Zambian miombo woodland the ECM of most trees were abundant at 10–15 cm depth, whereas the roots of grasses (which would be VAM) were mostly shallower. In another type of woodland in Zambia, however, this separation did not occur. In conifer forest in central Sweden the undergrowth shrubs, which form ericoid mycorrhizas, rooted predominantly in the surface humus layer, whereas the dominant tree *Pinus sylvestris* (ECM) had many fine roots in deeper mineral soil (Persson, 1980). *In Eucalyptus marginata* forests in Western Australia the ECM formed by certain fungi develop mainly in the litter, whereas other ECM and perhaps VAM on the same tree species are mainly in mineral soil (Reddell and Malajczuk, 1984).

If mycorrhizal links are important in nutrient cycling between plants, nutrients should be preferentially cycled between plants with the same mycorrhizal type. Eason (1987) grew the VAM grass *Molinia caerulea* in pots of heathland soil with two *Calluna vulgaris* (ericoid mycorrhizal). Each plant formed its normal mycorrhiza. ^{32}P was applied to the soil in some pots; ^{32}P uptake, expressed as counts per minute per mg shoot dry weight, was five times as great by *Molinia* as *Calluna*. In other pots one *Calluna* was fed ^{32}P through shoots, and later the whole shoot was removed, leaving the labelled roots in the soil. In these pots *Molinia* and the remaining *Calluna* plant acquired almost equal amounts of ^{32}P. This shows increased relative ability of *Calluna* to acquire phosphorus from dying roots of its own species. This could be due to transfer via mycorrhizal links, though there was no direct proof of mycorrhizal involvement since there were no non-mycorrhizal pots for comparison.

This section has shown that there is a limited amount of evidence supporting the hypothesis that mycorrhizal links favour the survival and growth of seedlings and small herbaceous plants in grassland and forest. But the evidence is at the moment little more than suggestive.

V. CONCLUSIONS

Evidence on almost every aspect of mycorrhizal links and their possible roles is inadequate; any conclusions must therefore be preliminary and tentative.

It is likely that mycorrhizal links between plants are common; but so far few species pairs have been examined for links. There is evidence that seedlings can become infected by forming mycorrhizal links with established plants, and growth of seedlings of some species is faster under these conditions. Transfer of ^{14}C from one plant to another via mycorrhizal links can occur, but it is not clear that net transfer of organic compounds is ever great enough to increase significantly growth or survival of the receiver plant. Transfer of ^{15}N and ^{32}P between plants is often markedly increased when the plants are mycorrhizal; however, most of the available evidence does not support net gain of mineral nutrients by one living plant from another. When roots die, the transfer of phosphorus from them to other plants is increased by VA mycorrhizal links, and the amounts of nutrients involved can be substantial.

Ecologists have for many years viewed relations between living green plants as primarily competitive, and the transfer of nutrients from one to another as occurring when one plant, or part of it, dies. I see no clear evidence that mycorrhizal links prevent these relationships from occurring or introduce fundamentally new interactions between plants. The evidence so far available suggests that mycorrhizal links can alter the relationships between plants, but that they do so by modifying competition and nutrient cycling rather than replacing them.

ACKNOWLEDGEMENTS

Unpublished results by K. Ritz and W. R. Eason were from research supported by the Natural Environment Research Council. I thank Dr D. J. Read and Dr D. M. Eissenstat for helpful suggestions and comments, and Dr K. Ritz and Dr D. J. Read for supplying the photographs for Fig. 1.

REFERENCES

Alexander, C. and Hadley, G. (1985). Carbon movement between host and mycorrhizal endophyte during the development of the orchid *Goodyera repens*. Br. *New Phytol.* **101**, 657–665.

Bethlenfalvay, G. J., Bayne, H. G. and Pacovsky, R. S. (1983). Parasitic and mutualistic associations between a mycorrhizal fungus and soybean: the effects of phosphorus on host plant–endophyte interactions. *Physiol. Plant.* **57**, 543–548.

Birch, C. P. D. (1986). Development of VA mycorrhizal infection in seedlings in semi-natural grassland turf. *Proc. First Europ. Symp. Mycorrhizas*, 233–237.

Björkman, E. (1960). *Monotropa Hypopitys* L.—an epiparasite on tree roots. *Physiol. Plant.* **13**, 308–327.

Brownlee, C., Duddridge, J. A., Malibari, A. and Read, D. J. (1983). The structure and function of mycelial systems of ectomycorrhizal roots with special reference to their role in forming inter-plant connections and providing pathways for assimilate and water transport. *Plant Soil* **71**, 433–443.

Buwalda, J. G. (1980). Growth of clover–ryegrass association with vesicular-arbuscular mycorrhizas. *NZ J. agric. Res.* **23**, 379–383.

Chiariello, N., Hickman, J. C. and Mooney, H. A. (1982). Endomycorrhizal role for interspecific transfer of phosphorus in a community of annual plants. *Science* **217**, 941–943.

Christie, P., Newman, E. I. and Campbell, R. (1978). The influence of neighbouring grassland plants on each others' endomycorrhizas and root-surface microorganisms. *Soil Biol. Biochem.* **10**, 521–527.

Clayton, J. S. and Bagyaraj, D. J. (1984). Vesicular-arbuscular mycorrhizas in submerged aquatic plants of New Zealand. *Aquat. Bot.* **19**, 251–262.

Cook, S. J. and Ratcliff, D. (1984). A study of the effects of root and shoot competition on the growth of green panic (*Panicum maximum* var. *trichoglume*) seedlings in an existing grassland using root exclusion tubes. *J. appl. Ecol.* **21**, 971–982.

Davis, R. M. and Menge, J. A. (1981). *Phytophthora parasitica* inoculation and intensity of vesicular-arbuscular mycorrhizae in citrus. *New Phytol.* **87**, 705–715.

Duddridge, J. A., Finlay, R. D., Read, D. J. and Söderstrom, B. (1988). The structure and function of the vegetative mycelium of ectomycorrhizal plants. III. Ultrastructural and autoradiographic analysis of inter-plant carbon distribution through intact mycelial systems. *New Phytol.*, in press.

Eason, W. R. (1987). *The cycling of phosphorus from dying roots including the role of mycorrhizas.* Ph.D. Thesis, University of Bristol.

Elliott, G. C., Lynch, J. and Lauchli, A. (1984). Influx and efflux of P in roots of intact maize plants. *Plant Physiol.* **76**, 336–341.

Finlay, R. D. and Read, D. J. (1986a). The structure and function of the vegetative mycelium of ectomycorrhizal plants. I. Translocation of ^{14}C-labelled carbon between plants interconnected by a common mycelium. *New Phytol.* **103**, 143–156.

Finlay, R. D. and Read, D. J. (1986b). The structure and function of the vegetative mycelium of ectomycorrhizal plants. II. The uptake and distribution of phosphorus by mycelial strands interconnecting host plants. *New Phytol.* **103**, 157–165.

Fitter, A. H. (1977). Influence of mycorrhizal infection on competition for phosphorus and potassium by two grasses. *New Phytol.* **79**, 119–125.

Fitter, A. H. (1985). Functioning of vesicular-arbuscular mycorrhizas under field conditions. *New Phytol.* **99**, 257–265.

Fleming, L. V. (1983). Succession of mycorrhizal fungi on birch: infection of seedlings planted around mature trees. *Plant Soil* **71**, 263–267.

Fleming, L. V. (1984). Effects of soil trenching and coring on the formation of ectomycorrhizas on birch seedlings grown around mature trees. *New Phytol.* **98**, 143–153.

Fox, F. M. (1983). Role of basidiospores as inocula of mycorrhizal fungi of birch. *Plant Soil* **71**, 269–273.

Francis, R. and Read, D. J. (1984). Direct transfer of carbon between plants connected by vesicular-arbuscular mycorrhizal mycelium. *Nature* **307**, 53–56.

Francis, R., Finlay, R. D. and Read, D. J. (1986). Vesicular-arbuscular mycorrhiza in natural vegetation systems. IV. Transfer of nutrients in inter- and intra-specific combinations of host plants. *New Phytol.* **102**, 103–111.

Gay, P. E., Grubb, P. J. and Hudson, H. J. (1982). Seasonal changes in the concentrations of nitrogen, phosphorus and potassium, and in the density of mycorrhiza, in biennial and matrix-forming perennial species of closed chalkland turf. *J. Ecol.* **70**, 571–593.

Graham, B. F. and Bormann, F. H. (1966). Natural root grafts. *Bot. Rev.* **32**, 255–292.

Grime, J. P., Mackey, J. M. L., Hillier, S. H. and Read, D. J. (1987). Floristic diversity in a model system using experimental microcosms. *Nature* **328**, 420–422.

Hall, I. R. (1978). Effects of endomycorrhizas on the competitive ability of white clover. *NZ J. agric. Res.* **21**, 509–515.

Harley, J. L. and Harley, E. L. (1987). A checklist of mycorrhiza in the British flora. *New Phytol.* (Suppl.) **105**, 1–102.

Harley, J. L. and Smith, S. E. (1983). *Mycorrhizal Symbiosis*. London: Academic Press.

Haystead, A. (1983). The efficiency of utilization of biologically fixed nitrogen in crop production systems. In: *Temperate Legumes* (Ed. by D. G. Jones and D. R. Davies), pp. 395–415. London: Pitman.

Heap, A. J. and Newman, E. I. (1980a). Links between roots by hyphae of vesicular-arbuscular mycorrhizas. *New Phytol.* **85**, 169–171.

Heap, A. J. and Newman, E. I. (1980b). The influence of vesicular-arbuscular mycorrhizas on phosphorus transfer between plants. *New Phytol.* **85**, 173–179.

Hirrel, M. C. and Gerdemann, J. W. (1979). Enhanced carbon transfer between onions infected with a vesicular-arbuscular mycorrhizal fungus. *New Phytol.* **83**, 731–738.

Hirrel, M. C., Mehravaran, H. and Gerdemann, J. W. (1978). Vesicular-arbuscular mycorrhizae in the Chenopodiaceae and Cruciferae: do they occur? *Can. J. Bot.* **56**, 2813–2817.

Högberg, P. and Piearce, G. D. (1986). Mycorrhizas in Zambian trees in relation to host taxonomy, vegetation type and successional patterns. *J. Ecol.* **74**, 775–785.

Jarvis, P. G. (1964). The adaptability to light intensity of seedlings of *Quercus petraea* (Matt.) Liebl. *J. Ecol.* **52**, 545–571.

Kessel, C. van, Singleton, P. W. and Hoben, H. J. (1985). Enhanced N-transfer from a soybean to maize via vesicular arbuscular mycorrhizal (VAM) fungi. *Plant Physiol.* **79**, 562–563.

Koide, R. (1985). The nature of growth depressions in sunflower caused by vesicular-arbuscular mycorrhizal infection. *New Phytol.* **99**, 449–462.

Kovacic, D. A., St John, T. V. and Dyer, D. I. (1984). Lack of vesicular-arbuscular mycorrhizal inoculum in a ponderosa pine forest. *Ecology* **65**, 1755–1759.

Lawley, R. A., Newman, E. I. and Campbell, R. (1982). Abundance of endomycorrhizas and root-surface microorganisms on three grasses grown separately and in mixtures. *Soil Biol. Biochem.* **14**, 237–240.

Ledgard, S. F., Freney, J. R. and Simpson, J. R. (1985). Assessing nitrogen transfer from legumes to associated grasses. *Soil. Biol. Biochem.* **17**, 575–577.

Linhart, Y. B. and Whelan, R. J. (1980). Woodland regeneration in relation to grazing and fencing in Coed Gorswen, North Wales. *J. appl. Ecol.* **17**, 827–840.

Mason, P. A., Wilson, J., Last, F. T. and Walker, C. (1983). The concept of succession in relation to the spread of sheathing mycorrhizal fungi on inoculated tree seedlings growing in unsterile soils. *Plant Soil* **71**, 247–256.

Medve, R. J. (1984). The mycorrhizae of pioneer species in disturbed ecosystems in western Pennsylvania. *Am. J. Bot.* **71**, 787–794.

Miller, R. M., Moorman, T. B. and Schmidt, S. K. (1983). Interspecific plant association effects on vesicular-arbuscular mycorrhiza occurrence in *Atriplex confertifolia*. *New Phytol.* **95**, 241–246.

Molina, R. and Trappe, J. M. (1982). Patterns of ectomycorrhizal host specificity and potential among Pacific Northwest conifers and fungi. *Forest Sci.* **28**, 423–458.

Morgan, M. A., Volk, R. S. and Jackson, W. A. (1973). Simultaneous influx and efflux of nitrate during uptake by perennial ryegrass. *Plant Physiol.* **51**, 267–272.

Nelsen, C. E. and Safir, G. R. (1982). Increased drought tolerance of mycorrhizal onion plants caused by improved phosphorus nutrition. *Planta* **154**, 407–413.

Newman, E. I. and Reddell, P. (1987). The distribution of mycorrhizas among families of vascular plants. *New Phytol.* **106**, 745–751.

Newman, E. I. and Reddell, P. (1988). Relationship between mycorrhizal infection and diversity in vegetation: evidence from the Great Smoky Mountains. *Functional Ecol.*, in press.

Newman, E. I. and Ritz, K. (1986). Evidence on the pathways of phosphorus transfer between vesicular-arbuscular mycorrhizal plants. *New Phytol.* **104**, 77–87.

Ocampo, J. A. (1986). Vesicular-arbuscular mycorrhizal infection of "host" and "non-host" plants: effect on the growth responses of the plants and competition between them. *Soil Biol. Biochem.* **18**, 607–610.

Ocampo, J. A., Martin, J. and Hayman, D. S. (1980). Influence of plant interactions on vesicular-arbuscular mycorrhizal infections. I. Host and non-host plants grown together. *New Phytol.* **84**, 27–35.

Parke, J. L., Linderman, R. G. and Black, C. H. (1983). The role of ectomycorrhizas in drought tolerance of Douglas-fir seedlings. *New Phytol.* **95**, 83–95.

Pendleton, R. L. and Smith, B. N. (1983). Vesicular-arbuscular mycorrhizae of weedy and colonizer plant species at disturbed sites in Utah. *Oecologia* **59**, 296–301.

Persson, H. (1980). Death and replacement of fine roots in a mature Scots pine stand. In: *Structure and Function of Northern Coniferous Forest—An Ecosystem Study* (Ed. by T. Persson). *Ecol. Bull.* (*Stockholm*) **32**, 251–260.

Rakhteyenko, I. N. (1958). The transfer of mineral nutrients from one plant to another through the interaction of their root systems. *Bot. Zhur.* **43**, 695–701. (In Russian.)

Read, D. J., Francis, R. and Finlay, R. D. (1985). Mycorrhizal mycelia and nutrient cycling in plant communities. In: *Ecological Interactions in Soil* (Ed. by A. H. Fitter), pp. 193–217. Oxford: Blackwell Scientific.

Read, D. J., Koucheki, H. K. and Hodgson, J. (1976). Vesicular-arbuscular mycorrhiza in natural vegetation systems. I. The occurrence of infection. *New Phytol.* **77**, 641–653.

Reddell, P. and Malajczuk, N. (1984). Formation of mycorrhizae by jarrah (*Eucalyptus marginata* Donn. ex Smith) in litter and soil. *Austr. J. Bot.* **32**, 511–520.

Reid, C. P. P. and Woods, F. W. (1969). Translocation of C^{14}-labeled compounds in mycorrhizae and its implications in interplant nutrient cycling. *Ecology* **50**, 179–187.

Ritz, K. (1984). *Phosphorus transfer between grassland plants.* Ph.D. Thesis, University of Bristol.

Ritz, K. and Newman, E. I. (1984). Movement of ^{32}P between intact grassland plants of the same age. *Oikos* **43**, 138–142.

Ritz, K. and Newman, E. I. (1985). Evidence for rapid cycling of phosphorus from dying roots to living plants. *Oikos* **45**, 174–180.

Ritz, K. and Newman, E. I. (1986). Nutrient transport between ryegrass plants differing in nutrient status. *Oecologia* **70**, 128–131.

Sims, P. L. and Singh, J. S. (1978). The structure and function of ten western North American grasslands. III. Net primary production, turnover and efficiencies of energy capture and water use. *J. Ecol.* **66**, 573–597.

Smith, S. E. (1966). Physiology and ecology of orchid mycorrhizal fungi with reference to seedling nutrition. *New Phytol.* **65**, 488–499.

Smith, S. E. (1967). Carbohydrate translocation in orchid mycorrhizas. *New Phytol.* **66**, 371–378.

Smith, S. E. (1980). Mycorrhizas of autotrophic higher plants. *Biol. Rev.* **55**, 475–510.

Snaydon, R. W. and Howe, C. D. (1986). Root and shoot competition between established ryegrass and invading grass seedlings. *J. appl. Ecol.* **23**, 667–674.

Sparling, G. P. and Tinker, P. B. (1978). Mycorrhizal infection in Pennine grassland. II. Effects of mycorrhizal infection on growth of some upland grasses on γ-irradiated soils. *J. appl. Ecol.* **15**, 951–958.

Theodorou, C. and Bowen, G. D. (1971). Effects of non-host plants on growth of mycorrhizal fungi of radiata pine. *Austr. For.* **35**, 17–22.

Tobiessen, P. and Werner, M. B. (1980). Hardwood seedling survival under plantations of Scotch pine and red pine in central New York. *Ecology* **61**, 25–29.

Tommerup, I. C. (1985). Inhibition of spore germination of vesicular-arbuscular mycorrhizal fungi in soil. *Trans. Br. Mycol. Soc.* **85**, 267–278.

Vogt, K. A., Grier, C. C. and Vogt, D. J. (1986). Production, turnover, and nutrient dynamics of above- and belowground detritus of world forests. *Adv. Ecol. Res.* **15**, 303–377.

Whitehead, D. C. (1970). The Role of Nitrogen in Grassland Productivity. Commonwealth Agricultural Bureaux, Farnham Royal.

Whittingham, J. and Read, D. J. (1982). Vesicular-arbuscular mycorrhiza in natural vegetation systems. III. Nutrient transfer between plants with mycorrhizal interconnections. *New Phytol.* **90**, 277–284.

Woods, F. W. (1970). Interspecific transfer of inorganic materials by root systems of woody plants. *J. appl. Ecol.* **7**, 481–486.

Woods, F. W. and Brock, K. (1964). Interspecific transfer of Ca-45 and P-32 by root systems. *Ecology* **45**, 886–889.

A Theory of Gradient Analysis

CAJO J. F. TER BRAAK AND I. COLIN PRENTICE

I.	Introduction	272
II.	Linear Methods	273
	A. Regression	277
	B. Calibration	278
	C. Ordination	278
	D. The Environmental Interpretation of Ordination Axes (Indirect Gradient Analysis)	280
	E. Constrained Ordination (Multivariate Direct Gradient Analysis)	281
III.	Non-Linear (Gaussian) Methods	282
	A. Unimodal Response Models	282
	B. Regression	285
	C. Calibration	285
	D. Ordination	286
	E. Constrained Ordination	286
IV.	Weighted Averaging Methods	287
	A. Regression	287
	B. Calibration	288
	C. Ordination	290
	D. Constrained Ordination	293
V.	Ordination Diagrams and Their Interpretation	295
	A. Principal Components: Biplots	295
	B. Correspondence Analysis: Joints Plots	296
	C. Redundancy Analysis	299
	D. Canonical Correspondence Analysis	300
VI.	Choosing the Method	302
	A. Which Response Model?	302
	B. Direct or Indirect?	304
	C. Direct Gradient Analysis: Regression or Constrained Ordination?	305
VII.	Conclusions	306
Acknowledgements		308
References		309
Appendix		313

ADVANCES IN ECOLOGICAL RESEARCH Vol. 18
ISBN 0-12-013918-9

I. INTRODUCTION

All species occur in a characteristic, limited range of habitats; and within their range, they tend to be most abundant around their particular environmental optimum. The composition of biotic communities thus changes along environmental gradients. Successive species replacements occur as a function of variation in the environment, or (analogously) with successional time (Pickett, 1980; Peet and Loucks, 1977). The concept of niche space partitioning also implies the separation of species along "resource gradients" (Tilman, 1982). Gradients do not necessarily have physical reality as continua in either space or time, but are a useful abstraction for explaining the distributions of organisms in space and time (Austin, 1985). Austin's review explores the interrelationships between niche theory and the concepts of ecological continua and gradients.

Our review concerns data analysis techniques that assist the interpretation of community composition in terms of species' responses to environmental gradients in the broadest sense. Gradient analysis *sensu lato* includes direct gradient analysis, in which each species' abundance (or probability of occurrence) is described as a function of measured environmental variables; the converse of direct gradient analysis, whereby environmental values are inferred from the species composition of the community; and indirect gradient analysis, *sensu* Whittaker (1967), in which community samples are displayed along axes of variation in composition that can subsequently be interpreted in terms of environmental gradients. There are close relationships among these three types of analysis. Direct gradient analysis is a *regression* problem—fitting curves or surfaces to the relation between each species' abundance or probability of occurrence (the response variable) and one or more environmental variables (the predictor variable(s)) (Austin, 1971). Inferring environmental values from species composition when these relationships are known is a *calibration* problem. Indirect gradient analysis is an *ordination* problem, in which axes of variation are derived from the total community data. Ordination axes can be considered as latent variables, or hypothetical environmental variables, constructed in such a way as to optimize the fit of the species data to a particular (linear or unimodal) statistical model of how species abundance varies along gradients (Ter Braak, 1985, 1987a). These latent variables are constructed without reference to environmental measurements, but they can subsequently be compared with actual environmental data if available. To these three well-known types of gradient analysis we add a fourth, *constrained ordination*, which has its roots in the psychometric literature on multidimensional scaling (Bloxom, 1978; De Leeuw and Heiser, 1980; Heiser, 1981). Constrained ordination also constructs axes of variation in overall community composition, but does so in such a way as to explicitly optimize the fit to supplied environmental data

(Ter Braak, 1986; Jongman *et al.*, 1987). Constrained ordination is thus a multivariate generalization of direct gradient analysis, combining aspects of regression, calibration and ordination. Table 1 gives an arbitrary selection of literature references, chosen simply to illustrate the wide range of ecological problems to which each of the four types of gradient analysis has been applied; the reader is also referred to Gauch (1982), who includes an extensive bibliography, and to Gittins (1985).

Standard statistical methods that assume linear relationships among variables exist for all four types of problems (regression, calibration, ordination and constrained ordination), but have found only limited application in ecology because of the generally non-linear, non-monotone response of species to environmental variables. Ecologists have independently developed a variety of alternative techniques. Many of these techniques are essentially heuristic, and have a less secure theoretical basis. These heuristic techniques can nevertheless give useful results, and can be understood as approximate solutions to statistical problems similar to those solved by standard methods, but formulated in terms of a *unimodal* (Gaussian or similar) response model instead of a linear one. We present here a theory of gradient analysis, in which the heuristic techniques are integrated with regression, calibration, ordination and constrained ordination as distinct, well-defined statistical problems.

The various techniques used for each type of problem are classified into families according to their implicit response model and the method used to estimate parameters of the model. We consider three such families (Table 2). First we treat the family of standard statistical techniques based on the linear response model, because these are conceptually the simplest and provide a basis for what follows, even though their ecological application is restricted. Second, we outline a family of somewhat more complex statistical techniques which are formal extensions of the standard linear techniques and incorporate unimodal (Gaussian-like) response models explicitly. Finally, we consider the family of heuristic techniques based on weighted averaging. These are not more complex than the standard linear techniques, but implicitly fit a simple unimodal response model rather than a linear one. Our treatment thus unites such apparently disparate data analysis techniques as linear regression, principal components analysis, redundancy analysis, Gaussian ordination, weighted averaging, reciprocal averaging, detrended correspondence analysis and canonical correspondence analysis in a single theoretical framework.

II. LINEAR METHODS

Species abundances may seem to change linearly through *short* sections of environmental gradients, so a linear response model may be a reasonable

Table 1
Selected applications of gradient analysis

Type of problem	Taxa	Environmental variables	Purpose of study
Regression			
Alderdice (1972)	Marine fish	Salinity, temperature	Defining ranges
Peet (1978)	Trees	Elevation, moisture, latitude	Biogeography
Wiens and Rotenberry (1981)	Birds	Vegetation structure	Niche characterization
Austin *et al.* (1984)	*Eucalyptus* spp.	Climatic indices	Habitat characterization
Bartlein *et al.* (1986)	Plant pollen types	Temperature, precipitation	Quaternary palaeoecology
Calibration			
Chandler (1970)	Benthic macro-invertebrates	Water pollution	Water quality management
Imbrie and Kipp (1971)	Foraminifera	Sea surface temperature	Palaeoclimatic reconstruction
Sládecek (1973)	Freshwater algae	Organic pollution	Ecological monitoring
Balloch *et al.* (1976)	Benthic macro-invertebrates	Water pollution	Ecological monitoring
Ellenberg (1979)	Terrestrial plants	Soil moisture, N, pH	Bioassay from vegetation
Van Dam *et al.* (1981)	Diatoms	pH	Acid rain effects
Böcker *et al.* (1983)	Terrestrial plants	Soil moisture, N, pH	Bioassay from vegetation
Bartlein *et al.* (1984)	Plant pollen types	Temperature, precipitation	Palaeoclimatic reconstruction
Battarbee (1984)	Diatoms	pH	Acid rain effects
Charles (1985)	Diatoms	pH	Acid rain effects
Atkinson *et al.* (1986)	Beetles	Summer temperature, annual	Palaeoclimatic reconstruction

Van der Aart and Smeenk–Enserink (1975)	Spiders	Microenvironmental features	Habitat characterization
Kooijman and Hengeveld (1979)	Beetles	Lutum content, elevation	Habitat characterization
Wiens and Rotenberry (1981)	Birds	Vegetation structure	Niche characterization
Prodon and Lebreton (1981)	Birds	Vegetation structure	Niche characterization
Kalkhoven and Opdam (1984)	Birds	Habitat and landscape features	Habitat characterization
Macdonald and Ritchie (1986)	Plant pollen types	Vegetation regions	Quaternary palaeoecology
Constrained ordination			
Webb and Bryson (1972)	Plant pollen types	Climate variables, airmass frequencies	Palaeoclimatic reconstruction
Gasse and Tekaia (1983)	Diatoms	pH classes	Palaeolimnology
Ås (1985)	Beetles	Vegetation types	Niche theory
Cramer and Hytteborn (1987)	Terrestrial plants	Time, elevation	Land uplift effects
Purata (1986)	Tropical trees	Successional boundary conditions	Study of secondary succession
Fängström and Willén (1987)	Phytoplankton	Physical/chemical variables	Environmental monitoring

[a]Excluding vegetation studies, where ordination is used routinely: see Gauch (1982) for a review.

Table 2

Classification of gradient analysis techniques by type of problem, response model and method of estimation.

	Linear response model	Unimodal response model	
Type of problem	Least-squares estimation	Maximum likelihood estimation	Weighted averaging estimation
Regression	Multiple regression	Gaussian regression	Weighted averaging of site scores (WA)
Calibration	Linear calibration; "inverse regression"	Gaussian calibration	Weighted averaging of species' scores (WA)
Ordination	Principal components analysis (PCA)	Gaussian ordination	Correspondence analysis (CA); detrended correspondence analysis (DCA)
Constrained ordination[a]	Redundancy analysis (RDA)[d]	Gaussian canonical ordination	Canonical correspondence analysis (CCA); detrended CCA
Partial ordination[b]	Partial components analysis	Partial Gaussian ordination	Partial correspondence analysis; partial DCA
Partial constrained ordination[c]	Partial redundancy analysis	Partial Gaussian canonical ordination	Partial canonical correspondence analysis; partial detrended CCA

[a] Constrained multivariate regression.
[b] Ordination after regression on covariables.
[c] Constrained ordination after regression on covariables = constrained partial multivariate regression.
[d] "Reduced-rank regression" = "PCA of y with respect to x".

basis for analysing quantitative abundance data spanning a narrow range of environmental variation.

A. Regression

If a plot of the abundance (y) of a species against an environmental variable (x) looks linear, or can easily be transformed to linearity, then it is appropriate to fit a straight line by linear regression. The formula $y = a + bx$ describes the linear relation, with a the intercept of the line on the y-axis and b the slope of the line, or regression coefficient. Separate regressions can be carried out for each of m species.

We are usually most interested in how the abundance of each species changes with a change in the environmental variable, i.e. in the slopes b_k (the index k refers to species k). If we first centre the data—by subtracting the mean of each species' abundances from the species data and the mean of the environmental values from the environmental data—the intercept disappears. Then if y_{ki} denotes the centred abundance of species k in the ith out of n sites, and x_i the centred environmental value for that site, the response model for fitting the straight lines becomes

$$y_{ki} = b_k x_i + e_{ki} \tag{1}$$

where e_{ki} is an error component with zero mean and variance v_{ki}. The standard estimator for the slope in equation (1) is

$$\tilde{b}_k = \sum_{i=1}^{n} y_{ki} x_i / s_x^2 \tag{2}$$

where $s_x^2 = \sum_{i=1}^{n} x_i^2$. This is the least-squares estimator, which is the best linear unbiased estimator when errors are uncorrelated and homogeneous across sites ($v_{ki} = v_k$). It is also the maximum likelihood (ML) estimator when the errors are normally distributed. The fitted lines can be used to predict the abundances of species in a site with a known value of the environmental variable simply by reading off the graph.

Species experience the effect of more than one environmental variable simultaneously, so more than one variable may be required to account for variation in species abundances. The joint effect of two or more environmental variables on a species can be analysed by multiple regression (see e.g. Montgomery and Peck, 1982). Standard computer packages are available to obtain least-squares (ML) estimates for the regression coefficients. Only when the environmental variables are uncorrelated will the partial regression coefficients be identical to the coefficients estimated by separate regressions using equation (1).

B. Calibration

We now turn to the inverse problem, calibration. When the relationship between the abundances of species and the environmental variable we are interested in is known, we can infer values of that environmental variable for new sites from the observed species abundances. If we took into account the abundance of only a single species, we could simply read off the graph, starting from a value on the vertical axis. However, another species may well give a different estimate. We therefore need a good and unambiguous estimator that combines the information from all m species. In terms of equation (1), the b_k are now assumed to be known and x_i is unknown. The role of the b_k and x_i have been interchanged. By interchanging their roles in equation (2) as well, we obtain

$$\tilde{x}_i = \sum_{k=1}^{m} y_{ki} b_k / s_b^2 \qquad (3)$$

where $s_b^2 = \sum_{k=1}^{m} b_k^2$. This is the least-squares estimator (and the ML-estimator) when the errors follow a normal distribution and are independent and homogeneous across species ($v_{ki} = v_i$).

A problem with equation (3) is that these conditions are likely to be unrealistic, because effects of other environmental variables can cause correlation between the abundances of different species even after the effects of the environmental variable of interest have been removed. Further, the residual variance v_{ki} may be different for different species. If this occurs, we also need to take the residual correlations and variances into account. In practice, the residual correlations and variances are estimated from the residuals of the regressions used for estimating the b_k's. Searching for the maximum of the likelihood with respect to x_i then leads to a general weighted least-squares problem (Brown, 1979; Brown, 1982) that can be solved by using standard algorithms.

Inferring values of more than one environmental variable simultaneously has been given surprisingly little attention in the literature. However, Williams (1959) and Brown (1982) derived the necessary formulae from the ML-principle (Cox and Hinkley, 1974).

C. Ordination

After having fitted a particular environmental variable to the species data by regression, we might ask whether another environmental variable would provide a better fit. For some species one variable may fit better, and for other species another variable. To get an overall impression we might judge the goodness-of-fit (explanatory power) of an environmental variable by the total regression sum of squares (Jongman *et al.*, 1987). The question then arises: what is the best possible fit that is theoretically obtainable with the straight line model of equation (1)?

This question defines an ordination problem, i.e. to construct the single "hypothetical environmental variable" that gives the best fit to the species data according to equation (1). This hypothetical environmental variable is termed the *latent variable*, or simply the (first) ordination axis. Principal components analysis (PCA) provides the solution to this ordination problem. In equation (1), x_i is then the score of site i on the latent variable, b_k is the slope for species k with respect to the latent variable (also called the species loading or species score) and the eigenvalue of the first PCA axis is equal to the goodness-of-fit, i.e. the total sum of squares of the regressions of the species abundances on the latent variable. PCA provides the least-squares estimates of the site and species scores: these estimates are also ML estimates if the errors are independently and normally distributed with constant variance ($v_{ki} = v$).

PCA is usually performed using a standard computer package, but several different algorithms can be used to do the same job. The following algorithm, known as the power method (Gourlay and Watson, 1973), makes the relationship between PCA and regression and calibration clear in a way that the usual textbook treatment, in terms of singular value decomposition of inner product matrices, does not; it also facilitates comparison with correspondence analysis, which we discuss later. The power method shows that PCA can be obtained by an alternating sequence of linear regressions and calibrations:

Step 1 Start with some (arbitrary) initial site scores $\{x_i\}$ with zero mean.
Step 2 Calculate new species scores $\{b_k\}$ by linear regression (equation (2)).
Step 3 Calculate new site scores $\{x_i\}$ by linear calibration (equation (3)).
Step 4 Remove the arbitrariness in scale by standardizing the site scores as follows: new x_i = old $x_i\sqrt{n}/s_x$, with s_x as defined beneath equation (2).
Step 5 Stop on convergence, i.e. when the newly obtained site scores are close to the site scores of the previous cycle of iteration, else go to Step 2.

The final scores do not depend on the initial scores.

The ordination problem for a two-dimensional linear model turns out to be relatively simple, compared with the regression and calibration problems. The solution does not need an alternating sequence of *multiple* regressions and calibrations, because the latent variables can always be chosen in such a way that they are uncorrelated; and if the latent variables are uncorrelated, then the multiple regressions and calibrations reduce to a series of separate linear regressions and calibrations. PCA provides the solution to the linear ordination problem in any number of dimensions; one latent variable is derived first, as in the one-dimensional case of equation (1), and the second latent variable can be obtained next by applying the same algorithm again but with one extra step—after Step 3, the trial scores are made uncorrelated with the first latent variable. On denoting the scores of the first axis by x_{il}, this orthogonalization is computed by

Step 3b Calculate $f = \Sigma_i x_i x_{i1}/n$,
Calculate new x_i = old $x_i - f x_{i1}$.

Further latent variables (ordination axes) may be derived analogously. As in the one-dimensional case, PCA provides the ML-solution to the multidimensional linear ordination problem if the errors are independently and normally distributed with constant variance across species and sites. Jolliffe (1986) reviews the theory and applications of PCA.

D. The Environmental Interpretation of Ordination Axes (Indirect Gradient Analysis)

In indirect gradient analysis the species data are first subjected to ordination, e.g. using PCA, to find a few major axes of variation (latent variables) with a good fit to the species data. These axes are then interpreted in terms of known variation in the environment, often by using graphical methods (Gauch, 1982). A more formal method for the latter step would be to calculate correlation coefficients between environmental variables and each of the ordination axes. This analysis is similar to performing a multiple regression of each separate environmental variable on the axes (Dargie, 1984), because the axes are uncorrelated. A joint analysis of all environmental variables can be carried out by multiple regression of each ordination axis on the environmental variables:

$$x_i = c_0 + \sum_{j=1}^{q} c_j z_{ji} \qquad (4)$$

in which x_i is the score of site i on that one ordination axis, z_{ij} denotes the value at site i of the jth out of q actual environmental variables, and c_j is the corresponding regression coefficient. For later reference, the error term in equation (4) is not shown. The multiple correlation coefficient R measures how well the environmental variables explain the ordination axis.

E. Constrained Ordination (Multivariate Direct Gradient Analysis)

Indirect gradient analysis, as outlined above, is a *two-step* approach to relate species data to environmental variables. A few ordination axes that summarize the overall community variation are extracted in the first step; then in the second step one may calculate weighted sums of the environmental variables that most closely fit each of these ordination axes. However, the environmental variables that have been studied may turn out to be poorly related to the first few ordination axes, yet may be strongly related to other, "residual" directions of variation in species composition. Unless the first few ordination axes explain a very high proportion of the variation, this residual variation can be substantial, and strong relationships between species and environment can potentially be missed.

In constrained ordination this approach is made more powerful by combining the two steps into one. The idea of constrained ordination is to search for a few weighted sums of environmental variables that fit the data of all species best, i.e. that give the maximum total regression sum of squares. The resulting technique, redundancy analysis (Rao, 1964; Van den Wollenberg, 1977), is an ordination of the species data in which the axes are constrained to be linear combinations of the environmental variables. These axes can be found by extending the algorithm of PCA described above with one extra step, to be performed directly after Step 3 (Jongman *et al.*, 1987):

Step 3a Calculate a multiple regression of the site scores $\{x_i\}$ on the environmental variables (equation (4)), and take as new site scores the fitted values of this regression.

The regression is thus carried out within the iteration algorithm, instead of afterwards. On convergence, the coefficients $\{c_j\}$ are termed canonical coefficients and the multiple correlation coefficient in Step 3a can be called the species–environment correlation.

Redundancy analysis is also known as reduced-rank regression (Davies and Tso, 1982), PCA of y with respect to x (Robert and Escoufier, 1976) and two-block mode C partial least-squares (Wold, 1982). It is intermediate between PCA and separate multiple regressions for each of the species: it is a constrained ordination, but it is also a constrained form of (multivariate) multiple regression (Davies and Tso, 1982; Israëls, 1984). By inserting equation (4) into equation (1), it can be shown that the "regression" coefficient of species k with respect to environmental variable j takes the simple form $b_k c_j$. With two ordination axes this form would be, in obvious notation, $b_{k1}c_{j1} + b_{k2}c_{j2}$. With two ordination axes, redundancy analysis thus uses $2(q + m) + m$ parameters to describe the species data, whereas the multiple regressions use $m(q + 1)$ parameters. One of the attractive features of redundancy analysis is that it leads to an ordination diagram that simultaneously displays (i) the main pattern of community variation as far as this variation can be explained by the environmental variables, and (ii) the main pattern in the correlation coefficients between the species and each of the environmental variables. We give an example of such a diagram later on.

Redundancy analysis is much less well known than canonical correlation analysis (Gittins, 1985; Tso, 1981), which is the standard linear multivariate technique for relating two sets of variables (in our case, the set of species and the set of environmental variables). Canonical correlation analysis is very similar to redundancy analysis, but differs from it in the assumptions about the error component: uncorrelated errors with equal variance in redundancy analysis and correlated normal errors in canonical correlation analysis (Tso, 1981; Jongman *et al.*, 1987). The most important practical difference is that redundancy analysis can analyse any number of species whereas in canonical correlation analysis the number of species (m) must be less than $n - q$ (Griffins, 1985:24); this restriction is often a nuisance.

Canonical variates analysis, or multiple discriminant analysis, is simply the special case of canonical correlation analysis in which the "environmental" variables are a series of dummy variables reflecting a single-factor classification of the samples. A similar restriction on the number of species thus also applies to canonical variates analysis. Redundancy analysis with dummy variables provides an alternative to canonical variates analysis, evading this restriction.

III. NON-LINEAR (GAUSSIAN) METHODS

A. Unimodal Response Models

Linear methods are appropriate to community analysis only when the species data are quantitative abundances (with few zeroes) and the range of environmental variation in the sample set is narrow. Alternative analytical methods can be derived from unimodal models.

A unimodal response model for one environmental variable can be obtained by adding a quadratic term (x_i^2) to the linear model, changing the response curve from a straight line into a parabola. But this quadratic model can predict large negative values, whereas species abundances are always zero or positive. A simple remedy for the problem of negative values is provided by the Gaussian response curve (Gauch and Whittaker, 1972) in which the *logarithm* of species abundance is a quadratic in the environmental variable:

$$\log y = b_0 + b_1 x + b_2 x^2$$
$$= a - \tfrac{1}{2}(x - u)^2/t^2 \qquad (5a)$$

where $b_2 < 0$ (otherwise the curve would have a minimum instead of a mode). The coefficients b_0, b_1 and b_2 are most easily interpreted by transformation to u, t and a (Fig. 1), u being the species' optimum (the value of x at the peak), t being its tolerance (a measure of response breadth or ecological amplitude), and a being a coefficient related to the height of the peak (Ter Braak and Looman, 1986).

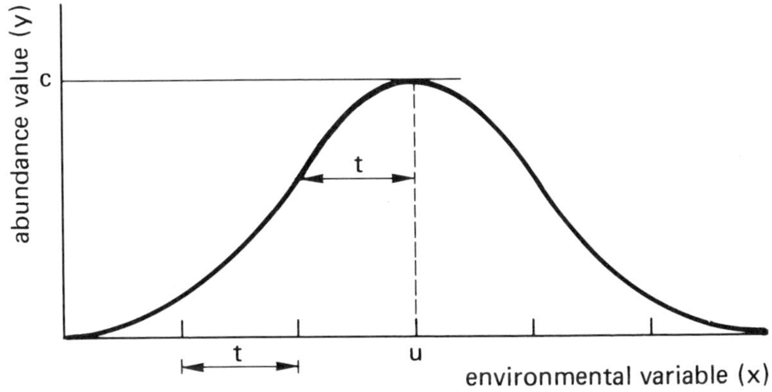

Fig. 1. A Gaussian curve displays a unimodal relation between the abundance value (y) of a species and an environmental variable (x). (u = optimum or mode; t = tolerance; c = maximum = exp (a)).

A closely related model can describe species data in presence–absence form. In analysing presence–absence data, we want to relate *probability of occurrence* (p) to environment. Probabilities are never greater than 1, so rather than using equation (5a) we use the Gaussian logit model,

$$\log\left(\frac{p}{1-p}\right) = b_0 + b_1 x + b_2 x^2 \qquad (5b)$$

which is very similar to the Gaussian model unless the peak probability is high (> 0·5); then equation (5b) gives a curve that is somewhat flatter on top. The coefficients b_0, b_1 and b_2 can be transformed as before into coefficients representing the species' optimum, tolerance and maximum probability value.

Although real ecological response curves are still more complex than implied by the Gaussian and Gaussian logit models, these models are nevertheless useful in developing statistical descriptive techniques for data showing mostly unimodal responses, just as linear models are useful in statistical analysis of data that are only approximately linear.

With two environmental variables, equations (5a) and (5b) become full quadratic forms with both square and product terms (Alderdice, 1972). For example, the Gaussian model becomes

$$\log y = b_0 + b_1 x_1 + b_2 x_1^2 + b_3 x_2 + b_4 x_2^2 + b_5 x_1 x_2 \tag{6}$$

If $b_2 + b_4 < 0$, and $4b_2 b_4 - b_5^2 > 0$ then equation (6) describes a unimodal surface with ellipsoidal contours (Fig. 2). If one of these conditions is not satisfied then equation (6) describes a surface with a minimum, or with a saddle point (e.g. Davison, 1983). Provided the surface is unimodal, its optimum (u_1, u_2) can be calculated from the coefficients in equation (6) by

$$\left.\begin{aligned} u_1 &= (b_5 b_3 - 2b_1 b_4)/d \\ u_2 &= (b_5 b_1 - 2b_3 b_2)/d \end{aligned}\right\} \tag{7}$$

where $d = 4b_2 b_4 - b_5^2$. When $b_5 \neq 0$, the optimum with respect to x_1, depends on the value of x_2; the environmental variables are then said to show interaction in their effect on the species. In contrast, when $b_5 = 0$ the optimum with respect to x_1 does not depend on the value of x_2 (no interaction) and equation (7) simplifies considerably (Ter Braak and Looman, 1986).

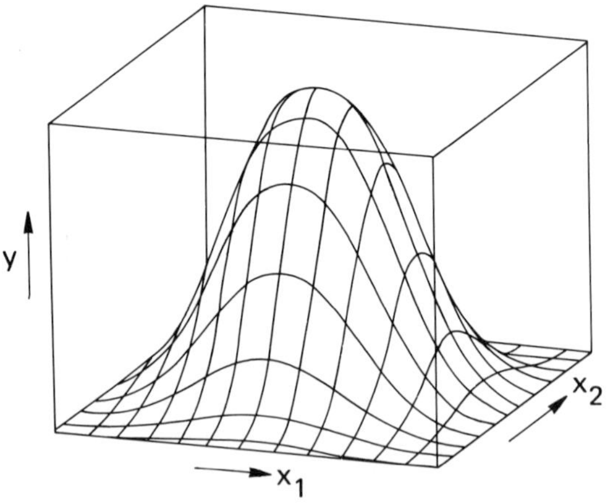

Fig. 2. A Gaussian surface displays a unimodal relation between the abundance value (y) of a species and two environmental variables (x_1 and x_2).

The unknown parameters of non-linear response models in the context of regression, calibration or ordination can (at least in theory) be estimated by the maximum likelihood principle, however difficult this may be in a particular situation. Usually iterative methods are required, and initial parameter values must be specified. The likelihood function may have local maxima, so that different sets of initial parameter values may result in different final estimates. It cannot be guaranteed that the global maximum has been found. Furthermore, all kinds of numerical problems may occur. However, the special cases of Gaussian and Gaussian logit response models do allow reasonably practical solutions, which we consider now.

B. Regression

The regression problems of fitting Gaussian or Gaussian logit curves or surfaces are relatively straightforward, since these models can be fitted by Generalized Linear Modelling (GLM: McCullagh and Nelder, 1983; Dobson, 1983). An elementary introduction to GLM directed at ecologists is provided by Jongman *et al.* (1987). GLM is more flexible than ordinary multiple regression because one can specify "link functions" and error distributions other than the normal distribution. For example, the Gaussian models of equations (5a) and (6) can be fitted with GLM to abundance data (which may include zeroes) by specifying the link function to be logarithmic and the error distribution to be Poissonian. The corresponding Gaussian logit models can be fitted with GLM to presence–absence data by specifying the link function to be logistic and the error distribution to be binomial-with-total-1. Alternatively, any statistical package that will do logit (= logistic) regression can be used to fit the Gaussian logit model. No initial estimates are needed and local maxima do not arise, so these techniques are quite practical for direct gradient analysis. For examples of the use of GLM in ecology see Austin and Cunningham (1981) and Austin *et al.* (1984).

The most common complications arise when the optimum for a species is estimated well outside the sampled range of environments, or if the fitted curve shows a minimum rather than a peak. These conditions suggest that the regression is ill-determined and that it might be better to fit a monotone curve by setting $b_2 = 0$ in equations (5); a statistical test can be used to determine whether this simplification is acceptable (Jongman *et al.*, 1987). Such cases are bound to arise in practice because any given set of samples will include some species that are near the edge of their range.

C. Calibration

The calibration problem of inferring environmental values at sites from species data and known Gaussian (logit) curves by ML is feasible by numerical optimization, but no computer programs are available at present that are easy to use (Jongman *et al.*, 1987). Local maxima may occur in the likelihood, when the tolerances of the species are unequal, and one needs to specify an initial estimate. The assumption of independence of species responses is required, but might not be tenable in practice; it remains to be studied how important this assumption is. Dependency among species could most obviously be caused by the effects of additional, unconsidered environmental variables, in which case the best remedy would be to identify these variables and include them in the analysis. Inferring the values of more than one environmental variable simultaneously on the basis of several Gaussian (logit) response surfaces is also possible in principle, but has not been done as far as we know.

D. Ordination

Ordination based on Gaussian (logit) curves aims to construct a latent variable such that these curves optimally fit the species data. This problem involves the ML estimation of site scores $\{x_k\}$ and the species' optima $\{u_k\}$, tolerances $\{t_k\}$ and maxima $\{a_k\}$, usually by an alternating sequence of Gaussian (logit) regressions and calibrations. This kind of ordination has been investigated by Gauch *et al.* (1974), Kooijman (1977), Kooijman and Hengeveld (1979), Goodall and Johnson (1982) and Ihm and Van Groenewoud (1975, 1984). The numerical methods required are computationally demanding, and in the general case, when the tolerances of the species are allowed to differ, the likelihood function typically contains many local maxima.

Kooijman (1977) and Goodall and Johnson (1982) reported numerical problems in their attempts to perform ML ordination using two-dimensional Gaussian-like models. A simple model with circular contours ($b_2 = b_4$ and $b_5 = 0$) may be amenable in practice, especially if b_2 is not allowed to vary among species (Kooijman, 1977). This model is equivalent to the "unfolding model" used by psychologists to analyse preference data (Coombs, 1964; Heiser, 1981; Davison, 1983; DeSarbo and Rao, 1984). But with more than two latent variables the Gaussian (logit) model with a second-degree polynomial as linear predictor contains so many parameters that it is likely to be difficult to get reliable estimates of them, even if all the interaction terms are dropped.

E. Constrained Ordination

The constrained ordination problem for Gaussian-like response models is to construct ordination axes that are also linear combinations of the environmental variables, such that Gaussian (logit) surfaces with respect to these axes optimally fit the data. As in redundancy analysis (Section II.E), the joint effects of the environmental variables on the species are "channelled" through a few ordination axes which can be considered as composite environmental gradients influencing species composition. Ter Braak (1986) refers to this approach as Gaussian canonical ordination, the word canonical being chosen by analogy with canonical correlation analysis. The estimation problem is actually simpler than in unconstrained Gaussian ordination, and is more easily soluble in practice because the number of parameters to be estimated is smaller: instead of n site scores one has to estimate q canonical coefficients. Meulman and Heiser (1984) have applied similar ideas in the context of non-metric multidimensional scaling. Gaussian canonical ordination can also be viewed as multivariate Gaussian regression with constraints on the coefficients of the polynomial (Ter Braak, 1988). In multivariate Gaussian regression each species has its own optimum in the q-dimensional space formed by the environmental variables; the constraints imposed in Gaussian canonical ordination amount to a requirement that these optima lie in a low-dimensional subspace. If the optima lie close to a plane then the most important species–environment relationships can be depicted graphically in an ordination diagram.

IV. WEIGHTED AVERAGING METHODS

Ecologists have developed alternative, heuristic methods that are simpler but have essentially the same aims as the methods of the previous section based on Gaussian-type models. Each method in the Gaussian family has a counterpart in the family of heuristic methods based on weighted averaging (WA). These methods have been used extensively, and even re-invented in different branches of ecology.

A. Regression

WA can be used to estimate species' optima with respect to known environmental variables. When a species shows a unimodal relationship with environmental variables, the species' presences will be concentrated around the peak of this function. One intuitively reasonable estimate of the optimum is the average of the values of the environmental variable over those sites in which the species is present. With abundance data, WA applies weights proportional to species abundance; absences still carry zero weight. The estimate of the optimum for species k is thus

$$\tilde{u}_k = \sum_{i=1}^{n} y_{ki} x_i / y_{k+} \tag{8}$$

where y_{ki} is from now onwards the abundance (*not* centred) or presence/absence (1/0) of species k at site i, y_{k+} is the species total ($y_{k+} = \Sigma_i y_{ki}$) and x_i is the value of the environmental variable at site i. As a follow-up to an investigation of the theoretical properties of this estimator (Ter Braak and Barendregt, 1986), Ter Braak and Looman (1986) showed by simulation of presence–absence data that WA estimates the optimum of a Gaussian logit curve as efficiently as the ML technique of Gaussian logit regression provided:

Condition 1a The site scores $\{x_i\}$ are equally spaced over the whole range of occurrence of the species along the environmental variable.

WA also proved to be only a little less efficient whenever the distribution of the environmental variable among the sites was reasonably homogeneous (rather than strictly equally spaced) over the whole range of species occurrences, or more generally for species with narrow ecological amplitudes. But the estimate of the optimum of a rare species may be imprecise, because the standard error of the estimate is inversely proportional to the square root of the number of occurrences. So for efficiency, we also need

Condition 1b The site scores $\{x_i\}$ are closely spaced in comparison with the species' tolerance.

B. Calibration

WA is also used in calibration, to estimate environmental values at sites from species' optima—which in this context are often called indicator values ("Zeigerwerte", Ellenberg, 1979) or scores (Whittaker, 1956). When species replace one another along the environmental variable of interest, i.e. have unimodal response functions with optima spread out along that variable, then species with optima close to the environmental value of a site will naturally tend to be represented at that site. Intuitively, to estimate the environmental value at a site, one can average the optima of the species that are present. With abundance data, the corresponding intuitive estimate is the weighted average,

$$\tilde{x}_i = \sum_{k=1}^{m} y_{ki} u_k / y_{+i} \tag{9}$$

where y_{+i} is the site total ($y_{+i} = \Sigma_k y_{ki}$).

Ter Braak and Barendregt (1986) showed that WA estimates the value x_i of a site as well as the corresponding ML techniques if the species show Gaussian curves and Poisson-distributed abundance values (or, for presence–absence data, show Gaussian logit curves), and provided:

Condition 2a The species' optima are equally spaced along the environmental variable over an interval that extends for a sufficient distance in both directions from the true value x_i;

Condition 3 The species have equal tolerances;

Condition 4 The species have equal maximum values.

These conditions amount to a "species packing model" wherein the species have equal response breadth and equal spacing (Whittaker *et al.*, 1973). The conditions may be relaxed somewhat (Ter Braak and Barendregt, 1986) without seriously affecting the efficiency of the WA-estimate. When the optima are uniformly distributed instead of being equally spaced, the efficiency is still high if the maximum probabilities of occurrence are small ($< 0{\cdot}5$). The species' maximum values may differ, but they must not show a trend along the environmental variable (for instance, leading to species-rich samples at one end of the gradient and species-poor samples at the other end). The efficiency of WA is less good if the tolerances substantially differ among species; a tolerance weighted version of WA, as suggested by Zelinka and Marvan (1961) and Goff and Cottam (1967), would be more efficient since it would give greater weight to species of narrower tolerance, which are more informative about the environment.

Under Conditions 2a–4 above, the standard error of the estimate of $\tilde{x}_i$ is approximately $t/\sqrt{y_{+i}}$, where t is the (common) species tolerance. For the weighted average to be practically useful, the number of species encountered in a site should therefore not be too small (not less than five). We therefore need the extra condition (cf. Section 5 in Ter Braak and Barendregt, 1986):

Condition 2b The species' optima must be closely spaced in comparison with their tolerances.

An alternative heuristic method of calibration is by "inverse regression". This is simply multiple linear regression of the environmental variable on the species abundances (Brown, 1982): the environmental variable is treated as if it were the response variable and the species abundances, possibly transformed, as predictor variables. The regression coefficients can be estimated from the training set of species abundances and environmental data, the resulting equations being applied directly to infer environmental values from further species abundance data. When applied to data on percentage composition, e.g. pollen spectra or diatom assemblages (Bartlein *et al.*, 1984; Charles, 1985), the method differs from WA calibration only in the way in which the species optima are estimated, since the linear combination of percentage values used to estimate the environmental value is by definition a weighted average of the regression coefficients.

C. Ordination

Hill (1973) turned weighted averaging into an ordination technique by applying alternating WA regressions and calibrations to a species-by-site data table. The algorithm of this technique of "reciprocal averaging" is similar to that given earlier for PCA:

Step 1 Start with arbitrary, but unequal, initial site scores $\{x_i\}$.
Step 2 Calculate new species scores $\{u_k\}$ by WA (equation (8)).
Step 3 Calculate new site scores $\{x_i\}$ by WA (equation (9)).
Step 4 Remove the arbitrariness in scale by standardizing the site scores by new $x_i = \{\text{old } x_i - z\}/s$ where $z = \Sigma_i y_{+i} x_i / \Sigma_i y_{+i}$ and

$$s^2 = \Sigma_i y_{+i}(x_i - z)^2 / \Sigma_i y_{+i} \quad (10)$$

Step 5 Stop on convergence, else go to Step 2.

As in PCA, the resulting site and species scores do not depend on the initial scores. The final scores produced by this reciprocal averaging algorithm form the first eigenvector or ordination axis of correspondence analysis (CA), an eigenvector technique that is widely used especially in the French-language literature (Laurec *et al.*, 1979; Hill, 1974). As with the power algorithm for PCA, the reciprocal averaging algorithm makes clear the relationship between CA and regression and calibration—this time, with WA regression and calibration. The method of standardization chosen in Step 4 is arbitrary, but chosen for later reference. On convergence, s in Step 4 is equal to the eigenvalue of the first axis, and lies between 0 and 1.

Correspondence analysis has many applications outside ecology. Nishisato (1980), Greenacre (1984) and Gifi (1981) provide a variety of different rationales for correspondence analysis, each adapted to a particular type of application. Heiser (1987) and Ter Braak (1985, 1987a) develop rationales for correspondence analysis that are particularly relevant to ecological applications.

Ter Braak (1985) showed that CA approximates ML Gaussian (logit) ordination under Conditions 1–4 listed above, i.e. under just these conditions for which WA is as good as ML-regression and ML-calibration. In practice CA can never be exactly equivalent to ML ordination, because Condition 1a implies that the range of site scores is broad enough to include the ranges of all of the species, whereas Condition 2a implies that there must be species with their optima situated beyond the edge of the range of site scores. These conditions cannot both be satisfied if the range of site scores is finite. As a result, CA shows an edge effect: the site scores near the ends of the axes become compressed relative to those in the middle (Gauch, 1982). This effect becomes less strong, however, as the range of site scores becomes wider and the spacing of the site scores and species scores becomes closer relative to the average species' tolerance.

Conditions 1–4 also disallow "deviant" sites and rare species. CA is sensitive to both (Hill, 1974; Feoli and Feoli Chiapella, 1980; Oksanen, 1983). This sensitivity may be useful in some applications, but is a nuisance if the aim is to detect major gradients. Deviant sites (and, possibly, the rarest species) should therefore ideally be removed from the data before analysis by CA.

As in PCA, further ordination axes can be extracted in CA by adding an extra step after Step 3, making the trial scores on the second axis uncorrelated with the (final) scores on the first axis. (In the calculation of f in Step 3b (see Section II.C) the sites are weighted proportional to the site total y_{+i}. This weighting is implicitly applied from now on.) However, there is a problem with the second and higher axes in CA. The problem is the well-known but hitherto not well-understood "arch effect" (Hill, 1974). If the species data come from an underlying one-dimensional Gaussian model the scores on the second ordination axis show a parabolic ("arch") relation with those of the first axis; if the species data come from a two-dimensional Gaussian model in which the true site and species scores are located homogeneously in a rectangular region in two-dimensional space (the extension to two dimensions of Conditions 1a and 2a), the scores of the second ordination axis lie not in a rectangle but in an arched band (Hill and Gauch, 1980). The arch effect arises because the axes are extracted sequentially in order of decreasing "variance". Suppose CA has succeeded in constructing a first axis, such that species appear one after the other along that axis as in a species packing model. Then a possible second axis is obtained by folding the first axis in the middle and bringing the ends together. This axis is a superposition of two species packing models, each with half the gradient length of the first axis. It is a candidate for becoming the second axis, because it has *no linear correlation* with the first CA-axis yet has as much as half the gradient length of the first axis (Jongman *et al.*, 1987). The folded axis by itself thus "explains" a part of the variation in the species data, even though when taken jointly with the first axis it contributes nothing. Even if there is a strong second gradient, CA will not associate it with the second axis if it separates the species less than a folded first axis. As a result of the arch effect, the two-dimensional CA-solution is generally not a good approximation to the ML-solution (two-dimensional Gaussian ordination).

Hill and Gauch (1980) developed detrended correspondence analysis (DCA) as a heuristic modification of CA designed to remedy both the edge effect and the arch effect. The edge effect is removed in DCA by non-linear rescaling of the axis. Assuming a species packing model with randomly distributed species' optima, Hill and Gauch (1980) noted that the variance of the optima of the species present at a site (the "within-site variance") is an estimate of the average response curve breadth of those species (they used the standard deviation as a measure of breadth, which is about equal to tolerance as we define it). Because of the edge effect, the species' curves before rescaling are narrower near the ends of the axis than in the middle, and the within-site variance is correspondingly smaller in sites near the ends of the axis than in sites in the middle. The rescaling therefore attempts to equalize the within-site variance at all points along the ordination axis by dividing the axis into small segments, expanding the segments with sites with small within-site variance, and contracting the segments with sites with large within-site variance. The site scores are then calculated as weighted averages of the species scores and the scores are standardized such that the within-site variance is equal to 1.

Hill and Gauch (1980) defined the length of the ordination axis to be the range of the site scores. This length is expressed in "standard-deviation units" (SD). The tolerance of the species' curves along the rescaled axis are close to 1, and each curve therefore rises and falls over about 4 SD. Sites that differ by 4 SD can thus be expected to have no species in common. Even if non-linear rescaling is not used, one can still set the average within-site variance of the species scores along a CA-axis equal to 1 by linear rescaling (Hill, 1979; Ter Braak, 1987b), so as to ensure that this useful interpretation of the length of the axis still approximately holds.

The arch effect, a more serious problem in CA, is removed in DCA by the heuristic method of "detrending-by-segments". This method ensures that at any point along the first ordination axis, the mean value of the site scores on subsequent axes is approximately zero. In order to achieve this, the first axis is divided into a number of segments and the trial site scores are adjusted within each segment by subtracting their mean after some smoothing across segments. Detrending-by-segments is built into the reciprocal averaging algorithm, and replaces Step 3b. Subsequent axes are derived similarly by detrending with respect to each of the existing axes.

DCA often works remarkably well in practice (Hill and Gauch, 1980; Gauch *et al.*, 1981). It has been critically evaluated in several recent simulation studies. Ter Braak (1985) showed that DCA gave a much closer approximation to ML Gaussian ordination than CA did, when applied to simulated data based on a two-dimensional species packing model in which species have identically shaped Gaussian surfaces and the optima and site scores are uniformly distributed in a rectangle. This improvement was shown to be mainly due to the detrending, not to the non-linear rescaling of axes. Kenkel and Orlóci (1986) found that DCA performed substantially better than CA when the two major gradients differed in length, but also noted that DCA sometimes "collapsed and distorted" CA results when there were: (a) few species per site, and (b) the gradients were long (we believe (a) to be the real cause of the collapse). Minchin (1987) further found that DCA can flatten out some of the variation associated with one of the underlying gradients. He ascribed this loss of information to an instability in the detrending-by-segments method. Pielou (1984, p. 197) warned that DCA is "overzealous" in correcting the "defects" in CA, and "may sometimes lead to the unwitting destruction of ecologically meaningful information". Minchin's (1987) results indicate some of the conditions under which such loss of information can occur.

DCA is popular among practical field ecologists, presumably because it provides an effective approximate solution to the ordination problem for a unimodal response model in two or more dimensions—given that the data are reasonably representative of sections of the major underlying environmental gradients. Two modifications might increase its robustness with respect to the problems identified by Minchin (1987). First, non-linear rescaling aggravates these problems; since the edge effect is not too serious, we advise against the routine use of non-linear rescaling. Second, the arch effect needs to be removed (as Heiser, 1987, also noted), but this can be done by a more stable, less "zealous" method of detrending which was also briefly mentioned by Hill and Gauch (1980): namely detrending-by-polynomials. Under the one-dimensional Gaussian model, it can be shown that the second CA-axis is a quadratic function of the first axis, the third axis is a cubic function of the first axis, and so on (Hill, 1974; Iwatsubo, 1984). Detrending-by-polynomials can be incorporated into the reciprocal averaging algorithm by extending Step 3b such that the trial scores are not only made uncorrelated with the previous axes, but are also made uncorrelated with polynomials of the previous axes. The limited experience so far suggests that detrending up to fourth-order polynomials should be adequate. In contrast with detrending-by-segments, the method of detrending-by-polynomials removes only specific defects of CA that are now theoretically understood.

D. Constrained Ordination

Just as CA/DCA is an approximation to ML Gaussian ordination, so is canonical correspondence analysis (CCA) an approximation to ML Gaussian canonical ordination (Ter Braak, 1986). CCA is a modification of CA in which the ordination axes are restricted to be weighted sums of the environmental variables, as in equation (4). CCA can be obtained from CA as redundancy analysis was obtained from PCA. An algorithm can be obtained by adding to the CA algorithm an extra multiple regression step. The only difference from Step 3a of redundancy analysis (see Section II.E) is that the sites must be weighted in the regression proportional to their site total y_{+i} (Ter Braak, 1986). CCA can also be obtained as the solution of an eigenvalue problem (Ter Braak, 1986). It is closely related to "redundancy analysis for qualitative variables" (Israëls, 1984) but has a different rationale and is applied to a different type of data.

In constrained ordination the constraints always become less strict as more environmental variables are included. If $q \geqslant n - 1$, then there are no real constraints, and CA and CCA become equivalent. As in CA, the edge effect in CCA is a minor problem that is best left untreated. Detrending may sometimes be required to remove the arch effect, i.e. to prevent CCA from selecting weighted sums of environmental variables that are approximately polynomials of previous axes. Detrending-by-segments does not work very well here for technical reasons; detrending-by-polynomials is better founded and more appropriate (see Appendix and Ter Braak, 1987b). However, the arch effect in CCA can be eliminated much more elegantly, simply by dropping superfluous environmental variables (Ter Braak, 1987a). Variables that are highly correlated with the "arched" axis (often the second axis) are the most likely to be superfluous. If the number of environmental variables is small enough for the relationship of individual variables to the ordination axes to be significant, the arch effect is not likely to occur at all.

CCA can be sensitive to deviant sites, but only when they are outliers with regard to both species composition and environment. When realistically few environmental variables are included, CCA is thus more robust than CA in this respect too.

CCA leads to an ordination diagram that simultaneously displays (a) the main patterns of community variations, as far as these reflect environmental variation, and (b) the main pattern in the weighted averages (not correlations as in redundancy analysis) of each of the species with respect to the environmental variables (Ter Braak, 1986, 1987a). CCA is thus intermediate between CA and separate WA calculations for each species. Geometrically, the separate WA calculations give each species a point in the q-dimensional space of the environmental variables, which indicates the centre of the species' distribution. CCA attempts to provide a low-dimensional representation of these centres; CCA is thus also a constrained form of WA, in which the weighted averages are restricted to lie in a low-dimensional subspace.

Like redundancy analysis, CCA can be used with dummy "environmental" variables to provide an ordination constrained to show maximum separation among pre-defined groups of samples. This special case of CCA is described, for example, by Feoli and Orlóci (1979) under the name of "analysis of concentration", by Greenacre (1984, Section 7.1) and by Gasse and Tekaia (1983).

V. ORDINATION DIAGRAMS AND THEIR INTERPRETATION

The linear ordination techniques (PCA and redundancy analysis) and the ordination techniques based on WA (CA/DCA and CCA) represent community data in substantially different ways. We focus on two-dimensional ordination diagrams, as these are the easiest to construct and to inspect, and illustrate the interpretation of each type of diagram with an example.

A. Principal Components: Biplots

PCA fits planes to each species' abundances in the space defined by the ordination axes. The species' point (b_{k1}, b_{k2}) may be connected with the origin (0,0) to give an arrow (Fig. 3). Such a diagram, in which sites are marked by points and species by arrows is called a "biplot" (Gabriel, 1971). There is a useful symbolism in this use of arrows: the arrow points in the direction of maximum variation in the species' abundance, and its length is proportional to this maximum rate of change. Consequently, species on the edge of the diagram (far from the origin) are the most important for indicating site differences; species near the centre are of minor importance. Ter Braak (1983) provides more detailed, quantitative rules for interpreting PCA ordination diagrams.

Van Dam *et al.* (1981) applied PCA to data consisting of diatom assemblages from 16 Dutch moorland pools, sampled in the 1920s and again in 1978, to investigate the impact of acidification on these shallow water bodies. Ten clearwater (non-humic) pools were situated in the province of Brabant and on the Veluwe and six brownwater (humic) pools in the province of Drenthe. Figure 3 displays the major variation in the data. The arrow of *Eunotia exigua* indicates that this species increases strongly along the first principal component: *E. exigua* is abundant in the recent Brabant and Veluwe samples, which lie on the right-hand side of the diagram, and rare in the remaining samples, which lie more to the left. The second axis accounts for some of the difference among the old and recent samples from Drenthe. These groups differ in the abundances of *Frustulia rhomboides* var. *saxonica*, *Tabellaria quadriseptata*, *Eunotia tenella*, *Tabellaria binalis*, and *Eunotia veneris*, as shown by the directions of the arrows for these species in Fig. 3. As *E. exigua* is acidobiontic and the first principal component is

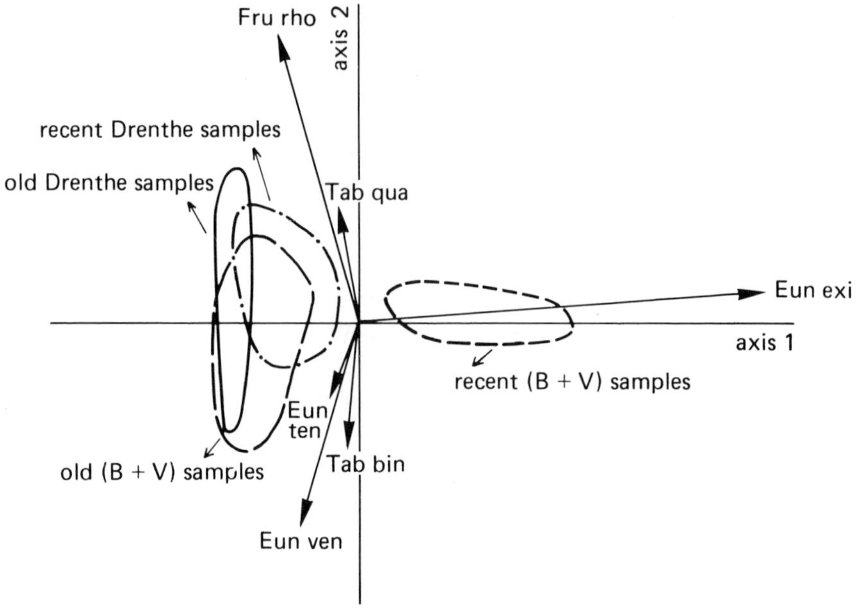

Fig. 3. Biplot based on principal components analysis of diatom assemblages from Dutch moorland pools (schematic after Van Dam *et al.* 1981). The arrows for the six most frequent species and the regions where different categories of samples lie jointly display the approximate community composition in each of the regions (old, *c.* 1920; recent, 1978; B + V, from the province of Brabant and the Veluwe). Abbreviations: Eun exi, *Eunotia exigua*; Eun ten, *Eunotia tenella*; Eun ven, *Eunotia veneris*; Fru rho, *Frustulia rhomboides* var. *saxonica*; Tab bin, *Tabellaria binalis*; Tab qua, *Tabellaria quadriseptata.*

strongly correlated with the sulphate concentration of the 1978 samples, this component clearly depicts the impact of acidification of the moorland pools in Brabant and the Veluwe (and to a smaller extent also in Drenthe). Thus Van Dam *et al.* (1981) used PCA to summarize the changes in diatom composition between the 1920s and 1978. PCA helped them to detect that the nature of the change differed among provinces, hence stressing the importance for diatoms of the distinction between clearwater and brownwater pools.

B. Correspondence Analysis: Joint Plots

In CA and DCA both sites and species are represented by points, and each site is located at the centre of gravity of the species that occur there. One may therefore get an idea of the species composition at a particular site by looking at "nearby" species points. Also, in so far as DCA approximates the fitting of

Gaussian (logit) surfaces (Fig. 2), the species points are approximately the optima of these surfaces; hence the abundance or probability of occurrence of a species tends to decrease with distance from its location in the diagram.

Figure 4 illustrates this interpretation of the species' points as optima in ordination space. DCA was applied to presence–absence data on 51 bird species in 526 contiguous, 100 m × 100 m grid-cells in an area with pastures and scattered woodlots in the Rhine valley near Amerongen, the Netherlands (Opdam *et al.*, 1984). Figure 4 shows the DCA scores of the 20 most frequent species by small circles, and the outline (dashed) of the region in which the scores for the grid-cells fall (the individual grid-cells are not shown, to avoid crowding). Opdam *et al.* (1984) interpreted the first axis, of length 5·6 SD, as a gradient from open to closed landscape and the second axis, of length 5·3 SD, as a gradient from wet to drier habitats.

In order to test the interpretation of species' scores as optima, we fitted a response surface for each species by logit regression using equation (6) with the first and the second DCA-axes as the predictor variables x_1 and x_2. For 13 of the 20 bird species, the fitted surface had a maximum. The optimum was calculated for each of these species by equation (7) and plotted as a triangle in Fig. 4. The fitted optima lie close to the DCA scores. The regression analysis also allowed us to estimate species' tolerances in ordination space: these are indicated in Fig. 4 by ellipses representing the region within which each species occurs with at least half of its maximum probability, according to the fitted surface.

The fitted surfaces for the remaining seven species had a minimum or saddle point, suggesting that their optima are located well outside the sampled range. For these species we fitted a "linear" logit surface by setting b_2, b_4 and b_5 in equation (6) to zero. The direction of steepest increase of each of the fitted surfaces is indicated in Fig. 4 by an arrow through the centroid of the site points; the beginning and end points of each arrow correspond to fitted probabilities of 0·1 and 0·9 respectively. As expected from our interpretation of DCA, these arrows point more or less in the same direction as the DCA scores of the corresponding species (Fig. 4).

In contrast to the PCA-diagram, the species points on the edge of the CA- or DCA-diagram are often rare species, lying there either because they prefer extreme (environmental) conditions, or (very often) because their few occurrences by chance happen to fall in sites with extreme conditions; one cannot decide between these possibilities without additional data. Such peripheral species have little influence on the analysis and it is often convenient not to display them at all. Furthermore, species near the centre of the diagram may be ubiquitous, unrelated to the ordination axes, bimodal, or in some other way not fitting a unimodal response model—or they may be genuinely specific with a habitat-optimum near the centre of the sampled

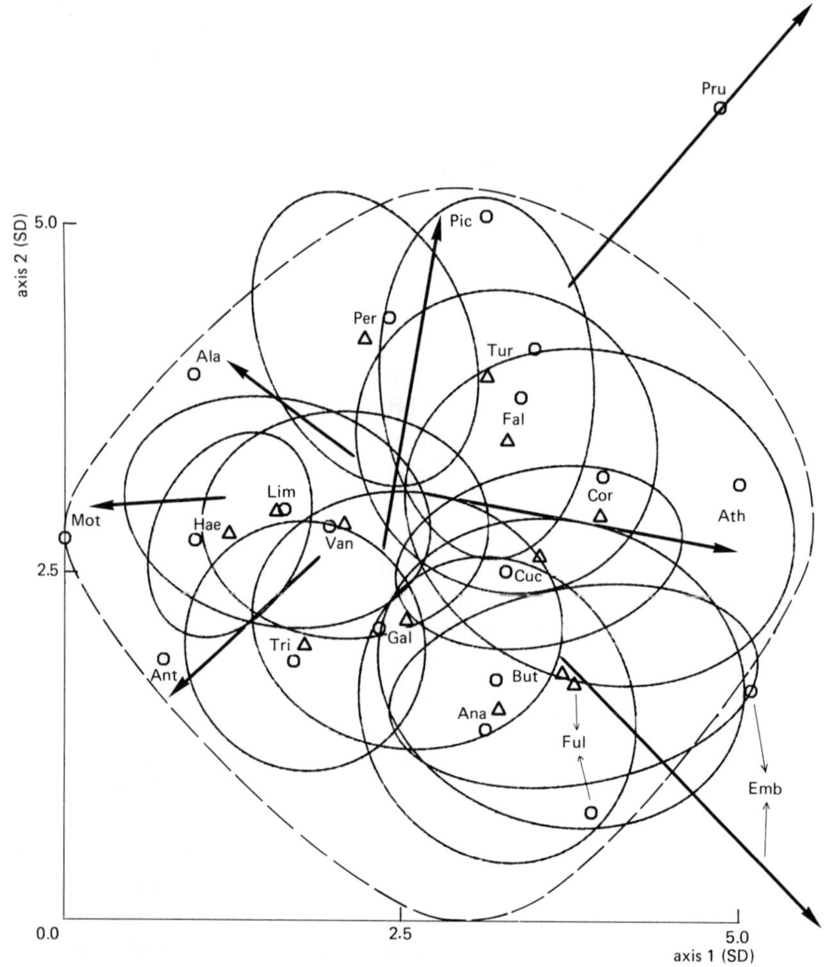

Fig. 4. Joint plot based on detrended correspondence analysis (DCA) of bird species communities in the Rhine valley near Amerongen, the Netherlands (data from Opdam *et al.*, 1984), displaying the major variation in bird species composition across the landscape. This plot shows the DCA-scores (○) of the 20 most frequent species and the region in which the samples fall (---). Also shown are optima (△) and lines of equal probability for the 13 species whose probability surfaces had clear maxima (as fitted by Gaussian logit regression), and arrows representing directions of increase for the seven species whose probability surfaces were monotonic. Abbreviations: Ala, *Alauda arvensis*; Ana, *Anas platyrhynchos*; Ant, *Anthus pratensis*; Ath, *Athene noctua*; But, *Buteo buteo*; Cor, *Corvus corone*; Cuc, *Cuculus canorus*; Emb, *Emberiza schoeniclus*; Fal, *Falco tinnunculus*; Ful, *Fulica atra*; Gal, *Gallinago gallinago*; Hae, *Haematopus ostralegus*; Lim, *Limosa limosa*; Mot, *Motacilla flava flava*; Per, *Perdix perdix*; Pic, *Pica pica*; Pru, *Prunella modularis*; Tri, *Tringa totanus*; Tur, *Turdus merula*; Van, *Vanellus vanellus*.

range of habitats. The correct interpretation may be found by the kind of secondary analysis shown in Fig. 4, or more straightforwardly just by plotting the species' abundances in ordination space.

C. Redundancy Analysis

In redundancy analysis sites are indicated by points, and both species and environmental variables are indicated by arrows whose interpretation is similar to that of the arrows in the PCA biplot. The pattern of abundance of each species among the sites can be inferred in exactly the same way as in a PCA biplot, and so may the direction of variation of each environmental variable. One may also get an idea of the correlations between species' abundances and environmental variables. Arrows pointing in roughly the same direction indicate a high positive correlation, arrows crossing at right angles indicate near-zero correlation, and arrows pointing in opposite directions indicate high negative correlation. Species and environmental variables with long arrows are the most important in the analysis; the longer the arrows, the more confident one can be about the inferred correlation. (It is assumed here that for the purpose of the ordination diagram the environmental variables have been standardized to zero mean and unit variance, so as to make the lengths of arrows comparable.) Jongman *et al.* (1987) provide more quantitative rules for interpreting the ordination diagrams derived from redundancy analysis.

The data we use to illustrate redundancy analysis were collected to study the relation between the vegetation and management of dune meadows on the island of Terschelling, The Netherlands (M. Batterink and G. Wijffels, unpublished). Figure 5 displays the main variation in the vegetation in relation to three environmental variables (thickness of the A1 horizon, moisture content of the soil and quantity of manuring). The arrows for *Poa trivialis* and *Elymus repens* make small angles with the arrow for manuring; these species are inferred to be positively correlated with manuring. *Salix repens* and *Leontodon autumnalis* have arrows pointing in directions roughly opposite to that of manuring, and are inferred to be negatively correlated with manuring. The former species are thus most abundant in the heavily manured meadows of standard farms (positioned at the top of the diagram), whereas the latter species are most abundant in the unmanured meadows (owned by the nature conservancy and positioned at the bottom of the diagram). The relationships of the species with moisture and thickness of the A1 horizon can be inferred in a similar way. The short arrows for *Bromus hordaceus* and *Sagina procumbens*, for example, indicate that their abundance is not so much affected by moisture, manure and thickness of the A1 horizon. Redundancy analysis can summarize the species–environment rela-

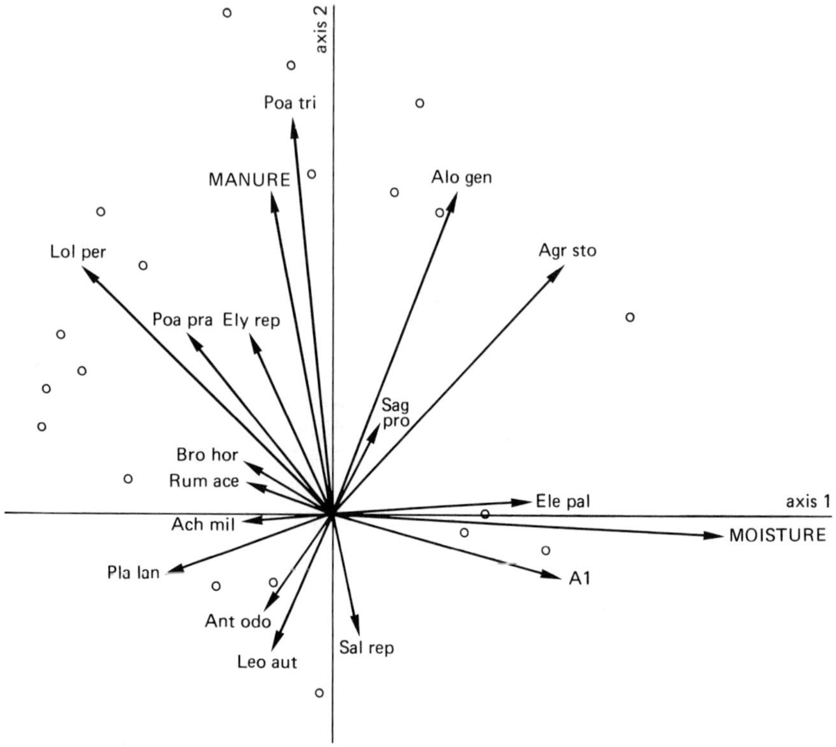

Fig. 5. Biplot based on redundancy analysis of vegetation with respect to three environmental variables (quantity of manure, soil moisture and thickness of the A1 horizon) in dune meadows (○) on the island of Terschelling, The Netherlands. The arrows for plant species and environmental variables display the approximate (linear) correlation coefficients between plant species and the environmental variables. Abbreviations: Ach mil, *Achillea millefolium*; Agr sto, *Agrostis stolonifera*; Alo gen, *Alopecurus geniculatus*; Ant odo, *Anthoxanthum odoratum*; Bro hor, *Bromus hordaceus*; Ele pal, *Eleocharis palustris*; Ely rep, *Elymus repens*; Leo aut, *Leontodon autumnalis*; Lol per, *Lolium perenne*; Pla lan, *Plantago lanceolata*; Poa pra, *Poa pratensis*; Poa tri, *Poa trivialis*; Rum ace, *Rumex acetosa*; Sag pro, *Sagina procumbens*; Sal rep, *Salix repens*.

tionships in such an informative way, because the gradients are short ($\approx$ 2SD: Ter Braak, 1987b).

D. Canonical Correspondence Analysis

In CCA, since species are assumed to have unimodal response surfaces with respect to linear combinations of the environmental variables, the species are

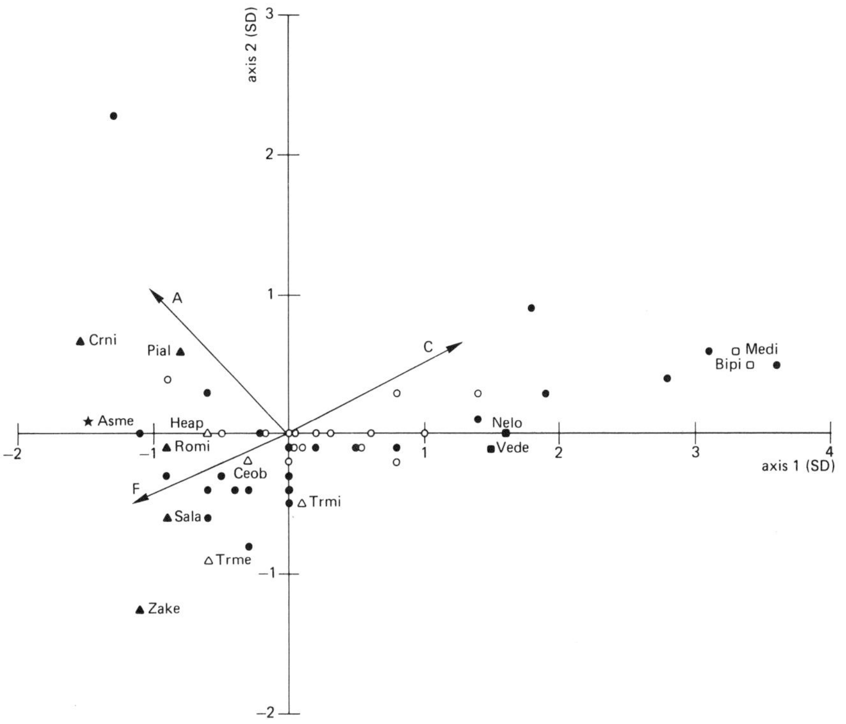

Fig. 6. Ordination diagram based on canonical correspondence analysis of successional plant communities with respect to three environmental variables (regrowth age A, length of cropping period C, and extent of forested perimeter F) on abandoned cultivation sites within Mexican tropical rain forest (Purata, 1986 and unpublished). ●, sites with environmental data; ○, sites added "passively" on the basis of floristic composition. The species shown are a selection among the 285 included in the analysis. □, denotes ruderals; ■, pioneer shrubs; △, pioneer trees; ▲, late-secondary canopy trees; and ★, an understorey palm. Abbreviations: Asme, *Astrocaryum mexicanum*; Bipi, *Bidens pilosa*; Ceob, *Cecropia obtusifolia*; Crni, *Croton nitens*; Heap, *Heliocarpus appendiculatus*; Medi, *Melampodium divaricatum*; Nelo, *Neurolaena lobata*; Pial, *Piper amalago*; Romi, *Robinsonella mirandae*; Sala, *Sapium lateriflorum*; Trme, *Trichospermum mexicanum*; Trmi, *Trema micrantha*; Vede, *Vernonia deppeana*; Zake, *Zanthoxylum kellermanii*.

logically represented by points (corresponding to their approximate optima in the two-dimensional environmental subspace), and the environmental variables by arrows indicating their direction and rate of change through the subspace.

Purata (1986, and unpublished results) applied CCA to plant species abundance data from 40 abandoned cultivation sites within Mexican tropical rain forest. Data were available for 24 of these sites on the regrowth age (A), the length of the cropping period in the past (C), and the proportion of the perimeter that had remained forested (F). These three variables were used as environmental variables in CCA. The remaining 16 sites were entered as "passive" sites, to be positioned with respect to the CCA axes according to their floristic composition in relation to the "active" sites.

Figure 6 illustrates the results. The first axis, with length 4·7 SD, was interpreted as an indicator of the general trend of secondary succession. The direction of the arrow for regrowth age shows that this trend runs broadly from right to left. The species' locations are consistent with their life-history characteristics: the trend of succession runs from ruderals (to the right), through pioneer shrubs and trees, to late-secondary canopy dominants and shade-tolerant understorey species (to the left). The directions of the other two arrows in relation to axis 1 show that a long cropping period delays succession, while an extensive forested perimeter accelerates succession. Axis 2 (3·0 SD) may (more speculatively) differentiate species whose establishment is favoured by the presence of mature forest around the site from those that simply require a long time to grow.

CCA also allows the computation of unconstrained, "residual" axes summarizing floristic variation that remains after the effect of the environmental variables has been taken out. In Purata's study, the successive eigenvalues of the first three (constrained) CCA axes were 0·49, 0·34 and 0·18. (There cannot be more constrained axes than there are environmental variables). The first residual axis gave an eigenvalue of 0·74, showing that at least as much floristic variation was *not* explained by the environmental variables. In our experience, terrestrial community data commonly give a residual eigenvalue as large as the first constrained eigenvalue, however carefully the environmental variables are chosen. Thus DCA and CCA tend to give different ordinations, and CCA is more powerful in detecting relationships between species composition and environment.

VI. CHOOSING THE METHODS

A. Which Response Model?

Regression methods can fit response models with a wide variety of shapes. The linear and Gaussian-like models are convenient starting points; more complex shapes can be fitted by adding further parameters, if the data are sufficiently detailed to support it. Other species may be used as additional explanatory variables if the specific aim is to detect species interactions (Fresco, 1982). The shapes of the response functions may be made even more general by applying Box–Cox transformations to the explanatory variables (Bartlein *et al.*, 1986) or still more general by fitting splines (Smith, 1979). Even with all these modifications, regression can still be done with standard packages for Generalized Linear Modelling.

After species response curves or surfaces have been fitted by regression, calibration based on the maximum likelihood principle can be used to make inferences about the environment from community data. If the surfaces fitted by regression have complex shapes, then calibration by *numerical* maximization of the likelihood may be problematic. But even then, if there are only a few environmental variables involved, the "most likely" combination of environmental values can be searched for on a grid across the environmental space (Atkinson *et al.*, 1986; Bartlein *et al.*, 1986). So the type of response model used in both regression and calibration should generally be guided by the characteristics and resolution of the data, and inspection of the data and the residuals after regression should show whether the model being used is adequate for the purpose.

In contrast to regression and calibration, the ordination problem requires the simultaneous estimation of large numbers of parameters and cannot be solved practically without some constraints on the structure one wants to fit. That these constraints may seem unduly restrictive simply shows that there are limits to what ordination can achieve. The number of ordination axes to be extracted must be small, and the type of response model must be restricted, in order to permit a solution. For example, it seems necessary to disregard the possibility of bimodal species distributions (Hill, 1977). Certainly bimodal distributions sometimes occur, but ordination has to assume that species "on average" have simple distributions—otherwise, the problem would be insoluble; the utility of ordination techniques depends on them being robust with respect to departures from the simple models on which they are based. The Gaussian model seems to be of the right order of complexity for ordination of ecological data, but the full second-degree model of equation (6) is already difficult to fit (Kooijman, 1977; Goodall and Johnson, 1982). The Gaussian model with circular contour lines and equal species tolerances, i.e. the unfolding model, might provide a good compromise between practical solubility and realism in ordination. Promising algorithms for unfolding are developed by Heiser (1987) and DeSarbo and Rao (1984). DCA provides a reasonably robust approximation to ML Gaussian ordination and requires far less computing time. Similarly, ML Gaussian canonical ordination is technically feasible, but CCA provides a practical and robust approximation to it.

Non-linear methods are appropriate if a reasonable number of species have their optima located within the data set. If the gradient length is reduced to less than about 3 SD, the approximations involved in WA become worse and ultimately (if the gradient length is less than about 1·5 SD) the methods yield poor results because most species are behaving monotonically over the observed range. Thus if the community variation is within a narrow range, the linear ordination methods—PCA and redundancy analysis—are appropriate. If the community variation is over a wider range, non-linear ordination methods—including DCA and CCA—are appropriate.

B. Direct or Indirect?

Direct gradient analysis allows one to study the part (large or small) of the variation in community composition that can be explained by a particular set of environmental variables. In indirect gradient analysis attention is first focused on the major pattern of variation in community composition; the environmental basis of this pattern is to be established later. If the relevant environmental data are to hand, the direct approach—either fitting separate response surfaces by regression for each major species, or analysing the overall patterns of the species–environment relationship by constrained ordination—is likely to be more effective than the traditional indirect approach. However, indirect gradient analysis does have the advantage that no prior hypothesis is needed about what environmental variables are relevant. One does not need to measure the environmental variables in advance, and one can use informal field knowledge to help interpret the patterns that emerge—hence the emphasis in the literature on ordination as a technique for "hypothesis generation", the implication being that experimental or more explicit statistical approaches can be used for subsequent hypothesis testing. This distinction is not hard and fast, but it does draw attention to the strengths and limitations of indirect gradient analysis.

In Section V.D, we showed in passing how an indirect gradient analysis can be carried out *after* a direct gradient analysis in order to summarize the community variation that remains after known effects have been removed. When the known environmental variables are not the prime object of study, they are called concomitant variables (Davies and Tso, 1982) or covariables. It would be convenient to solve for the residual (unconstrained) axes without having to extract all the constrained axes first. Fortunately, this is straightforward. In the iterative algorithm for PCA and CA, one simply extends Step 3b such that the trial scores are not only made uncorrelated with any previous axis (if present) but are also made uncorrelated with all specified covariables (see Appendix for details.) In this way the effects of the covariables are partialled out from the ordination; hence the name "partial ordination". The theory of "partial components analysis" and "partial correspondence analysis", as we call these extensions of PCA and CA, is given by Gabriel (1978, theorem 3) and Ter Braak (1988), respectively. Swaine and Greig-Smith (1980) used partial components analysis to obtain an ordination of within-plot vegetation change in permanent plots. Partial correspondence analysis, or its detrended form, would be more appropriate if the gradients were long.

C. Direct Gradient Analysis: Regression or Constrained Ordination?

Whether to use constrained ordination (multivariate direct gradient analysis) instead of a series of separate regressions (the traditional type of direct gradient analysis) depends on whether or not there is any advantage in analysing all the species simultaneously. Both constrained and unconstrained ordination assume that the species react to the *same* composite gradients of environmental variables, while in regression a separate composite gradient is constructed for each species. Regression can therefore allow more detailed descriptions and more accurate prediction and calibration, if properly carried out (with due regard to its statistical assumptions) and if sufficient data are available. However, ecological data that are collected over a large range of habitat variation require non-linear models, and building good non-linear models by regression is demanding in time and computation. In CCA the composite gradients are linear combinations of environmental variables and the non-linearity enters through a unimodal response model with respect to a few composite gradients, taken care of in CCA by the procedure of weighted averaging. Constrained ordination is thus easier to apply, and requires less data, than regression; it provides a summary of the species–environment relationship, and we find it most useful for the exploratory analysis of large data sets.

Constrained ordination can also be carried out *after* regression, in order to relate the residual variation to other environmental variables. This type of analysis, called "partial constrained ordination", is useful when the explanatory (environmental) variables can be subdivided in two sets, a set of covariables—the effects of which are not the prime object of study—and a further set of environmental variables whose effects are of particular interest. For example, in the illustration of Section V.C, the study was initiated to investigate differences in vegetation among dune meadows that were exploited under different management regimes (standard farming, biodynamical farming, nature management, among others). Standard CCA showed systematic differences in vegetation among management regimes. A further question is then whether these differences can be fully accounted for by the environmental variables moisture, quantity of manure and thickness of the A1 horizon, whose effects are displayed in Fig. 5, or whether the variation that remains after fitting the three environmental variables (three constrained ordination axes) is systematically related to management regimes. This question can be tackled using partial constrained ordination, with the three environmental variables as covariables, and a series of dummy variables (for each of the management regimes) as the variables of interest.

Technically, partial constrained ordination can be carried out by any computer program for constrained ordination. The usual environmental variables are replaced by the residuals obtained by regressing each of the variables of interest on the covariables (see Appendix). Davies and Tso (1982) gave the theory behind partial redundancy analysis; Ter Braak (1988) derived partial canonical correspondence analysis as an approximation to "partial Gaussian canonical ordination".

Partial constrained ordination has the same essential aim as Carleton's (1984) residual ordination, i.e. to determine the variation in the species data that is uniquely attributable to a particular set of environmental variables, taking into account the effects of other (co-) variables; however, Carleton's method is somewhat less powerful, being based on a pre-existing DCA which may already have removed some of the variation of interest. Partial constrained ordination is, by contrast, a true direct gradient analysis technique which seems promising, e.g. for the analysis of permanent plot data (effects of time, with location and/or environmental data as covariables), and a variety of other applications in which effects of particular environmental variables are to be sorted out from the "background" variation imposed by other variables.

VII. CONCLUSIONS

Regression, calibration, ordination and constrained ordination are well-defined statistical problems with close interrelationships. Regression is the tool for investigating the nature of individual species' response to environment, and calibration is the tool for (later) inferring the environment from species composition at an individual site. Both tools come in various degrees of complexity. The simplest are linear and WA regression and calibration. The linear methods are applicable over short ranges of environment, where species' abundance appears to vary monotonically with variation in the environment. The WA methods are applicable over wider ranges of environment; WA regression is a crude method to estimate each species' optimum, and WA calibration just averages the optima of the species that are present.

WA works with presence–absence data. If abundances are available, they provide the weights. These WA techniques can be shown to give approximate estimates of the species' optima and environmental values when the species' response surfaces (the relationships between the species' abundance, or probability of occurrence, and the environmental variables) are Gaussian (or for probabilities, Gaussian–logit) in form. Gaussian regression and calibration are also possible, but the WA techniques are simpler and are approximations to the Gaussian methods.

These simple tools are suitable when there are many species of interest and the exact form of the response surface is not critical, and they are very easy to use. If the form of the response surfaces *is* critical, more complex models can be fitted using Generalized Linear Modelling (for regression) and maximum likelihood techniques (for calibration). These more complex tools are becoming important in the theoretical study of species–environment relationships (Austin, 1985) and environmental dynamics (Bartlein *et al.*, 1986). Naturally, they require skilled users who are aware of their statistical assumptions, limitations and pitfalls.

Ordination and constrained ordination can be related to the simpler methods of regression and calibration. Ordination is the tool for exploratory analysis of community data with no prior information about the environment. Constrained ordination is the equivalent tool for the analysis of community variation in relation to environment. Both implicitly assume a common set of environmental variables and a common response model for all of the species. (Without these simplifying assumptions, they could not work; such major simplifications of data can only be achieved at the expense of some realism.) The basic ordination techniques are PCA and CA. PCA constructs axes that are as close as possible to a linear relationship with the species. These axes can be found by a converging sequence of alternating linear regressions and calibrations. Each axis after the first is obtained by partialling out linear relationships with the previous axis. CA is mathematically related to PCA, but has a very different effect. CA axes can be found by a converging sequence of WA regressions and calibrations. In CA, axes after the first are obtained analogously with PCA; in DCA they are obtained by removing all trends, linear or non-linear, with respect to previous axes. CA suffers from the arch effect, which DCA eliminates. DCA is a reasonably robust approximation to Gaussian ordination, in which the axes are constructed so that the species response curves with respect to the axes are Gaussian in form. Gaussian ordination is feasible but not convenient. DCA is much more practical. But there are problems with the detrending, and the method can break down when the connections between sites are too tenuous. Some modifications—including an improved method of detrending—may improve DCA's robustness; alternatively, some forms of nonmetric multidimensional scaling may be more robust (Kenkel and Orlóci, 1986; Minchin, 1987).

Constrained ordination methods have the added constraint that the ordination axes must be linear combinations of environmental variables. This constraint can be implemented as an extra multiple regression step in the general iterative ordination algorithm. PCA then becomes redundancy analysis (a more practical alternative to canonical correlation), Gaussian ordination becomes Gaussian canonical ordination, and CA becomes CCA (Table 2). The constraint makes Gaussian canonical ordination somewhat more stable than its unconstrained equivalent, but still CCA provides a much more practical alternative. All these constrained methods are most powerful if the number of environmental variables is small compared to the number of sites. Then the constraints are much stronger than in normal ordination, and the common problems of ordination (such as the arch effect, the need for detrending and the sensitivity to deviant sites) disappear.

Often, community–environment relationships have been explored by "indirect gradient analysis"—ordination, followed by interpretation of the axes in terms of environmental variables. But if the environmental data are to hand, constrained ordination ("multivariate direct gradient analysis") provides a more powerful means to the same end. Hybrid (direct/indirect) analyses are also possible. In partial ordination and partial constrained ordination, the analysis works on the variation that remains after the effects of particular environmental, spatial or temporal "covariables" have been removed.

The choice between linear and non-linear ordination methods is not a matter of personal preference. Where gradients are short, there are sound statistical reasons to use linear methods. Gaussian methods break down, and edge effects in CA and related techniques become serious; the representation of species as arrows becomes appropriate. As gradient lengths increase, linear methods become ineffective (principally through the "horseshoe effect", which scrambles the order of samples along the first axis as well as creates a meaningless second axis); Gaussian methods become feasible, and CA and related techniques become effective. The representation of species as points, representing their optima, becomes informative. The range 1·5–3 SD for the first axis represents a "window" over which both PCA and CA/DCA, or both redundancy analysis and CCA, can be used to good effect.

ACKNOWLEDGEMENTS

We thank Dr M. P. Austin, Dr P. J. Bartlein, Professor L. C. A. Corsten,

J. A. Hoekstra, Dr P. Opdam and Dr H. van Dam for comments on the manuscript. Our collaboration was supported by a Netherlands Science Research Council (ZWO) grant to I.C.P. and a Swedish Natural Science Research Council (NFR) grant to the project "Simulation Modelling of Natural Forest Dynamics". We also thank Dr S. E. Purata V. for supplying unpublished results.

REFERENCES

Alderdice, D. F. (1972). Factor combinations: responses of marine poikilotherms to environmental factors acting in concert. In: *Marine Ecology* (Ed. by O. Kinne), Vol. 1, Part 3, pp. 1659–1722. New York: John Wiley.

Ås (1985). Biological Community patterns in insular environments. *Acta Unit. Ups.* **792:** 1–55.

Atkinson, T. C., Briffa, K. R., Coope, G. R., Joachim, M. J. and Perry, D. W. (1986). Climatic calibration of coleopteran data. In: *Handbook of Holocene Palaeoecology and Palaeohydrology* (Ed. by B. E. Berglund), pp. 851–858. Chichester: John Wiley.

Austin, M. P. (1971). Role of regression analysis in plant ecology. *Proc. ecol. Soc. Austr.* **6,** 63–75.

Austin, M. P. (1985). Continuum concept, ordination methods, and niche theory. *Ann. Rev. Ecol. Syst.* **16,** 39–61.

Austin, M. P. and Cunningham, R. B. (1981). Observational analysis of environmental gradients. *Proc. ecol. Soc. Austr.* **11,** 109–119.

Austin, M. P., Cunningham, R. B. and Fleming, P. M. (1984). New approaches to direct gradient analysis using environmental scalars and statistical curve-fitting procedures. *Vegetatio* **55,** 11–27.

Balloch, D., Davies, C. E. and Jones, F. H. (1976). Biological assessment of water quality in three British rivers: the North Esk (Scotland), the Ivel (England) and the Taf (Wales). *Wat. Pollut. Control* **75,** 92–114.

Bartlein, P. J., Webb, T. III and Fleri, E. (1984). Holocene climatic changes in the Northern Midwest: pollen-derived estimates. *Quat. Res.* **22,** 361–374.

Bartlein, P. J., Prentice, I. C. and Webb, T. III (1986). Climatic response surfaces from pollen data for some eastern North American taxa. *J. Biogeogr.* **13,** 35–57.

Battarbee, R. W. (1984). Diatom analysis and the acidification of lakes. *Phil. Trans. Roy. Soc. London Ser. B* **305,** 451–477.

Bloxom, B. (1978). Constrained multidimensional scaling in N spaces. *Psychometrika* **43,** 397–408.

Böcker, R., Kowarik, I. and Bornkamm, R. (1983). Untersuchungen zur Anwendung der Zeigerwerte nach Ellenberg. *Verh. Ges. Oekol.* **11,** 35–56.

Brown, G. H. (1979). An optimization criterion for linear inverse estimation. *Technometrics* **21,** 575–579.

Brown, P. J. (1982). Multivariate calibration. *J. Roy. statist. Soc. B* **44,** 287–321.

Carleton, T. J. (1984). Residual ordination analysis: a method for exploring vegetation environment relationships. *Ecology* **65,** 469–477.

Chandler, J. R. (1970). A biological approach to water quality management. *Wat. Pollut. Control* **69,** 415–421.

Charles, D. F. (1985). Relationships between surface sediment diatom assemblages and lakewater characteristics in Adirondack lakes. *Ecology* **66,** 994–1011.

Coombs, C. H. (1964). *A Theory of Data.* New York: John Wiley.

Cox, D. R. and Hinkley, D. V. (1974). *Theoretical Statistics*. London: Chapman and Hall.

Cramer, W. and Hytteborn, H. (1987). The separation of fluctuation and long-term change in vegetation dynamics of a rising sea-shore. *Vegetatio* **69,** 155–167.

Dargie, T. C. D. (1984). On the integrated interpretation of indirect site ordinations: a case study using semi-arid vegetation in southeastern Spain. *Vegetatio* **55,** 37–55.

Davies, P. T. and Tso, M. K.-S. (1982). Procedures for reduced-rank regression. *Appl. Statist.* **31,** 244–255.

Davison, M. L. (1983). *Multidimensional Scaling*. New York: John Wiley.

De Leeuw, J. and Heiser, W. (1980). Multidimensional scaling with restrictions on the configuration. In: *Multivariate Analysis-V* (Ed. by P. R. Krishnaiah), pp. 501–522. Amsterdam: North-Holland.

DeSarbo, W. S. and Rao, V. R. (1984). GENFOLD2: a set of models and algorithms for the general unfolding analysis of preference/dominance data. *J. Class.* **1,** 147–186.

Dobson, A. J. (1983). *Introduction to Statistical Modelling*. London: Chapman and Hall.

Ellenberg, H. (1979). Zeigerwerte der Gefässpflanzen Mitteleuropas. *Scripta Geobotanica* **9,** 1–121.

Fängström, I. and Willén, E. (1987). Clustering and canonical correspondence analysis of phytoplankton and environment variables in Swedish lakes. *Vegetatio* **71,** 87–95.

Feoli, E. and Feoli Chiapella, L. (1980). Evaluation of ordination methods through simulated coenoclines: some comments. *Vegetatio* **42,** 35–41.

Feoli, E. and Orlóci, L. (1979). Analysis of concentration and detection of underlying factors in structured tables. *Vegetatio* **40,** 49–54.

Fresco, L. F. M. (1982). An analysis of species response curves and of competition from field data: some results from heath vegetation. *Vegetatio* **48,** 175–185.

Gabriel, K. R. (1971). The biplot graphic display of matrices with application to principal component analysis. *Biometrika* **58,** 453–467.

Gabriel, K. R. (1978). Least squares approximation of matrices by additive and multiplicative models. *J. Roy. statist. Soc. B* **40,** 186–196.

Gasse. F. and Tekaia, F. (1983). Transfer functions for estimating paleoecological conditions (pH) from East African diatoms. *Hydrobiologia* **103,** 85–90.

Gauch, H. G. (1982). *Multivariate Analysis in Community Ecology*. Cambridge: Cambridge Univ. Press.

Gauch, H. G. and Whittaker, R. H. (1972). Coenocline simulation. *Ecology* **53,** 446–451.

Gauch, H. G., Chase, G. B. and Whittaker, R. H. (1974). Ordination of vegetation samples by Gaussian species distributions. *Ecology* **55,** 1382–1390.

Gauch, H. G., Whittaker, R. H. and Singer, S. B. (1981). A comparative study of nonmetric ordinations. *J. Ecol.* **69,** 135–152.

Gifi, A. (1981). *Nonlinear Multivariate Analysis*. Department of Data Theory, University of Leiden, Leiden.

Gittins, R. (1985). *Canonical Analysis. A Review With Applications in Ecology*. Berlin: Springer-Verlag.

Goff, F. G. and Cottam, G. (1967). Gradient analysis: The use of species and synthetic indices. *Ecology* **48,** 793–806.

Goodall, D. W. and Johnson, R. W. (1982). Non-linear ordination in several dimensions: a maximum likelihood approach. *Vegetatio* **48,** 197–208.

Gourlay, A. R. and Watson, G. A. (1973). *Computational Methods for Matrix Eigen Problems*. New York: John Wiley.

Greenacre, M. J. (1984). *Theory and Applications of Correspondence Analysis*. London: Academic Press.

Heiser, W. J. (1981). *Unfolding Analysis of Proximity Data*. Thesis, University of Leiden, Leiden.

Heiser, W. J. (1987). Joint ordination of species and sites: the unfolding technique. In: *Developments in Numerical Ecology* (Ed. by P. Legendre and L. Legendre), pp. 189–221. Berlin: Springer-Verlag.

Hill, M. O. (1973). Reciprocal averaging: an eigenvector method of ordination. *J. Ecol.* **61,** 237–249.

Hill, M. O. (1974). Correspondence analysis: a neglected multivariate method. *Appl. Statist.* **23,** 340–354.

Hill, M. O. (1977). Use of simple discriminant functions to classify quantitative phytosociological data. In: *First International Symposium on Data Analysis and Informatics* (Ed. by E. Diday, L. Lebart, J. P. Pagès and R. Tomassone), Vol. 1, pp. 181–199. Chesnay: INRIA.

Hill, M. O. (1979). *DECORANA—A FORTRAN Program for Detrended Correspondence Analysis and Reciprocal Averaging*. Section of Ecology and Systematics, Cornell University, Ithaca, New York.

Hill, M. O. and Gauch, H. G. (1980). Detrended correspondence analysis: an improved ordination technique. *Vegetatio* **42,** 47–58.

Ihm, P. and Van Groenewoud, H. (1975). A multivariate ordering of vegetation data based on Gaussian type gradient response curves. *J. Ecol.* **63,** 767–777.

Ihm, P. and Van Groenewoud, H. (1984). Correspondence analysis and Gaussian ordination. *COMPSTAT Lectures* **3,** 5–60.

Imbrie, J. and Kipp, N. G. (1971). A new micropaleontological method for quantitative paleoclimatology: application to a late Pleistocene Caribbean core. In: *The Late Cenozoic Glacial Ages* (Ed. by K. K. Turekian), pp. 71–181. New Haven, CT: Yale University Press.

Israëls, A. Z. (1984). Redundancy analysis for qualitative variables. *Psychometrika* **49,** 331–346.

Iwatsubo, S. (1984). The analytical solutions of eigenvalue problem in the case of applying optimal scoring method to some types in data. In: *Data Analysis and Informatics 3* (Ed. by E. Diday *et al.*), pp. 31–40. Amsterdam: North-Holland.

Jolliffe, I. T. (1986). *Principal Component Analysis*. Berlin: Springer-Verlag.

Jongman, R. H. G., Ter Braak, C. J. F. and Van Tongeren, O. F. R. (1987). *Data Analysis in Community and Landscape Ecology*. Wageningen: Pudoc.

Kalkhoven, J. and Opdam, P. (1984). Classification and ordination of breeding bird data and landscape attributes. In: *Methodology in Landscape Ecological Research and Planning* (Ed. by J. Brandt and P. Agger), Vol. 3, Theme 3, pp. 15–26. Roskilde: Roskilde Universitetsforlag GeoRue.

Kenkel, N. C. and Orlóci, L. (1986). Applying metric and nonmetric multidimensional scaling to ecological studies: some new results. *Ecology* **67,** 919–928.

Kooijman, S. A. L. M. (1977). Species abundance with optimum relations to environmental factors. *Ann. Syst. Res.* **6,** 123–138.

Kooijman, S. A. L. M. and Hengeveld, R. (1979). The description of a non-linear relationship between some carabid beetles and environmental factors. In: *Contemporary Quantitative Ecology and Related Econometrics* (Ed. by G. P. Patil, and M. L. Rosenzweig), pp. 635–647. Fairland MD: International Co-operative Publishing House.

Laurec, A., Chardy, P., de la Salle, P. and Rickaert, M. (1979). Use of dual structures in inertia analysis: ecological implications. In: *Multivariate Methods in Ecological Work* (Ed. by L. Orlóci, C. R. Rao and W. M. Stiteler), pp. 127–174. Fairland MD: International Co-operative Publishing House.

McCullagh, P. and Nelder, J. A. (1983). *Generalized Linear Models.* London: Chapman and Hall.

Macdonald, G. M. and Ritchie, J. C. (1986). Modern pollen spectra from the western interior of Canada and the interpretation of Late Quaternary vegetation development. *New Phytol.* **103,** 245–268.

Meulman, J. and Heiser, W. J. (1984). Constrained multidimensional scaling: more directions than dimensions. *COMPSTAT 1984*, pp. 137–142. Vienna: Physica-Verlag.

Minchin, P. (1987). An evaluation of the relative robustness of techniques for ecological ordination. *Vegetatio* **69,** 89–107.

Montgomery, D. C. and Peck, E. A. (1982). *Introduction to Linear Regression Analysis.* New York: John Wiley.

Nishisato, S. (1980). *Analysis of Categorical Data: Dual Scaling and Its Applications.* Toronto: University of Toronto Press.

Oksanen, J. (1983). Ordination of boreal heath-like vegetation with principal component analysis, correspondence analysis and multidimensional scaling. *Vegetation* **52,** 181–189.

Opdam, P. F. M., Kalkhoven, J. T. R. and Phillippona, J. (1984). *Verband tussen Broedvogelgemeenschappen en Begroeiing in een Landschap bij Amerongen.* Wageningen: Pudoc.

Peet, R. K. (1978). Latitudinal variation in southern Rocky Mountain forests. *J. Biogeogr.* **5,** 275–289.

Peet, R. K. and Loucks, O. L. (1977). A gradient analysis of southern Wisconsin forests. *Ecology* **58,** 485–499.

Pickett, S. T. A. (1980). Non-equilibrium coexistence of plants. *Bull. Torrey bot. Club* **107,** 238–248.

Pielou, E. C. (1984). *The Interpretation of Ecological Data.* New York: John Wiley.

Prodon, R. and Lebreton, J.-D. (1981). Breeding avifauna of a Mediterranean succession: the holm oak and cork oak series in the eastern Pyrenees, 1. Analysis and modeling of the structure gradient. *Oikos* **37,** 21–38.

Purata, S. E. (1986). Studies on secondary succession in Mexican tropical rain forest. *Acta Univ. Ups.* Comprehensive Summaries of Uppsala Dissertations from the Faculty of Science 19. Stockholm: Almqvist and Wiksell International.

Rao, C. R. (1964). The use and interpretation of principal components analysis in applied research. *Sankhya A* **26,** 329–358.

Robert, P. and Escoufier, Y. (1976). A unifying tool for linear multivariate statistical methods: the RV-coefficient. *Appl. Statist.* **25,** 257–265.

Sládecek, V. (1973). System of water quality from the biological point of view. *Arch. Hydrobiol. Beiheft* **7,** 1–218.

Smith, P. L. (1979). Splines as a useful and convenient statistical tool. *Am. Stat.* **33,** 57–62.

Swaine, M. D. and Greig-Smith, P. (1980). An application of principal components analysis to vegetation change in permanent plots. *J. Ecol.* **68,** 33–41.

Ter Braak, C. J. F. (1983). Principal components biplots and alpha and beta diversity. *Ecology* **64,** 454–462.

Ter Braak, C. J. F. (1985). Correspondence analysis of incidence and abundance data: properties in terms of a unimodal response model. *Biometrics* **41,** 859–873.

Ter Braak, C. J. F. (1986). Canonical correspondence analysis: a new eigenvector technique for multivariate direct gradient analysis. *Ecology* **67,** 1167–1179.

Ter Braak, C. J. F. (1987a). The analysis of vegetation-environment relationships by canonical correspondence analysis. *Vegetatio* **69,** 69–77.

Ter Braak, C. J. F. (1987b). *CANOCO—a FORTRAN Program for Canonical Community Ordination by [Partial] [Detrended] [Canonical] Correspondence Analysis, Principal Components Analysis and Redundancy Analysis (Version 2.1).* Agriculture Mathematics Group, Wageningen.

Ter Braak, C. J. F. (1988). Partial canonical correspondence analysis. In: *Classification Methods and Related Methods of Data Analysis* (Ed. by H. H. Bock) Amsterdam: North-Holland, pp. 551–558.

Ter Braak, C. J. F. and Barendregt, L. G. (1986). Weighted averaging of species indicator values: its efficiency in environmental calibration. *Math. Biosci.* **78,** 57–72.

Ter Braak, C. J. F. and Looman, C. W. N. (1986). Weighted averaging, logistic regression and the Gaussian response model. *Vegetatio* **65,** 3–11.

Tilman, D. (1982). *Resource Competition and Community Structure.* Princeton: Princeton University Press.

Tso, M. K-S. (1981). Reduced-rank regression and canonical analysis. *J. Roy. statist. Soc. B* **43,** 183–189.

van der Aart, P. J. M. and Smeenk-Enserink, N. (1975). Correlations between distribution of hunting spiders (Lycosidae, Ctenidae) and environmental characteristics in a dune area. *Neth. J. Zool.* **25,** 1–45.

van den Wollenberg, A. L. (1977). Redundancy analysis. An alternative for canonical correlation analysis. *Psychometrika* **42,** 207–219.

van Dam, H., Suurmond, G. and Ter Braak, C. J. F. (1981). Impact of acidification on diatoms and chemistry of Dutch moorland pools. *Hydrobiologia* **83,** 425–459.

Webb, T., III and Bryson, R. A. (1972). Late- and postglacial climatic change in the northern Midwest, USA: quantitative estimates derived from fossil spectra by multivariate statistical analysis. *Quat. Res.* **2,** 70–115.

Whittaker, R. H. (1956). Vegetation of the Great Smoky Mountains. *Ecol. Monogr.* **26,** 1–80.

Whittaker, R. H. (1967). Gradient analysis of vegetation. *Biol. Rev.* **49,** 207–264.

Whittaker, R. H., Levin, S. A. and Root, R. B. (1973). Niche, habitat and ecotope. *Am. Natur.* **107,** 321–338.

Wiens, J. A. and Rotenberry, J. T. (1981). Habitat associations and community structure of birds in shrubsteppe environments. *Ecol. Monogr.* **51,** 21–41.

Williams, E. J. (1959). *Regression Analysis.* New York: John Wiley.

Wold, H. (1982). Soft modeling. The basic design and some extensions. In: *Systems Under Indirect Observation. Causality–Structure–Prediction* (Ed. by K. G. Jöreskog and H. Wold), Vol. 2, pp. 1–54. Amsterdam: North-Holland.

Zelinka, M. and Marvan, P. (1961). Zür Präzisierung der biologischen Klassifikation der Reinheit fliessender Gewässer. *Arch. Hydrobiol.* **57,** 389–407.

APPENDIX

A general iterative algorithm can be used to carry out the linear and weighted-averaging methods described in this review. The algorithm is essentially the one used in the computer program CANOCO (Ter Braak, 1987b). It operates on response variables, each recording the abundance or presence/absence of a species at various sites, and on two types of explanatory variables: environmental variables and covariables. By environmental variables we mean here explanatory variables of prime interest, in contrast with covariables which are "concomitant" variables whose effect is to be removed. When all three types of variables are present, the algorithm describes how to obtain a *partial constrained ordination*. The other linear and WA techniques are all special cases, obtained by omitting various irrelevant steps.

Let $Y = [y_{ki}]$ $(k = 1, \ldots, m;\ i = 1, \ldots, n)$ be a species-by-site matrix containing the observations of m species at n sites $(y_{ki} \geqq 0)$ and let $Z_1 = [z_{1li}]$ $(l = 0, \ldots, p;\ i = 1, \ldots, n)$ and $Z_2 = [z_{2ji}]$ $(j = 1, \ldots, q;\ i = 1, \ldots, n)$ be covariable-by-site and environmental variable-by-site matrices containing the observations of p covariables and q environmental variables at the same n sites, respectively. The first row of Z_1, with index $l = 0$, is a row of 1's, which is included to account for the intercept in equation (4). Further, denote the species and site scores on the sth ordination by $\mathbf{u} = [u_k]$ $(k = 1, \ldots, m)$ and $\mathbf{x} = [x_i]$ $(i = 1, \ldots, n)$, the canonical coefficients of the environmental variables by $\mathbf{c} = [c_j]$ $(j = 1, \ldots, q)$ and collect the site scores on the $(s - 1)$ previous ordination axes as rows of the matrix A. If detrending-by-polynomials is in force (Step A10), then the number of rows of A, s_A say, is greater than $s - 1$. In the algorithm we use the assign statement ": = ", for example $a := b$ means "a is assigned the value b". If the left-hand side of the assignment is indexed by a subscript, it is assumed that the assignment is made for all permitted subscript values: the subscript k will refer to species $(k = 1, \ldots, m)$, the subscript i to sites $(i = 1, \ldots, n)$ and the subscript j to environmental variables $(j = 1, \ldots, q)$.

Preliminary Calculations

P1 Calculate species totals $\{y_{k+}\}$, site totals $\{y_{+i}\}$ and the grand total y_{++}. If a linear method is required, set

$$r_k := 1, \quad w_i := 1, \quad w_i^* := \frac{1}{n} \tag{A.1}$$

and if a weighted averaging method is required, set

$$r_k := y_{k+}, \quad w_i := y_{+i}, \quad w_i^* := y_{+i}/y_{++} \tag{A.2}$$

P2 Standardize the environmental variables to zero mean and unit variance. For environmental variable j calculate its mean $\bar{z}$ and variance v

$$\bar{z} := \Sigma_i w_i^* z_{2ji}, \quad v := \Sigma_i w_i^* (z_{2ji} - \bar{z})^2 \tag{A.3}$$

and set $z_{2ji} := (z_{2ji} - \bar{z})/\sqrt{v}$

P3 Calculate for each environmental variable j the residuals of the multiple regression of the environmental variables on the covariables, i.e.

$$\mathbf{c}_j^* := (Z_1 W Z_1')^{-1} Z_1 W \mathbf{z}_{2j} \tag{A.4}$$

$$\tilde{\mathbf{z}}_{2j} := \mathbf{z}_{2j} - Z_1' \mathbf{c}_j^* \tag{A.5}$$

where $\mathbf{z}_{2j} = (z_{2j1}, \ldots, z_{2jn})'$, $W = \text{diag}\,(w_1, \ldots, w_n)$ and $\mathbf{c}_j^*$ is the $(p+1)$–vector of the coefficients of the regression of $\mathbf{z}_{2j}$ on Z_1. Now define $\tilde{Z}_2 = [\tilde{z}_{2ji}]$ $(j = 1, \ldots, q,\ i = 1, \ldots, n)$.

Iteration Algorithm

Step A0 Start with arbitrary, but unequal site scores $\mathbf{x} = [x_i]$. Set $x_i^0 = x_i$

Step A1 Derive new species scores from the site scores by

$$u_k := \sum_i y_{ki} x_i / r_k \tag{A.6}$$

Step A2 Derive new site scores $\mathbf{x}^* = [x_i^*]$ from the species scores

$$x_i^* := \sum_k y_{ki} u_k / w_i \tag{A.7}$$

Step A3 Make $\mathbf{x}^* = [x_i^*]$ uncorrelated with the covariables by calculating the residuals of the multiple regression of $\mathbf{x}^*$ on Z_1:

$$\mathbf{x}^* := \mathbf{x}^* - Z_1'(Z_1 W Z_1')^{-1} Z_1 W \mathbf{x}^* \tag{A.8}$$

Step A4 If $q \leqslant s_A$, set $x_i := x_i^*$ and skip Step A5.

Step A5 If $q > s_A$, calculate a multiple regression of $\mathbf{x}^*$ on $\tilde{Z}_2$

$$\mathbf{c} := (\tilde{Z}_2 W \tilde{Z}_2')^{-1} \tilde{Z}_2 W \mathbf{x}^* \tag{A.9}$$

and take as new site scores the fitted values:

$$\mathbf{x} := \tilde{Z}_2' \mathbf{c} \tag{A.10}$$

Step A6 If $s > 1$, make $\mathbf{x} = [x_i]$ uncorrelated with previous axes by calculating the residuals of the multiple regression of $\mathbf{x}$ on A:

$$\mathbf{x} := \mathbf{x} - A'(AWA')^{-1}AW\mathbf{x} \tag{A.11}$$

Step A7 Standardized $\mathbf{x} = [x_i]$ to zero mean and unit variance by

$$\bar{x} := \sum_i w_i^* x_i, \quad \sigma^2 := \sum_i w_i^*(x_i - \bar{x})^2$$
$$x_i := (x_i - \bar{x})/\sigma \tag{A.12}$$

Step A8 Check convergence, i.e. if

$$\sum_i w_i^*(x_i^0 - x_i)^2 < 10^{-10} \tag{A.13}$$

goto Step A9, else set $x_i^0 := x_i$ and goto Step A1.

Step A9 Set the eigenvalue λ equal to σ in (A.12) and add $\mathbf{x} = [x_i]$ as a new row to the matrix A.

Step A10 If detrending-by-polynomials is required, calculate polynomials of $\mathbf{x}$ up to order 4 and first-order polynomials of $\mathbf{x}$ with the previous ordination axes,

$$x_{2i} := x_i^2, \quad x_{3i} := x_i^3, \quad x_{4i} := x_i^4, \quad x_{(b)i} := x_i a_{bi} \tag{A.14}$$

where a_{bi} are the site scores of a previous ordination axis ($b = 1, \ldots, s - 1$). Now perform for each of the $(s + 2)$-variables in (A.14) the Steps A3–A6 and add the resulting variables as new variables to the matrix A.

Step A11 Set $s := s + 1$ and goto Step A0 if required and if further ordination axes can be extracted, else stop.

At convergence, the algorithm gives the solution with the greatest real value of λ to the following transition formulae [where $R = \text{diag}(r_1, \ldots, r_m)$ and $W = \text{diag}(w_1, \ldots, w_n)$ and where the notation B^0 is used to denote $B'(BWB')^{-1}BW$, the projection operator on the row space of a matrix B in the metric defined by the matrix W]:

$$\mathbf{u} = R^{-1}Y\mathbf{x} \tag{A.15}$$
$$\mathbf{x}^* = (I - \tilde{Z}_1^0)W^{-1}Y'\mathbf{u} \tag{A.16}$$
$$\mathbf{c} = (\tilde{Z}_2 W \tilde{Z}_2')^{-1}\tilde{Z}_2 W \mathbf{x}^* \tag{A.17}$$
$$\lambda\mathbf{x} = (I - A^0)\tilde{Z}_2'\mathbf{c} \tag{A.18}$$

The tilde above Z_2 is there as a reminder that the original environmental variables were replaced by residuals of a regression on Z_1 in (A.5), i.e. in terms of the original variables

$$\tilde{Z}_2' = (I - Z_1^0)Z_2' \tag{A.19}$$

Remarks

(1) Note that u_k in the algorithm takes the place of b_k in Section II.

(2) Special cases of the algorithm are: constrained ordination: $p = 0$; partial ordination: $q = 0$; (unconstrained) ordination: $p = 0$, $q = 0$; linear calibration and weighted averaging: $p = 0$, $q = 1$; (partial) multiple regression: $m = 1$. The corresponding transition formulae follow from (A.15)–(A.18) with the proviso that, if $q = 0$, Z_2 in (A.19) must be replaced by the $n \times n$ identity matrix and generalized matrix inverses are used. Note that, if $p = 0$, Z_1 is a $1 \times n$ matrix containing 1's; Z_1 renders the centring of the species data in the linear methods in Section II redundant.

(3) The standardization in P2 removes the arbitrariness in the units of measurement of the environmental variables, and makes the canonical coefficients comparable among each other, but does not influence the values of λ, $\mathbf{u}$ and $\mathbf{x}$ to be obtained in the algorithm.

(4) Step A6 simplifies to Step 3b of the main text if the rows of A are W-orthonormal. The steps A3–A6 form a single projection of $\mathbf{x}^*$ on the column space of $(I - A^0)\tilde{Z}_2'$ if and only if A defines a subspace of the row space of $\tilde{Z}_2$. As each ordination axis defines such a subspace, this is trivially so without detrending. The method of detrending-by-polynomials as defined in Step A10, ensures that A defines also a subspace of $\tilde{Z}_2$ if detrending is in force. The transition formulae (A.15)–(A.18) define an eigenvalue equation of which all eigenvalues are real non-negative (Ter Braak, 1987b).

(5) If a particular scaling of the biplot or the joint plot is wanted, the ordination axes may require linear rescaling. With linear methods one can choose between a Euclidean distance biplot and a covariance biplot, which focus on the approximate Euclidean distances between sites and correlations among species, respectively (Ter Braak, 1983). With weighted averaging methods it is customary to use the site scores $\mathbf{x}^*$ (A.16) and the species scores $\mathbf{u}$ (A.15) to prepare an ordination diagram after a linear rescaling so that the average within-site variance of the species scores is equal to 1 (cf. Section IV.C), as is done in DECORANA (Hill, 1979) and CANOCO (Ter Braak, 1987b).

Index

Abies magnifica, 50–51
sunfleck activity variations, 15
Abundance of species, *see* Gradient analysis
Acclimatory responses, 37–41, 53–54
seasonal variation, 40–41
spatial variation, 38–40
Acer pseudoplatanus, 264
Acer rubrum, 264
Acer saccharum, 22–23
Acleris variana, see Black-headed budworm
Adirondack lake, 140
Aegopodium podagraria, 34
Agrypon flaveolatum, 199
Alocasia macrorrhiza
gas exchange, 21–22
induction response, 26–28
photosynthetic dynamics, 29, 39
Ambrosia pollen, 94, 121
Analysis of concentration, 295
Anemone raddeana, 40
Aphids, 211
Aquatic systems
age and accumulation, 89, 90–91
bioturbation, 80, 124
cation exchange, 74
complexation with organic matter, 74, 76
diatom flora, *see* phytoplankton
heavy metals cycling and redistribution in, 73–82
interactions in water column, 74–76
phytoplankton, 74, 76, 77, 94
redistribution by plants, 80
redox potentials, 77–78, 86
retention rates, 80, 82
sediment, 75, 76–82
acidification of lakes, 78, 80
Antarctic lake, 154
decrease in heavy metal concentration, 78–80
diagenetic modification, 77, 116
focussing, 77
N.Atlantic Ocean, 147–148
normalization, 102–103
organic degradation, 77
profiles, 79
Arabis hirsuta, 255, 257
Argyrodendron
induction state, 33
stomatal response, 32
Argyrodendron peralatum, 23, 24
Arnica species, 15
A. cordifolia
acclimatory responses, 38–39
distributions, 51
leaf temperature and transpiration, 37
photosynthesis, 22
stomatal response, 31–32
water-use efficiency, 34
A. latifolia
distributions, 51
leaf temperature and transpiration, 37
photosynthesis, 22
water-use efficiency, 34
Aster acuminatus
acclimatory responses, 39

Aster acuminatus—continued
growth, 46–47
reproduction, 49, 50
Astrocaryum mexicanum, 49
Atriplex confertifolia, 248
Autumnal moth, 200

Bacillus thuringiensis, 210, 222, 225
Baltic Sea, 112, 117, 133
Betula pendula, 248
Betula pubescens, 247, 248
Big Heath Bog, Maine, 125
Biological controls, 199, 225
Biological monitoring, 68
Biotic communities, *see* Gradient analysis
Bioturbation, 80, 124
Black-headed budworm, 183, 187, 201
Blanc, Mont, 117, 136
Blanket mires, 83, 84, 85, 112
Brassica oleracea, 260
Bromus hordeaceus, 299
Brownian motion transfer, 70
particle diameters, 70, 71
Bryophytes, 68, 82, 83, 92
Budworms, *see* Black-headed budworm: Eastern *and* Western spruce budworm
Bupalus piniarius, see Pine looper

Cabbage butterfly, 198, 211
Caesium-137 dating, 97, 98
Calibration, 272, 274
linear methods, 278
non-linear methods, 285–286
response model choice, 303
weighted averaging methods, 288–290
Calluna vulgaris, 83, 265
Calvin cycle intermediates, 42
Canonical correlation analysis, 282
Canonical correspondence analysis (CCA), 293–295
interpretation, 300–302
joint plots, 296–299
Carbon gain
transport through mycorrhizal links, 250–253
see also Photosynthesis
Castanea pollen, 94
Cation exchange, 74, 83, 84–85
Cecropia, 54
Chamaecyparis lawsoniana, 251
Chazdon, R. L., 1–63
Chelation processes, 84, 86
Choristoneura fumiferana, see Eastern spruce budworm
Choristoneura occidentalis, see Western spruce budworm
Circaea lutetiana, 36–37
Claoxylon sandwicense
growth, 44–45
heat damage, 36
induction response, 27
photosynthesis, 22
photosynthetic dynamics, 28–29
Clarkia rubicunda, 247
Climate release hypothesis, 206–208
Cloud effects, 12, 15, 17, 34
Concentration, analysis of, 295
Coniferous forests, sunflecks, 14–15
Constance, lake, 117, 133
Constrained ordination, *see* Ordination
Cordyline rubra, 21
Cornus florida, 52
Crustal enrichment, 102–103, 104
Cyzenia albicans, 199

Daphnia, 226, 227
Defoliation, 196, 197, 200
impact on forest, 201–202
Dendrolimus pini, 185
Detrended correspondence analysis (DCA), 292–293
joint plots, 296–299
Diagenesis, 77, 116
Diatom stratigraphy, 94
Dipteryx panamensis, 18
growth, 45–46
Distribution of species
canopy gaps and, 48, 50–51
vertical, 48, 51–52
Don Juan Pond, Antarctica, 154
Douglas-fir, water-use efficiency, 34
Douglas-fir tussock moth, 182, 183, 187, 191–192
parasitoid mortality, 197
see also Tussock moths

Dragonflies, 228
Draved Mose, Denmark, 112, 113, 136

Eastern spruce budworm, 182, 184, 202
  DDT and 224–225
  fecundity, 196
  predator exclusion, 226
  see also Spruce budworm
Ecological continua and gradients, *see* Gradient analysis
Ectomycorrhizas (ECM), *see* Mycorrhizal links
Elymus repens, 299
Ennerdale Water, 112
Epirrita, see Oporinia autumnata
Eriophorum vaginatum, 83
Erythronium americanum, 40
Eucalyptus marginata, 265
Eunotia exigua, 295
Eunotia tenella, 295
Eunotia veneris, 295
Euphorbia forbesii
  growth, 44–45
  heat damage, 36
  induction response, 27
  photosynthesis, 22
  photosynthetic dynamics, 28–29

Fagus grandifolia, 31
Fall webworm, 183
False henlock looper, 183, 187
Featherbed Moss, Derbyshire, 107, 111
Festuca ovina, 247, 255, 262–263
Fish-eye camera, 11–12
Flin Flon smelting complex, Manitoba, 130, 132
Forest lepidoptera, *see* Population cycles of forest lepidoptera
Forest tent caterpillars, 182, 184, 193, 207
Fossil fuels, 132–143
Fragaria vesca, 50
Fragaria virginiana
  acclimatory responses, 38, 40–41
  carbon gain, 23
  photosynthesis, 28
  reproduction, 50
Fraxinus americana, 264
Fraxinus excelsior, 264
Frustilia rhomboides var. *saxonica,* 295
Fungus, *see* Mycorrhizal links

Generalized linear modelling, 285
Geochemical monitoring, *see* Heavy metal pollution
Geomagnetic dating, 97, 99
Germination, sunflecks and, 42–44
Glenshieldaig, 110, 112
Glycine max, 254
Gordano Valley, Bristol, 107
Gradient analysis, 271–317
  calibration, *see* Calibration
  canonical correlation analysis, 282
  canonical correspondence analysis (CCA), 293–295, 296–302
  classification of techniques, 276
  conclusions, 306–308
  constrained ordination, *see* Ordination
  detrended correspondence analysis, 292–293, 296–299
  direct analysis, 304–306
  indirect analysis, 280, 304–305
  linear models, 272, 273–282
  model choice
    direct or indirect analysis, 304–305
    regression or constrained ordination, 305–306
    response model, 302–304
  multivariate direct analysis, *see* Ordination, constrained
  non-linear (Gaussian) methods
    regression, 285
    unimodal response models, 282–283
  ordination, *see* Ordination
  partialling of covariables, 304–305
  principal component analysis (PCA), 279–280
  redundancy analysis, 281–282
    interpretation of plots, 299–300
  regression, 272, 277–278, 285, 287–288, 305–306
  selected applications, 274–275
  weighted averaging methods
    calibration, 288–290

Gradient analysis—*continued*
constrained ordination, 293–295
ordination, 290–292
regression, 287–288
Grana stacking, 31, 42
Granulosis virus, 214
Grasshoppers, 211
Grassington Moor, W.Yorks, 109, 116
Gravitational settling, 70
Gypsy moth, 201, 202

Heavy metal pollution, 65–177
accumulation rates, 90, 91
age, 89
conclusions, 154–157
cycling and recycling, *see* Aquatic systems: Peat ecosystems
dating techniques, 91–99
biostratigraphic, 94
geomagnetic, 97, 99
pollen analysis, 94
radiometric techniques, 94, 96–97
stratigraphic, 92, 94
deposition from atmosphere
Brownian motion transfer, 70, 71
dry, 70
gravitational settling, 70
impaction, 70
nucleation, 72
scavenging, 72
turbulent transfer, 70
wet, 72–73
discriminative assessment, 101–106
crustal enrichment, 102–103, 104
inflection point, 101
physico-chemical partitioning, 105–106
stable lead isotope analysis, 103, 105
fossil fuels, coal consumption decline, 117, 132–143
heavy metal definition, 68–69
heavy metal flux rates, 100
history
from Roman times, 107–116
industrial revolution
Europe, 110–116
N.America, 119–127
twentieth century
Europe, 114–115, 116–119
N.America, 127–129
ice deposits, 88–89, 147, 148–154
interpretation of records, 99–106
monitoring
biological, 68
development of, 67–68
pollution definition, 69
present situation, 143–154
N.Atlantic Ocean sediments, 147–148
polar ice deposits, 147, 148–154
quantitative assessment, 100–101
reduction, 117
Scandinavia, 112–113, 115
sources, 130–143
dispersal from point source, 130–132
fossil fuels, 132–143
motor vehicles, 132–143
multiple sources, 132–143
tall stacks, 130, 140
theoretical considerations, 69–106
Hemispherical photographs, *see* Fish-eye camera
Hepatica acutiloba, 41
Heracleum lanatum, 36
Heterocampa guttivitta, see Saddled prominent
Holcus lanatus, 250, 263
Hopea pedicellata, 44
Humic substances, 83, 84–85
Hyacinthoides (Scilla) non-scripta, 46
Hyloicus pinastri, 185
Hyphal links, *see* Myocorrhizal links
Hyphantria cunea, see Fall webworm
Hypothetical environment variable, 279

Ice and snow
age and accumulation, 89, 90–91
Antarctic lake sediments, 154
crustal enrichment, 102
heavy metals in
cycling and redistribution, 88–89
polar ice deposits, 147, 148–154
quantitative assessment, 100
Impaction, 70

Impatiens parviflora
growth, 46
water-stress, 34
Industrial revolution pollution
Europe, 110–116
N.America, 119–127
Inflection point, 101
Intersection plane, 52
Isothermal remnant magnetism (IRM), 99

Kotaochalia junodi, 183

Lactarius pubescens, 248
Lambdina fiscella lugubrosa, see Western hemlock loopers
Larch budmoth, 183, 185, 186, 193, 199–200
Bacillus thuringiensis, 225
disease susceptibility, 214
dispersal of, 203
foliage quality and population, 210
parasitoid mortality, 197
Lead, *see* Heavy metal pollution
Lead–210 dating, 96–97
Leaf movement, 35–36, 53
Leaf temperatures, 6, 24, 36–37
Leccinium species, 248
Lecythis ampla, 18
growth, 45–46
Leontodon autumnalis, 299
Lepidoptera, *see* Population cycles of forest lepidoptera
Lichens, 68
Light acclimation, *see* Acclimatory responses
Livett, Elizabeth A., 65–177
Lolium perenne, 247, 249, 255–61
Lolium rigidum, 253
Lomond, loch, 112, 116
Loopers, *see* False *and* Western hemlock looper: Pine looper
Lymantria dispar, 183
Lymantria fumida, 183
Lymantria ninayi, 188

Magnetism, *see* Geomagnetic dating
Malacosoma californicum pluviata, see Western tent caterpillar
Malacosoma disstria, see Forest tent caterpillar
Mamestra brassicae, 215
Mercurialis perennis, 36
Mercury
enrichments, 117
focussing, 77
see also Heavy metal pollution
Molinia caerulea, 265
Monotropa hypopitys, 250–251, 262
Mont Blanc, 117, 136
Moor House, Cumbria, 107, 110, 116
Motor vehicle pollution source, 132–143
Multivariate direct gradient analysis, *see* Ordination, constrained
Mycorrhizal links, 243–70
arterial hyphae, 245
competitive effects, 256, 261, 262–263
ectomycorrhizas (ECM), 244, 247
evidence for, 245–249
infection, greater or rapid, 249–250
roles in ecosystem, 261–265
seedling establishment, 261, 262
transport between plants
C-containing substances, 250–253
mineral nutrients
dying roots, 258–261
living plants, 253–258
tracer experiments, 253–255
vegetation with mixed links, 264–265
vesicular-arbuscular mycorrhizas (VAM), 243–244, 247
Myers, Judith H., 179–242

Natural remnant magnetism (NRM), 99
Neagh, lough, 112, 116
Neodiprion autumnalis, see Pine sawfly
Nepytia freemani, see False hemlock looper
Net radiometers, 9
Newman, E. I., 243–270
Niche theory, 272
Nuclear polyhedrosis virus (NPV), 213–216
Nucleation, 72

Oak species, 264
  oak-chestnut, 248
Ochroma, 54
Operophtera, 182, 183
Operophtera brumata, see Winter moth
Oporinia autumnata, 182, 202
Orchids, 250
Ordination, 272–3, 275
  constrained, 272–3, 275
    choice of, 305–306
    linear methods, 281
    non-linear methods, 286–287
    weighted averaging methods, 293–295
  diagrams and interpretation, 295–302
    correspondence analysis, 296–299, 300–302
    PCA biplots, 295–296
    redundancy analysis, 299–300
  first ordination axis, 279–280
  linear methods, 278–280
  non-linear methods, 286
  response model choice, 303
  weighted averaging methods, 290–292
Orygia pseudotsugata, see Douglas-fir tussock moth
Oxalis oregana, 35–36, 53

Pachysandra terminalis, 41
Panolia flammea, 185
Parasitoids, 196–199, 224, 228
Parthenocissus quinquefolia, 40
Partitioning, physico-chemical, 105–106
Peat ecosystems, 82–88
  age and accumulation, 89, 90–91
  binding ability, 85
  biomass cycling, 83
  blanket mires, 83, 84, 85, 112
  bryophytes, 82, 83, 92
  cation exchange, 83, 84–85
  chelation processes, 84, 96
  depletion and enrichment horizons, 86, 87
  heavy metal–peat associations, 82–85
  humic substances, 83, 84–85
  hummock, 86
  moss growth increments, 92, 94
  ombrotrophic, 85, 89
  raised mires, 84, 85
  redistribution of heavy metals, 85–88
  relocation by plants, 86, 88
  saturated or hollow peat, 86
  sheet erosion, 83
  water table, 86
Pelargonium, 33
Phaseolus, 39–40
Photoelectric cells, 9
Photographic techniques, 11–12
Photoinhibition, 35–36, 54
Photon flux density, 17–18, 19, 24
Photosynthesis, 41–42
  acclimatory responses, 37–41
  carbon gain, 20–25, 24, 28–31
    computer simulations, 23–25
    field studies, 21–23
    lightfleck length and, 28–29
  dynamic responses, 25, 28–29, 31
  grana stacking, 31, 42
  induction, 25–28, 54
  leaf induction state, 33
  light utilization efficiency, 29
  low light compensation points, 22, 23
  photosynthetic efficiency, 29, 31, 42
  post-illumination, 28, 29, 31, 42
  readiness, 27–28, 33
  RuBP production, 31
  stomatal response, 31–32
    conductance, 21, 24
    water-stress and, 33–35, 54
  time lags, 28, 31–32
  triose-phosphates, 31
Photosynthetically active radiation (PAR), 1
Physico-chemical partitioning, 105–106
Phytoplankton, 74, 76, 77
Picea abies, 247, 251
Picea sitchensis, 247
Pieris brassicae, 213
Pieris rapae, see Cabbage butterfly
Pine looper, 183, 185
  biological controls, 199
  fecundity, 196
  parasitoid mortality, 197
Pine sawfly, 212
Pinus cembra, 203–204
Pinus contorta, 247, 251
Pinus patula, 188
  population change patterns, 189–193
Pinus ponderosa, 264

Pinus radiata, 264
Pinus resinosa, 264
light variations, 14
Pinus sylvestris, 247, 251, 264, 265
light variations, 14
Piper genus, 39
P. aequale
distributions, 51
seed germination, 44
P. amalago, 51
P. auritum, 44
P. hispidum
distributions, 51
water-use efficiency, 34–35
P. treleaseanum, 22
P. umbellatum, 44
Plantago erecta, 247
Plantago lanceolata, 245, 246, 247, 255–261
Poa trivialis, 299
Pollen analysis, 94
Population cycles of forest lepidoptera, 179–242
biological controls, 199, 225
climate release hypothesis, 206–208
cycles in other species, 226–228
cyclic and non-cyclic populations, 199–201
cyclic species characteristics, 188–189
decline beginnings, 193–194
defoliation, 196, 197, 200, 201–202
disease susceptibility, 212–216 224, 229–231
dispersal, 203
drought and, 211
evidence for, 182–188
fecundity, 195–196
food quality
plant deterioration, 208–211
plant stress, 211–212
in tropics, 187–188
insect quality variation
genetic, 202–204
qualitative, 204–206
mathematical models, 216–223
multiple species, 220–223
single species, 217–220
mortality agents, 196
parasitoids, 196–199, 224, 228
periodicity of outbreaks, 182–184
periodicity ranges, 186–187
predator exclusion, 226
self-regulation, 223
synchronous and incongruent fluctuations, 186
time lags, 217, 219
yoke deposition and offspring, 204, 206
Prentice, I. C., 271—317
Principal component analysis (PCA), 279–280
biplots, 295–6
Pyranometers, 9

Quercus species, 264
Q. castanea, 248

Radiocarbon dating, 96
Radiometric dating techniques, 94, 96–97
caesium-137 dating, 97, 98
lead-210 dating, 96–97
radiocarbon dating, 96
thorium-232 dating, 97
Ramets, 46–47
Red fir, *see Abies magnifica*
Red grouse, 227–228
Redox potentials, 77–78, 86
Redundancy analysis, *see* Gradient analysis
Redwood forests, sunflecks, 15
Regression, 272, 274
linear methods, 277–278
model choice, 302–303, 305–306
non-linear methods, 285
weighted averaging method, 287–288
Reproductive behaviour (plants)
size variation and, 49
sunflecks and, 48, 49–50
vegetation and sexual, 50
Respiration, dark 40, 41
Ringinglow Bog, S.Yorks, 109, 110, 111, 112, 133
RuBP, 40
Rubus chamaemorus, 86, 88

Saddled prominent, 183, 211

Sagina procumbens, 299
Salix repens, 299
Saplings, growth, 44–46
Scandinavia, 112–113, 115
Scavenging, 72
  in water column, 76
Sediments, *see* Aquatic systems
Seed germination, 42–44
Seedlings, acclimatory responses, 39
Self-regulation, 223, 227
Shea Sisters Lake, Antarctica, 154
Shetland, 110
Snow, *see* Ice and snow
Snowshoe hares, 227
Solar radiation
  diffuse, 8, 11, 15–16, 32, 44
  direct-beam, 8, 11
  photon flux density (PFD), 9, 10, 17–18, 19, 24
  wavelength variations, 6, 7, 9
    acclimatory responses and, 39
  see also Sunflecks
Solidago flexicaulis, 40
Species abundances, *see* Gradient analysis
Species packing model, 289
Sphagnum, 88
  S. fuscum, 91
Spruce budworm, 191, 193, 194
  see also Eastern *and* Western spruce budworm
Stable lead isotope analysis, 103, 105, 132–133, 136, 137
Stacks, tall, 130, 140
Stomatal conductance, *see* Photosynthesis
Stomatal response, *see* Photosynthesis
Stratigraphic dating techniques, 92, 94
  biogenic laminations, 92
  varves, 92, 124, 126
Sudbury smelters, Ontario, 130
Suess effect, 96
Suillus bovinus, 247
Sunflecks
  acclimatory responses, 37–41, 53–54
  activity, 13–20
  as resource, 3–7
  canopy structure, 19–20, 47
  cloud effects, 12, 15, 17, 34
  coniferous forests, 14–15
  definition habitat dependent, 12–13
  distribution of species, 50–52
  dynamic responses, *see* Photosynthesis
  establishment, 42–44
  future research, 53–54
  growth
    tree seedlings and saplings, 44–46
    understorey species, 46–47
  importance of, 52–53
  induction responses, *see* Photosynthesis
  leaf temperatures, 6, 24, 36–37
  light variation
    spatial, 5, 6
    temporal, 5
  measurement
    area-survey techniques, 8–9, 16
    instantaneous sensor measurements, 9–11
    photographic techniques, 11–12
    sampling intervals, 9, 11
  penumbra, 12, 20
  photoinhibition, 35–36, 54
  photosynthesis, *see* Photosynthesis
  photosynthetically active radiation (PAR), 1
  reproductive behaviour, 48, 49–50
  respiration, dark, 40, 41
  seasonal variations, 13–14
  seed germination, 42–44
  temperate deciduous forests, 13–14
  tropical evergreen forests, 15–19
  utilization constraints, 32–37
    leaf induction state, 33
    leaf temperature, 36–37
    photoinhibition, 35–36
    transpiration, 36–37
    water-use efficiency, 33–35
  utilization determinants, 25–32, 54
  water status, 6, 33–35
  see also Solar radiation
Tabellaria binalis, 295
Tabellaria quadriseptata, 295
Temperate deciduous forests, sunflecks in, 13–14
Tent caterpillars
  parasitoid mortality, 198
  see also Forest *and* Western tent caterpillar

Ter Braak, C. J. F., 271–317
Thelephora terrestris, 251
Thorium-232 dating, 97
Toona australis
induction response, 27
photosynthetic dynamics, 29, 39
Transpiration, 36–37
Tree seedlings, growth, 44–46
Tree top disease, 215
Trifolium repens, 263
Trifolium subterraneum, 253
Trillium grandiflorum, 40
Trillium ovatum, 35
Triose-phosphates, 31
Tropical evergreen forests, 15–19
Turbulent transfer, 70
Tussock moths, 193
drought stress on, 212
population changes, 191–192
predator exclusion, 226
see also Douglas-fir tussock moth

Unimodal response models, 282–283
United States, heavy metal pollution
East Coast, 119
Great Lakes region, 119–122
NW coastal region, 123
regional baselines, 124–127
Southern California, 123–124

Varves, 92
biogenic, 124, 126
Vesicular-arbuscular mycorrhizas (VAM), *see* Mycorrhizal links
Viola blanda, 32
Volcanic emissions, 102

Wastwater, 112
Water status, 6
Water-use efficiency, 33–35
Western hemlock loopers, 183, 187
Western spruce budworm, 183, 187
defoliation and, 200
parasitoid mortality, 197, 198
predator exclusion, 226
see also Spruce budworm
Western tent caterpillar, 182, 190, 192–193, 193
disease, 215
drought stress on, 212
fecundity, 195–196
foliage quality and population, 210
parasitoid mortality, 198
qualitative variation in, 204–206
Windermere, lake, 112, 116
Winter moths, 183
fecundity, 196
parasitoid mortality, 199

Zamia skinneri, 49
Zea mays, 254
Zeiraphera diniana, see Larch budmoth
Zinc, in solid sediments, 79–81
Zunacetha annulata, 187–188

Advances in Ecological Research Volumes 1–17

Cumulative List of Titles

Aerial heavy metal pollution and terrestrial ecosystems, **11,** 218
Analysis of processes involved in the natural control of insects, **2,** 1
Ant-plant-homopteran interactions, **16,** 53
Biological strategies of nutrient cycling in soil systems, **13,** 1
Bray-Curtis ordination: an effective strategy for analysis of multivariate ecological data, **14,** 1
Communities of parasitoids associated with leafhoppers and planthoppers in Europe, **17,** 282
The decomposition of emergent macrophytes in fresh water, **14,** 115
Developments in ecophysiological research on soil invertebrates, **16,** 175
The distribution and abundance of lake-dwelling Triclads—towards a hypothesis, **3,** 1
The dynamics of aquatic ecosystems, **6,** 1
The dynamics of field population of the pine looper, *Bupalus piniarius* L. (Lep., Geom.), **3,** 207
Earthworm biotechnology and global biogeochemistry, **15,** 379
Ecological aspects of fishery research, **7,** 114
Ecological conditions affecting the production of wild herbivorous mammals on grasslands, **6,** 137
Ecological implications of dividing plants into groups with distinct photosynthetic production capabilities, **7,** 87
Ecological studies at Lough Ine, **4,** 198
Ecological studies at Lough Hyne, **17,** 115
The ecology of the Cinnabar moth, **12,** 1
Ecology of coarse woody debris in temperate ecosystems, **15,** 133
Ecology, evolution and energetics: a study in metabolic adaptation, **10,** 1
Ecology of fire in grasslands, **5,** 209
The ecology of pierid butterflies: dynamics and interactions, **15,** 51
The ecology of serpentine soils, **9,** 255
Ecology, systematics and evolution of Australian frogs, **5,** 37
The effects of modern agriculture, nest predation and game management on the population ecology of partridges (*Perdix perdix and Alectoris rufa*), **11,** 2
El Niño effects on Southern California kelp forest communities, **17,** 243
Energetics, terrestrial field studies and animal productivity, **3,** 73
Energy in animal ecology, **1,** 69
Estimating forest growth and efficiency in relation to canopy leaf area, **13,** 327
The evolutionary consequences of interspecific competition, **12,** 127
Forty years of genecology, **2,** 159
The general biology and thermal balance of penguins, **4,** 131

Heavy metal tolerance in plants, **7,** 2
Human ecology as an interdisciplinary concept: a critical inquiry, **8,** 2
Industrial melanism and the urban environment, **11,** 373
Integration, identity and stability in the plant association, **6,** 84
Isopods and their terrestrial environment, **17,** 188
Landscape ecology as an emerging branch of human ecosystem science, **12,** 189
Litter production in forests of the world, **2,** 101
Mathematical model building with an application to determine the distribution of Dursban® insecticide added to a simulated ecosystem, **9,** 133
The method of succesive approximation in descriptive ecology, **1,** 35
Nutrient cycles and H^+ budgets of forest ecosystems, **16,** 1
Pattern and process in competition, **4,** 1
Phytophages of xylem and phloem: a comparison of animal and plant sap-feeders, **13,** 135
The population biology and turbellaria with special reference to the freshwater triclads of the British Isles, **13,** 235
Population cycles in small mammals, **8,** 268
Population regulation in animals with complex life-histories: formulation and analysis of a damselfly model, **17,** 1
Predation and population stability, **9,** 1
The pressure chamber as an instrument for ecological research, **9,** 165
Principles of predator-prey interaction in theoretical experimental and natural population systems, **16,** 249
The production of marine plankton, **3,** 117
Production, turnover, and nutrient dynamics of above- and belowground detritus of world forests, **15,** 303
Quantitative ecology and the woodland ecosystem concept, **1,** 103
Realistic models in population ecology, **8,** 200
Renewable energy from plants: bypassing fossilization, **14,** 57
Rodent long distance orientation ("homing"), **10,** 63
Secondary production in inland waters, **10,** 91
The self-thinning rule, **14,** 167
A simulation model of animal movement patterns, **6,** 185
Soil arthropod sampling, **1,** 1
Stomatal control of transpiration: Scaling up from leaf to region, **15,** 1
Studies on the cereal ecosystem, **8,** 108
Studies on grassland leafhoppers (Auchenorrhyncha, Homoptera) and their natural enemies, **11,** 82
Studies on the insect fauna on Scotch Broom *Sarothamnus scoparius* (L.) Wimmer, **5,** 88
A synopsis of the pesticide problem, **4,** 75
Theories dealing with the ecology of landbirds on islands, **11,** 329
Throughfall and stemflow in the forest nutrient cycle, **13,** 57
Towards understanding ecosystems, **5,** 1
The use of statistics in phytosociology, **2,** 59
Vegetation, fire and herbivore interactions in heathland, **16,** 87
Vegetational distribution, tree growth and crop success in relation to recent climate change, **7,** 177
The zonation of plants in freshwater lakes, **12,** 37